Applications of Small Unmanned Aircraft Systems

Best Practices and Case Studies

T0315378

Applications of Small Unmanned Aircraft Systems

Best Practices and Case Studies

Edited by
J.B. Sharma

CRC Press
Taylor & Francis Group
Boca Raton London New York

CRC Press is an imprint of the
Taylor & Francis Group, an **informa** business

CRC Press
Taylor & Francis Group
6000 Broken Sound Parkway NW,
Suite 300 Boca Raton, FL
33487-2742

First issued in paperback 2022

© 2020 by Taylor & Francis Group, LLC
CRC Press is an imprint of Taylor & Francis Group, an Informa business

No claim to original U.S. Government works

ISBN 13: 978-1-03-247515-8 (pbk)
ISBN 13: 978-0-367-19924-1 (hbk)

DOI: 10.1201/9780429244117

Library of Congress Cataloging-in-Publication Data

Names: Sharma, J. B. (Jitendra Bal), editor.
Title: Applications of small unmanned aircraft systems : best practices and case studies / edited by J.B. Sharma.
Description: Boca Raton, Florida : CRC Press/Taylor & Francis Group, 2019.
| Includes bibliographical references and index. | Summary: "Advances in high spatial resolution mapping capabilities and the new rules established by the Federal Aviation Administration for the operation of Small Unmanned Aircraft Systems have provided new opportunities to acquire aerial data at a lower cost and more safely versus other methods. Also, sUAS can access hazardous or inaccessible areas during disaster events and provide rapid response when needed. This is the first book that brings together the best practices of sUAS applied to a broad range of issues in high spatial resolution mapping projects. The case studies included in this book will help readers understand and develop sUAS based projects"-- Provided by publisher.
Identifiers: LCCN 2019024677 | ISBN 9780367199241 (hardback) | ISBN 9780429244117 (ebook)
Subjects: LCSH: Aerial photography in geography. | Drone aircraft--Scientific applications. | Aeronautics in geodesy. | Aerial photogrammetry.
Classification: LCC G142 .A67 2019 | DDC 526.028/4--dc23
LC record available at https://lccn.loc.gov/2019024677

Visit the Taylor & Francis Web site at
http://www.taylorandfrancis.com

and the CRC Press Web site at
http://www.crcpress.com

I dedicate this book to my family, to my friends, and to the continued splendor of Mother Nature

Contents

Foreword

The recent advancement of small Unmanned Aerial Systems (sUAS) into every aspect of our lives is truly a positive, disruptive technology. It has changed how we view everything. The use of sUAS has put the power in our hands to quickly and effectively view our world from above in a way that has never before been possible. Now, anyone can do it. It is amazing.

In this book, Dr J.B. Sharma has compiled a state-of-the-art collection of chapters that represents the breadth of the impact that sUAS are having today while providing many recommendations and thoughts that will take us into the future. It is an invaluable read for anyone wishing to grasp the fundamentals of sUAS use for a variety of applications. Each chapter provides information that is not readily available elsewhere with the authors sharing successes, failures, and recommendations that would not be found in peer-reviewed journal papers. Those new to the field of sUAS will benefit greatly from the basic information and applications shared in this book. Those who have used sUAS will appreciate the insights and depth of knowledge shared by the authors. This book is timely and valuable, and Dr Sharma should be commended for bringing together such a valuable resource.

Russell G. Congalton, PhD
Professor of Remote Sensing & Geographic Information Systems
Department of Natural Resources & the Environment
University of New Hampshire
Durham, New Hampshire

Preface

Humans have always looked up to the sky since time immemorial. Gazing at the wonders of the starlit nights have provided inspiration for the development of mathematics, science, technology, and indeed civilization. However, the ability to look back upon our own self on our own home planet is very new and no more than two centuries old. The advent of hot-air balloons, aircraft, and satellites have begun to provide insights into our global civilization and upon the impact of humanity on our home planet. Databases of land, oceanic, and atmospheric systems have been created by satellite observations and their global interactions are beginning to be understood. Remote sensing has developed into a robust discipline in the past century and has been the domain of those who specialized in geospatial science and technology.

Humans have also always aspired to fly like a bird and to be able to see a transcendent view of themselves; to be able to look back upon oneself from the outside. This capability has just become available to all as a result of sensor and hardware miniaturization, advances in computing, and the internet. Small Unmanned Aircraft Systems (sUAS) are now available to everybody, including children, and can fit in the palm of your hand. They can be tethered to smartphones serving as controllers and everybody can see themselves and their proximity from the outside, just like a bird can. This is a disruptive technology that is in its beginning phases. It can be harnessed to benefit humanity and our planet in many unprecedented ways.

sUAS technology is inherently local and fuses both space and place in a very meaningful manner. It makes remote sensing personal. It allows for a near real time geospatial awareness capability that can inform governance, planning, safety, and efficiency that smart spaces of the 21st century will need. Regulations are being created at levels ranging from national, state, county, city, and institution level to protect privacy, public safety, and airspace safety. These nested regulations need to be adhered to for any sUAS operations with a commercial or educational import. Not many curricular resources are currently available for sUAS based remote sensing applications. The high spatial, spectral, radiometric, and temporal resolution afforded by sUAS brings with it challenges that involve sensors, software, algorithms, computing, data warehousing, and analytics. This opens up the space for the creation of new courses, curricula, and research aimed at the development of sUAS applications and is an opportunity for universities worldwide. This book seeks to address this need.

This book has a primary focus on sUAS-based remote sensing applications. It begins with the consideration of positional accuracy of the imagery collected and the thematic accuracy of the maps produced. Optimal workflows are beginning to be formulated for particular sUAS applications and replicability is a critical issue for effective mapping applications; an in-depth discussion is presented on this issue. A curricular template for sUAS called a DACUM has been developed to inform curriculum and programs design in sUAS applications. An overview of the United States Federal Government sUAS applications over many different areas and conducted for the public good is presented. This is followed by case studies of sUAS applications to wildlife conservation, natural conservation, cultural monitoring, riverine hydrology, machine learning, educational issues, thermal audits, and forest monitoring from below the canopy. The reader is invited to explore this compendium of sUAS-based remote sensing case studies written by foremost experts in this area.

It is hoped that this treatment of sUAS mapping principles and projects gives the reader insights and resources to develop their own local, highly spatially detailed applications that enrich communities.

MATLAB® is a trademark of The MathWorks, Inc. and is used with permission. The MathWorks does not warrant the accuracy of the text or exercises in this book. This book's use or discussion of MATLAB® software or related products does not constitute endorsement or sponsorship by The MathWorks of a particular pedagogical approach or particular use of the MATLAB® software.

MATLAB® is a registered trademark of The MathWorks, Inc. For product information, please contact:

The MathWorks, Inc.
3 Apple Hill Drive
Natick, MA, 01760-2098 USA
Tel: 508-647-7000
Fax: 508-647-7001
E-mail: info@mathworks.com
Web: www.mathworks.com

Acknowledgments

Special acknowledgment is given to the University of North Georgia Foundation for supporting my research in sUAS applications through the Eminent Scholar of Teaching and Learning Endowed Chair. The AmericaView Consortium is dedicated to the advancement of remote sensing education, research and outreach, and deserves a special mention. This book is a product of the collegiality vested in AmericaView as most of the chapter authors are fellow colleagues of this consortium. The participation of all the scholars in this book project and their support is most appreciated.

Editor

Dr Jitendra Bal 'J.B.' Sharma is a professor and the assistant department head of Physics at the University of North Georgia (UNG). He is also affiliated with the Lewis F. Rogers Institute for Environmental Spatial Analysis. He has held the Eminent Scholar of Teaching and Learning Endowed Chair since 2009 and has been a faculty member in the University System of Georgia for 35 years. He has a BS in physics from Jacksonville State University and an MS in physics from the University of Georgia. He has a PhD in geography from the University of Georgia with a specialization in Remote Sensing and Geographic Information Science. He was elected as the fellow of the American Association of Physics Teachers in 2018. He was awarded the American Society of Photogrammetry and Remote Sensing (ASPRS) Leidos-Estes Teaching Excellence Award in 2015 and was the Carnegie-CASE 1999 Georgia Professor of the Year. He teaches courses in physics, remote sensing, and image processing. He is very interested in physics and remote sensing education and in fostering societal literacy in these disciplines. His research interests are in environmental physics, high spatial resolution remote sensing, geographic object-based image analysis (GEOBIA), 3D digital fabrication, and small Unmanned Aerial Systems (sUAS). He lives in Gainesville, Georgia with his wife, has two grown-up sons and enjoys sailing, woodworking, gardening, reading, and hiking.

Contributor List

Amr Abd-Elrahman
University of Florida
Gainesville, Florida, United States

Qassim Abdullah
Woolpert, Inc.
Canonsburg, Pennsylvania, United States

Matthew Belt
Idaho State University
Pocatello, Idaho, United States

Sergio Bernardes
University of Georgia
Athens, Georgia, United States

Gregory K. Brown
University of Wyoming
Laramie, Wyoming, United States

Kristy Bly
World Wildlife Fund
Washington, D.C., United States

Thomas Calton
Idaho State University
Pocatello, Idaho, United States

James B. Campbell
Virginia Tech
Blacksburg, Virginia, United States

Russell G. Congalton
University of New Hampshire
Durham, New Hampshire, United States

Katie Corcoran
Orbital Sidekick, Inc.
San Francisco, California, United States

David Cotten
University of Georgia
Athens, Georgia, United States

Donna M. Delparte
Idaho State University
Pocatello, Idaho, United States

Brian Duran
University of North Georgia
Dahlonega, Georgia, United States

Benjamin T. Fraser
University of New Hampshire
Durham, New Hampshire, United States

Brandon S. Gellis
University of Wyoming
Laramie, Wyoming, United States

Cari Goetcheus
University of Georgia
Athens, Georgia, United States

Barry Haack
George Mason University
Fairfax, Virginia, United States

Richard Ham
University of Arkansas
Fayetteville, Arkansas, United States

Lance Hundt
University of North Georgia
Dahlonega, Georgia, United States

Thomas Jordan
University of Georgia
Athens, Georgia, United States

Bandana Kar
Oak Ridge National Laboratory
Oak Ridge, Tennessee, United States

Michael Kinsey
Fort Belknap
Fort Belknap Reservation, Montana,
    United States

Tao Liu
Oak Ridge National Laboratory
Oak Ridge, Tennessee, United States

Marguerite Madden
University of Georgia
Athens, Georgia, United States

Salvatore Manfreda
University of Basilicata
Potenza, Italy

John A. McGee III
Virginia Tech
Blacksburg, Virginia, United States

J. Zachary Miller
University of North Georgia
Dahlonega, Georgia, United States

Jana Müllerová
Institute of Botany of the Czech Academy
  of Sciences
Průhonice, Czechia

Riadh Munjy
California State University - Fresno
Fresno, California, United States

Sarah Olimb
World Wildlife Fund
Washington, D.C., United States

Kristen Olson
Vertical-Access Inc. and Cornell University
Ithaca, New York, United States

Alonso Pizarro
University of Basilicata
Potenza, Italy

Bruce Quirk
USGS - Retired
Virginia, United States

Silvano Fortunato Dal Sasso
University of Basilicata
Potenza, Italy

Ramesh Sivanpillai
University of Wyoming
Laramie, Wyoming, United States

Travis Stone
Idaho State University
Pocatello, Idaho, United States

Flavia Tauro
University of Tuscia
Viterbo, Italy

Jason A. Tullis
University of Arkansas
Fayetteville, Arkansas, United States

Malcolm Williamson
University of Arkansas
Fayetteville, Arkansas, United States

1

sUAS Data Accuracy in Photogrammetric Workflows

Qassim Abdullah and Riadh Munjy

CONTENTS

1.1 Source of Errors in UAS Photogrammetry

The collection of UAS imagery for mapping applications involves several issues impacting the positional accuracy of the data collected. The awareness and quantification of these errors is very important for developing mapping products that accurately represent the imaged terrain. These foundational issues that introduce positional errors in UAS imagery are articulated in the following sections.

1.1.1 Datum and Coordinate Systems

Errors introduced from the vertical and horizontal datum are frequent and a serious occurrence of error during the map making process. Often, users are either mistakenly using the wrong datum or its version or realization. *Realization of a datum means that positional data can be viably displayed in a coordinate system based on that datum.* Here in the United States, there are several versions for the North American Datum of 1983 (NAD83) and several geoid models used with the North American Vertical Datum of 1988 (NAVD88). Using the wrong datum or confusing its version in the ground control networks or the airborne Global Positioning System (GPS) data may introduce positional errors ranging from millimeters to meters in the derived geospatial products.

1.1.2 Flight Planning and Execution

Flight planning is a very important consideration in successful sUAS mapping operations. There are many issues that contribute to accurate mapping products and these are articulated as follows.

Flight Pattern: Flight patterns are optimized for minimizing flight time, wind direction, and topography (Anders et al. 2013). Windy conditions consume more battery energy than calm conditions. Flight patterns can be vertical, oblique, or circular (Chiabrando et al. 2017). Flights in opposite directions can contribute to a better estimation of the principal point when self-calibration is applied in the block adjustment (Gerke and Przybilla 2016). It is advisable to add extra frames at the end of each flight line since poor image quality results during turn-about (Zajkowski and Snyder 2016).

Imagery Overlap: Certain imagery overlap is required to achieve accurate and reliable products. To take advantage of using Structure from Motion (SfM) algorithms, the forward overlap must be increased. Forward overlap of 80% and side overlap of 60% to 70% provide better accuracy and reduce image mismatches (Nasrullah 2016).

Flying Altitude: The imagery product's positional accuracy is influenced by the accuracy of the sensor orientation determination during the aerial triangulation process. Accordingly, the accuracy of the geospatial products decreases as the flying height increases assuming the same camera and lens were used during these flights. Lowering the flight altitude demands a longer acquisition time and may increase building lean effect (Nasrullah 2016). Adding images with different altitudes (20% higher) may improve the block accuracy (Yanagi 2016; Hoffman et al. 2016), which will contribute to better focal length estimation when self-calibration is being used.

Corridor mapping: For corridor mapping, the accuracy depends on the number and distribution of the Ground Control Points (GCPs), the accuracy of the GCPs, and the accuracy of the directly measured camera positions using the Global Navigation Satellite System (GNSS). Measurements of camera angles using an Inertial Measurement Unit (IMU) may contribute to better accuracy (Rehak et al. 2013).

Ground Sampling Distance (GSD): The Ground Sampling Distance (GSD) is a function of flying height, camera focal length, and camera sensor size. The GSD has a strong influence on the accuracy achieved in photogrammetric mapping (Rumpler et al. 2017). Besides the GSD, there is an important parameter related to the resolving power of the camera lens. It is the Ground Resolved Distance (GRD). The GRD has substantial impact on matching results and the achieved accuracy. The GRD is highly correlated with lens quality, f-stop, motion blur, and GSD (Nasrullah 2016). It is recommended to use a small aperture lens to minimize the image blur (Nasrullah 2016).

Ground Control and Check Points: Ground control points are required to achieve high accuracy as current direct geo-referencing technology and algorithms do not provide such accuracy. The accuracy of the image products is influenced by the quality of the GCPs marker (panel) size and shape, number, and distribution of GCPs, and their accuracy. The number of GCPs and their distribution is not well defined in the literature. While some suggest five control points (Reshetyuk and Mårtensson 2016; Agüera-Vega et al. 2017), others suggest seven or eight GCPs (Gillan et al. 2017).

Based on the authors' research findings, five to nine GCPs are adequate to achieve high accuracy for small and medium size projects. The shape of the targets that have been used are either squares from 20 cm to 40 cm (Bendea et al. 2007; Brutto et al. 2014), or circular (Fras et al. 2016). The size of the target depends on the GSD. It is recommended that the image size of the GCPs be 6 to 10 pixels (James et al. 2017; Yusoff et al. 2017). A 17-cm diameter circle with 5-cm white square in the center was also suggested in the literature (Gillan et al. 2017). Black circular targets on a white background are found to result in good performance (Fras et al. 2016; Zajkowski and Snyder 2016).

GCPs are surveyed in most cases with GNSS equipment using the Real Time Kinematic (RTK) mode, either with respect to a nearby base station or with differential corrections. Such an RTK survey usually results in horizontal accuracy of 1–2 cm and vertical accuracy of 2–3 cm (Benassi et al. 2017). The disadvantage of this mode of surveying is the need to continuously track the rover which makes it unsuitable for some cases in urban areas (Nasrullah 2016).

To verify the image product accuracy, the accuracy of check points should be at least three times better than that of the targeted validation results (Mesas-Carrascosa et al. 2014). This requirement was also found in the "ASPRS Positional Accuracy Standards for Digital Geospatial Data." Terrestrial laser scanning has been used as ground truth with an accuracy of 2–4 mm and GSD as low as 1.5 mm (Reshetyuk and Mårtensson 2016; Rumpler et al. 2017).

1.2 Lens Quality and Lens Distortion

Camera systems have a tremendous impact on the achieved photogrammetric accuracy. Since almost every camera system used with UAS is a non-metric camera, it is highly recommended to use a camera with firm lens elements that maintains its calibration (Rosnell and Honkavaara 2012). Cameras with retracted lenses in shutdown should be avoided (Rosnell and Honkavaara 2012). The magnitude of the camera lens distortion is large in small and mobile cameras. They display complex patterns where the distortion is not a monotonic function of radial distance from the center of the image (Carbonneau and Dietrich 2016). Combined with multiple images, it will propagate to form a doming or dishing effect (Carbonneau and Dietrich 2016).

Many consumer grade cameras have image stabilization features to detect and correct camera movement during image acquisition by moving the camera lens elements. In a UAS environment, the stability of camera calibration is a prerequisite for precise photogrammetric mapping. It is recommended to disable image stabilization in UAS photography (Gillan et al. 2017).

Optical sensors that are commonly used in current digital cameras are either Charge Coupled Devices (CCD) or Complementary Metal-Oxide-Semiconductors (CMOS), while mirrorless cameras are becoming increasingly popular, mainly because of their light weight (Georgopoulos et al. 2016). CCD registers are appropriately light-shielded; such CCDs provide an electronic global shutter without additional complexity (Vautherin et al. 2016). CCDs are mostly found in consumer grade electronics, but recently CMOS sensors have been widely used (Vautherin et al. 2016).

The UAS camera shutter plays an important role in mapping accuracies. A variety of mechanical and electronic shutters are used, among those are the following:

Electronic Rolling Shutter: With this shutter, the image is captured by scanning across the scene either vertically or horizontally. The entire image is not recorded at the same time which will be an issue for moving UASs, causing image skew and smear. This type of shutter is often found in cameras that use a CMOS sensor or video (Vautherin et al. 2016). The effect of this type of shutter can be mathematically modeled in photogrammetric software such as Pix4D and Metashape (formerly Photoscan) by AgiSoft.

Electronic Global Shutters: With this type of shutter, the entire frame is captured at the same time. It is usually found with cameras that use a CCD or CMOS sensor used in high-speed cameras (Vautherin et al. 2016). The S.O.D.A. camera used on the eBee UAS by SensFly uses this type of shutter and it has been shown that it can outperform cameras with a rolling shutter (Vautherin et al. 2016).

Mechanical Shutter: Mechanical shutters are found in most cameras that have a non-removable lens. They are usually located inside the lens and consist of two curtains. The first curtain opens to start the exposure, and the other curtain ends the exposure (Vautherin et al. 2016).

Image Blur: Image blur is usually caused by fast UAS motion or vibration; this in turn degrades the quality of the acquired imagery. Images with high image blur would detect fewer matching features using image correlation and can reduce the accuracy of a bundle solution (Nasrullah 2016). Multi-copters can minimize image blur if they hover during the image exposure (Zajkowski and Snyder 2016). Small image blur does not affect the accuracy of the block if the control points can be detected and successfully measured and there are sufficient numbers of control points (Nasrullah 2016). Image blur can be minimized also by having a high exposure rate with short exposure time (1/2000 seconds) (Rosnell and Honkavaara 2012).

1.2.1 Geo-Location Auxiliary Systems

The camera geo-location and orientation in the sUAS imaging process is very important for positional accuracy of the image data. This is dependent on several interlinked systems which are elaborated upon as follows:

GPS: Camera positioning has a large impact on the block accuracy. If the UAS is equipped with a GPS receiver, the GPS-based geo-location information for each image can be captured and encoded in the image EXIF header or saved in a separate file (Benassi et al. 2017). There are two types of GPS receivers used with UASs. These are:

Single Frequency (L1) Receivers: These are used to establish coarse positioning, which in many cases is not accurate enough to achieve mapping accuracy since single frequency GPS systems have a precision that is in the order of meters and not in the order of centimeters.

Dual Frequency (L1 and L2) Receivers: Some UASs have a high-end receiver on board for GNSS L1 and L2 signals, with corrections sent from the master station to achieve the desired accuracy (Benassi et al. 2017). Those systems are called Real Time Kinematic

(RTK) capable. The main advantage of RTK-enabled UASs is to minimize the number of GCPs needed and to enable real time mapping capability. As GNSS kinematic surveys might suffer from false ambiguity fixing that can cause systematic positioning errors (Benassi et al. 2017), the Post Processed Kinematic (PPK) technique can be used instead. With the PPK mode, the raw GPS data collected are processed and corrected after the flight is completed using the GPS data from a base station on site, which yields higher precision. Some systems offer a Direct Mapping Solution for Unmanned Aerial Vehicles (DMS-UAS) using an integrated GNSS/INS system, which enables direct geo-referencing (Reshetyuk and Mårtensson 2016); thus, considerably reducing the number of GCPs (Skaloud et al. 2014).

Inertial Measurement Unit (IMU): Some UASs have a low-end IMU which measures the camera orientation angle. In most cases, these low accuracy angles are not used during the data processing of the imagery. Camera orientation angles from accurate IMU can be used during the data processing to reduce the number of ground control points needed for the imagery processing.

1.3 Block Adjustment Utilizing the Structure from Motion Algorithm

Block adjustment is a very important photogrammetric process in which the individual images acquired are orthomosaiced into a seamless image factoring in both the camera internal and exterior imaging parameters and terrain morphology. The Structure from Motion (SfM) algorithm is widely used for sUAS block adjustment because it provides greater flexibility and high-quality results (Chiabrando et al. 2013). SfM is a computer vision technique which, for reconstructing the geometry of the scene, solves the camera interior and exterior parameters without providing known 3D point positions (Agüera-Vega et al. 2017).

The majority of software cited in the literature uses the Python script interface to establish an automated process for the implementation of the SfM technique (El Meouche et al. 2016). Otherwise, the user in principle has very little control over the data processing. It is another example of a black box approach (Reshetyuk and Mårtensson 2016), which shows increasing robustness and visually appealing 3D models (Rumpler et al. 2017). The SfM-based software does not provide a thorough statistical analysis of results, but a mean re-projection and the RMS from check points are given in the report. The lack of details means that it is not possible to understand fully the underlying sources of error (James et al. 2017; Cramer 2013). Careful selection of initial settings, combined with a thorough check and analysis of the output, are prerequisites for obtaining the best possible results in UAS photogrammetry (Reshetyuk and Mårtensson 2016). Another drawback is that manual digitizing of GCPs and check points is required (Rumpler et al. 2017).

It is highly recommended that after the bundle adjustment the positional residuals are analyzed. For all images showing large differences, the standard deviation should be increased, and the bundle adjustment repeated. This process is iterated until the residuals of all remaining images with high residuals are below a predefined threshold (Daakir et al. 2015).

Camera Self-Calibration: In general, an Unmanned Aircraft System (UAS) uses consumer grade digital cameras, which are classified as non-metric cameras. They require self-calibration to evaluate the camera interior parameters.

The parameters in camera self-calibration are usually focal length (f), principal point coordinates (cx, cy), radial distortion parameters (k1, k2, k3), affinity and non-orthogonal (b1, b2), and tangential distortion (p1, p2) (Gerke and Przybilla 2016). Another parameter that can be included in the self-calibration is GNSS drift to eliminate the average time delay between exposure time and GNSS time (Reshetyuk and Mårtensson 2016). When using non-metric cameras, the physical stability of these parameters is unknown (Gerke and Przybilla 2016). The estimation of focal length might be correlated with camera Z position, especially for flat terrain (Gerke and Przybilla 2016). A dense pattern of highly accurate GCPs with good spatial distribution and inclusion of oblique imagery for rotary wings and cross strips at different altitudes are recommended for successful self-calibration (Benassi et al. 2017).

Self-calibration in some SfM workflows is performed before the measurement of the GCPs (Reshetyuk and Mårtensson 2016). The result of the bundle adjustment will be in an arbitrary coordinate system which can be adjusted using 3D conformal transformation using the GCPs. Using this approach, large errors are reported, especially in the Z-component. Instead, when GCPs were introduced into the bundle adjustment, self-calibration errors were reduced by a factor of three and they became even smaller when oblique images were used (Anders et al. 2013). Those oblique images provide an image block geometry which also supports self-calibration (Gerke and Przybilla 2016).

It is generally agreed that, unless a pre-calibration is performed on-site, just prior or just after the flight, a robust self-calibration procedure is better for non-metric cameras (Benassi et al. 2017; Cramer 2013; Eisenbeiss 2011). Some research has shown that using pre-calibration is comparable to on-site calibration (Reshetyuk and Mårtensson 2016). For reduced GCP distributions, pre-calibrated cameras should be used, but the calibration needs to be better than the required network precision, and this may be difficult to achieve with non-metric cameras (James et al. 2017). For dense GCPs with good spatial distribution, self-calibration bundle block adjustments are likely to outperform adjustments carried out with pre-calibrated camera models (James et al. 2017).

1.3.1 Strengths and Weaknesses of UAS-Based Photogrammetry

Despite the fact that low-cost cameras and auxiliary systems are used in the UAS photogrammetric workflow, we are witnessing a great deal of success when it comes to quality and accuracy of the imagery products. Such success is attributed to the following factors:

High-Resolution Imagery and Low Altitude: High-resolution imagery is usually acquired from lower altitude and therefore resulting in lower positional errors in the products. UAS-based imagery is usually flown from altitudes ranging from 70 ft. to 400 ft. Above Ground Level (AGL). Positional errors caused by the errors in camera orientation angles are much smaller for imagery acquired from a UAS flown at 100 ft. AGL as compared to imagery from manned aircraft flown at 6,000 ft. AGL; see Table 1.1. In addition, the high-resolution imagery helps the image matching algorithms and enhances the quality of match points and points cloud.

Excessive Overlap and Image Redundancy: The increased image overlap associated with UAS flights increases the redundancy and therefore enhances the reliability figure in the photogrammetric solution. Increased image overlap coupled with SfM workflows result in high-quality image matching and high-quality point clouds.

TABLE 1.1

Effect of Flying Altitude on Products Accuracy

	Horizontal Error in X or Y (ft.)	
Flying Altitude AGL (ft.)	**ft.**	**cm**
100.0	0.007	0.22
150.0	0.011	0.33
200.0	0.015	0.44
400.0	0.029	0.89
3000.0	0.218	6.65
6000.0	0.436	13.30

Capable Processing Software: A new generation of software has been introduced to the market to specifically deal with the non-metric nature of UAS-based imagery. Such software has accelerated the adoption of UAS for mapping by the geospatial industry. Without such software, UAS business growth would have been sluggish.

Small Size Projects: Small UASs are usually utilized for small projects resulting in fewer images. Such a small number of images reduces the amount of errors and observations that need to be modeled during the bundle block adjustment. In addition, fewer images translate to better GCP/image ratio assuming ground control points are used in processing the data.

As for the weakness in UAS photogrammetry, it is all about the non-metric nature of the consumer grade cameras used with small UAS. Consumer grade cameras are not stable, and they do not hold their internal geometry firmly. In other words, the internal relationship between the lens focal length, the principal point location, and the CCD array are changing during changing operational conditions. It is almost impossible to obtain accurate products from UAS imagery without utilizing a camera self-calibration approach during the bundle adjustment.

1.4 Image Data Processing and Analysis of Results

The following sections elaborate on the important issues in analyzing the results of the processed UAS imagery.

1.4.1 Bundle Adjustment Report Analysis

It is important for the person who is performing the data processing to pay close attention to the summary of the results after the orthomosaicing of the imagery, as follows:

Fit to Ground Controls: Good ground control fit in the adjustment is crucial to achieving an accurate orthophoto product. Figure 1.1 illustrates the ground control fit as reported in the quality report produced by Pix4D software. One needs to evaluate not only the "RMS" values but the values of "Mean" as it is sometimes indicative of the presence of biases in the ground controls coordinates or fit.

Camera Self-Calibration Magnitudes: It is good practice to evaluate the magnitude of the corrections the software applies to the internal camera parameters after each block adjustment. Figure 1.2 illustrates the effect of different numbers and configurations of ground control points on the calibrated camera parameters as reported by Pix4D software. Figure 1.2 clearly shows that adding four ground control points to the adjustment changed the initial value of the focal length by 437 microns. Such large adjustment of the focal length is either due to inaccurate initial value of the

GCP Name	Accuracy XY/Z [US survey foot]	Error X [US survey foot]	Error Y [US survey foot]	Error Z [US survey foot]	Projection Error [pixel]	Verified/Marked
3 (3D)	0.215/0.215	0.001	-0.065	-0.023	1.534	26 / 26
4 (3D)	0.215/0.215	0.086	-0.147	-0.087	2.165	25 / 25
13 (3D)	0.215/0.215	0.072	0.032	0.211	1.760	30 / 32
14 (3D)	0.215/0.215	-0.029	0.031	0.043	1.575	36 / 36
23 (3D)	0.215/0.215	0.016	0.108	0.046	1.420	40 / 40
25 (3D)	0.215/0.215	-0.101	0.090	-0.111	1.614	26 / 26
26 (3D)	0.215/0.215	0.038	0.272	-0.246	1.272	14 / 14
GCP20 (3D)	0.215/0.215	-0.041	-0.055	0.175	1.804	30 / 30
GCP22 (3D)	0.215/0.215	-0.033	-0.239	-0.236	1.882	21 / 21
Mean [US survey foot]		0.000847	0.003074	-0.025300		
Sigma [US survey foot]		0.055726	0.142201	0.152513		
RMS Error [US survey foot]		0.055732	0.142234	0.154597		

FIGURE 1.1
Processing quality report.

	Focal Length	Principal Point x	Principal Point y	R1	R2	R3	T1	T2
Initial Values	4114.286 [pixel] 16.000 [mm]	3000.000 [pixel] 11.667 [mm]	2000.000 [pixel] 7.778 [mm]	0.000	0.000	0.000	0.000	0.000
Scenario C 4 GCPs								
Optimized Values	4001.811 [pixel] 15.563 [mm]	3024.497 [pixel] 11.762 [mm]	1994.082 [pixel] 7.755 [mm]	-0.065	0.082	0.008	-0.001	-0.000
Uncertainties (Sigma)	2.477 [pixel] 0.010 [mm]	0.344 [pixel] 0.001 [mm]	0.422 [pixel] 0.002 [mm]	0.000	0.001	0.001	0.000	0.000
Scenario E 7 GCPs								
Optimized Values	4015.186 [pixel] 15.615 [mm]	3024.273 [pixel] 11.761 [mm]	1992.625 [pixel] 7.749 [mm]	-0.063	0.082	0.009	-0.001	-0.000
Uncertainties (Sigma)	2.307 [pixel] 0.009 [mm]	0.344 [pixel] 0.001 [mm]	0.408 [pixel] 0.002 [mm]	0.000	0.001	0.001	0.000	0.000
Scenario G 13 GCPs								
Optimized Values	4017.736 [pixel] 15.625 [mm]	3024.343 [pixel] 11.761 [mm]	1991.995 [pixel] 7.747 [mm]	-0.063	0.082	0.010	-0.001	-0.000
Uncertainties (Sigma)	2.250 [pixel] 0.009 [mm]	0.342 [pixel] 0.001 [mm]	0.401 [pixel] 0.002 [mm]	0.000	0.001	0.001	0.000	0.000
Scenario A 28 GCPs								
Optimized Values	4016.885 [pixel] 15.621 [mm]	3024.327 [pixel] 11.761 [mm]	1992.283 [pixel] 7.748 [mm]	-0.063	0.082	0.010	-0.001	-0.000
Uncertainties (Sigma)	2.262 [pixel] 0.009 [mm]	0.340 [pixel] 0.001 [mm]	0.402 [pixel] 0.002 [mm]	0.000	0.001	0.001	0.000	0.000

FIGURE 1.2
Effect of ground control points on camera self-calibration.

focal length or due to the low accuracy of the airborne GPS or the combination of both. Once the first set of GCPs is used in the adjustment, changing the number of GCPs in subsequent adjustments has a less dramatic effect on the computed focal length.

1.5 Evaluating Errors According to ASPRS Accuracy Standards

Products' accuracy should be evaluated according to the new "ASPRS Positional Accuracy Standards for Digital Geospatial Data." The new standard calls for the following relationships in the accuracy of the product, the ground control points, and the aerial triangulation.

1.5.1 Accuracy Requirements for Aerial Triangulation and INS-Based Sensor Orientation of Digital Imagery

Accuracy of aerial triangulation designed for digital planimetric data (orthoimagery and/or digital planimetric map) only is specified as follows:

$$\text{RMSE}_{x(AT)} \text{ or } \text{RMSE}_{y(AT)} = \tfrac{1}{2} * \text{RMSE}_{x(Map)} \text{ or } \text{RMSE}_{y(Map)} \tag{1.1}$$

$$\text{RMSE}_{z(AT)} = \text{RMSE}_{x(Map)} \text{ or } \text{RMSE}_{y(Map)} \text{ of Orthoimagery} \tag{1.2}$$

Accuracy of aerial triangulation designed for elevation data, or planimetric data (orthoimagery and/or digital planimetric map) and elevation data production is specified as follows:

$$\text{RMSE}_{x(AT)}, \text{RMSE}_{y(AT)} \text{ or } \text{RMSE}_{z(AT)}$$
$$= \tfrac{1}{2} * \text{RMSE}_{x(Map)}, \text{RMSE}_{y(Map)} \text{ or } \text{RMSE}_{z(DEM)} \tag{1.3}$$

1.5.2 Accuracy Requirements for Ground Control Used for Aerial Triangulation

Accuracy of ground control designed for planimetric data (orthoimagery and/or digital planimetric map) production only is given as follows:

$$\text{RMSE}_{x} \text{ or } \text{RMSE}_{y} = 1/4 * \text{RMSE}_{x(Map)} \text{ or } \text{RMSE}_{y(Map)} \tag{1.4}$$

$$\text{RMSE}_{z} = 1/2 * \text{RMSE}_{x(Map)} \text{ or } \text{RMSE}_{y(Map)} \tag{1.5}$$

Accuracy of ground control designed for elevation data, or planimetric data and elevation data production is given as follows:

$$\text{RMSE}_{x}, \text{RMSE}_{y} \text{ or } \text{RMSE}_{z} = 1/4 * \text{RMSE}_{x(Map)}, \text{RMSE}_{y(Map)} \text{ or } \text{RMSE}_{y(DEM)} \tag{1.6}$$

According to the above requirements, for a UAS mission intended to produce an orthorectified mosaic and digital surface model with horizontal and vertical accuracy of RMSE = 4-cm, the following accuracy figures need to be specified for the project:

Ground Control Horizontal Accuracy ($RMSE_{x,y}$) = 1-cm

Ground Control Vertical Accuracy ($RMSE_z$) = 1-cm

Aerial Triangulation Horizontal Accuracy ($RMSE_{x,y}$) = 2-cm

Aerial Triangulation Vertical Accuracy ($RMSE_z$) = 2-cm

As for the check points required to verify the products' accuracy, the new standards call for an independent source of higher accuracy for check points that is at least three times more accurate than the required accuracy of the geospatial data set being tested. Therefore, the accuracy of the check points for the previous mission is:

Check Points Horizontal Accuracy ($RMSE_{x,y}$) = 1.33-cm

Check Points Vertical Accuracy ($RMSE_z$) = 1.33-cm

1.6 UAS Project Data Accuracy Examples

The positional accuracy achieved by sUAS data depends on many factors. Image resolution, camera shutter type, and the terrain texture in the scene are significant factors (Harwin and Lucieer 2012). Errors in measurements and systematic errors propagate through to the block adjustment final results affecting scaling, rotation, and translation. In addition, investigations have shown systematic doming deformations between the GCPs (Carbonneau and Dietrich 2016; Tonkin and Midgley 2016). Accuracy is highly dependent on the amount and distribution of the GCPs used (Nasrullah 2016). On the other hand, a low number of GCPs without accurate initial camera positions results in a decreased accuracy by about a factor of six (Nasrullah 2016). Time synchronization problems between the time of exposure and the event marker in the GNSS file can lead to large deviations in camera position (3 cm) which will have noticeable impacts on the overall block accuracy (Eling et al. 2015).

To assess the capability of the SenseFly eBee fixed-wing UAS with a global shutter camera, a 320 m × 320 m control field was established (Munjy and Lopez 2018). The control points were placed in an approximately 40 m grid pattern totaling 81 control points. The terrain of the area is rolling hills with sparse vegetation, structures, and roads. The points were surveyed to 1-cm horizontal and 0.3-cm vertical accuracy both at one sigma confidence level. The horizontal positioning was established using static GPS while differential levelling was used to establish the vertical positioning of the ground control points. The UAS flight produced 293 images (18 strips) with 70% forward lap and 80% side lap. The average Ground Sampling Distance (GSD) was 2.45 cm and the average flying height was 114 m AGL. The dataset was processed with Pix4D with self-calibration parameters (f, xp, yp, K1, K2, K3, P1, and P2). The GPS image positions included in the adjustment had an average a priori sigma of 2-cm H, 2.5-cm V. For images with large external orientation residuals greater than 3-sigma, their a priori sigma values were modified, and the data were reprocessed to eliminate constraints

TABLE 1.2

Accuracy of UAS-Based Adjustments

# of check points	# of GCP	RMSE (cm)				Avg (cm)			
		X	Y	Z	XY	X	Y	Z	XY
80	0	1.25	1.47	8.11	1.37	0.72	1.17	−7.98	1.76
75	5	1.04	0.92	1.30	0.98	−0.33	0.14	0.48	1.23
73	7	0.90	0.98	1.24	0.94	−0.15	0.23	0.58	1.18
71	9	0.94	0.98	1.17	0.96	−0.11	0.13	0.35	1.22

on the aero-triangulation solution. Results with different control schemes are shown in Table 1.2.

With no ground control points, the adjustment showed a larger z-bias due to the fact that Pix4D has no option to input the GPS antenna offset. Results show that the adjustment stabilizes with five control points with RMSE-XY = 0.92 cm and RMSE-Z = 1.3 cm.

Many UAS camera systems are non-metric cameras and most have a CMOS sensor with an electronic rolling shutter. Each scanline of the acquired image with a rolling shutter is exposed at a different time, which induces distortion when the camera is in motion. Pix4D and Metashape have the option to mathematically model this distortion and to analyze the rolling shutter effect. These software packages are also very effective in removing this systematic error against a control field with many check points.

Images were acquired using the "Ricoh GR Digital 3" camera mounted on a Trimble Gatewing X100 fixed-wing UAS with a single frequency (L1) GPS receiver. An average GSD of 5 cm was achieved from 160-m Above Ground Level (AGL) and images were captured with 80% forward overlap and 60% side overlap.

When rolling shutter is off, obvious systematic errors in the results are clearly shown in Figure 1.3a (planimetric errors for rolling shutter model off) and Figure 1.3b (vertical error contour map for rolling shutter model off) even when using 15 control points. The vertical error shows a doming effect with peaks and depressions as displayed by the contour map in Figure 1.3b.

Using the rolling shutter model will minimize the planimetric and vertical errors as shown in Figure 1.4a (planimetric errors for rolling shutter model on) and Figure 1.4b (vertical errors contour map for rolling shutter model on). A summary of the two solutions is shown in Table 1.3. It is worth noting that the average errors when the rolling shutter is off are not large. This is due to the fact that errors are canceling each other as shown in Figure 1.3a.

Finally, UAS operators need to understand that utilizing UAS for mapping practices requires a thorough knowledge and appreciation of the photogrammetric process and the factors effecting such processes. Factors such as camera quality and stability, the sophistication of the mathematical modeling utilized within the image processing software, quality, number, and distribution of the ground control points, and the quality of the GPS-based camera positions play a significant role in determining the derived product quality and accuracy. In order to understand the fundamental knowledge for effective sUAS mapping, the professional communities of photogrammetrists and mapping scientists are a good resource to seek guidance and advice. Getting involved with The American Society of Photogrammetry and Remote Sensing (ASPRS) and its certification programs is a great place to start building the foundational knowledge, to conduct sustained, safe, and effective sUAS operations.

FIGURE 1.3
a, b Positional errors with rolling shutter model on and rolling shutter model off.

FIGURE 1.4
a, b Planimetric errors for rolling shutter model on and vertical errors contour map for rolling shutter model on.

TABLE 1.3

Rolling Shutter off

15gcp	cm	15gcp	cm
CP	82	CP	82
GCP	15	GCP	15
RMSEx	10.58	RMSEx	2.33
RMSEy	9.13	RMSEy	2.4
RMSEz	6.33	RMSEz	3.44
Std_dx	10.56	Std_dx	2.33
Std_dy	9.15	Std_dy	2.4
Std_dz	6.37	Std_dz	3.41
absavg_dx	9.17	absavg_dx	1.92
absavg_dy	7.87	absavg_dy	1.75
absavg_dz	4.9	absavg_dz	2.75
avg_dx	1.37	avg_dx	0.27
avg_dy	0.79	avg_dy	0.19
avg_dz	−0.12	avg_dz	0.6
maxdx	18.71	maxdx	5.27
maxdy	16.46	maxdy	5.1
maxdz	12.77	maxdz	8.44
mindx	−20.21	mindx	−5.23
mindy	−19.29	mindy	−9.25
mindz	−16.98	mindz	−6.74

References

Agüera-Vega, F., Carvajal-Ramírez, F., and Martínez-Carricondo, P. (2017). Assessment of Photogrammetric Mapping Accuracy Based on Variation Ground Control Points Number Using Unmanned Aerial Vehicle. *Measurement* 98, 221–227.

Anders, N., Masselink, R., Keesstra, S., and Suomalainen, J. (2013). High-Res Digital Surface Modeling using Fixed-Wing UAS-Based Photogrammetry. *Geomorphometry*, O.2.1–O.2.4. Retrieved from http://geomorphometry.org/system/files/Anders2013geomorphometry.pdf

Benassi, F., Dall'Asta, E., Diotri, F., Forlani, G., Morra di Cella, U., Roncella, R., and Santise, M. (2017). Testing Accuracy and Repeatability of UAS Blocks Oriented with GNSS-Supported Aerial Triangulation. *Remote Sensing* 9, 172–194. doi:10.3390/rs9020172

Bendea, H., Chiabrando, F., Tonolo, F.G., and Marenchino, D. (2007). Mapping of Archaeological Areas using a Low-Cost UAS the Augusta Bagiennorum Test Site. Proceedings from XXI International CIPA Symposium. Athens, Greece.

Brutto, M.L., Garraffa, A., and Meli, P. (2014). UAS Platforms for Cultural Heritage Survey: First Results. *ISPRS-Annals of the Photogrammetry, Remote Sensing and Spatial Information Sciences* II-5, 227. doi:10.5194/isprsannals-II-5-227-2014.173

Carbonneau, P.E., and Dietrich, J.T. (2016). Cost-Effective Non-metric Photogrammetry from Consumer-Grade sUAS: Implications for Direct Georeferencing of Structure from Motion Photogrammetry. *Earth Surface Processes and Landforms* 42, 473–486. doi:10.1002/esp.4012

Chiabrando, F., Lingua, A., and Piras, M. (2013). Direct Photogrammetry Using UAS: Tests and First Results. *ISPRS-International Archives of the Photogrammetry, Remote Sensing and Spatial Information Sciences* XL-1/W2, 81–86. doi:10.5194/isprsarchives-XL-1-W2-81-2013

Chiabrando, F., Lingua, A., Maschio, P., and Losé, L.T. (2017). The Influence of Flight Planning and Camera Orientation in UAS's Photogrammetry. A Test in the Area of Rocca San Silvestro (LI), Tuscany. *ISPRS-International Archives of the Photogrammetry, Remote Sensing and Spatial Information Sciences* XLII-2/W3, 163–170.

Cramer, M. (2013). The UAS@LGL BW Project-A NMCA Case Study. Proceedings from Photogrammetric Week 2013, 165–179. Retrieved from www.ifp.uni-stuttgart.de/publications/phowo13/150Cramer.pdf

Daakir, M., Pierrot-Deseilligny, M., Bosser, P., Pichard, F., and Thom, C. (2015). UAS Onboard Photogrammetry and GPS Positioning for Earthworks. *ISPRS-International Archives of the Photogrammetry, Remote Sensing and Spatial Information Science* XL-3/W3, 293–298. doi:10.5194/isprsarchives-XL-3-W3-293-2015

Eisenbeiss, H., and Sauerbier, M. (2011). Investigation of UAV Systems and Flight Modes for Photogrammetric Application. *The Photogrammetric Record* 26(136), 400–421. Retrieved from https://onlinelibrary.wiley.com/toc/14779730/2011/26/136

El Meouche, R., Hijazi, I., Poncet, P., Abunemeh, M., and Rezoug, M. (2016). UAS Photogrammetry Implementation to Enhance Land Surveying, Comparisons and Possibilities. *ISPRS-International Archives of the Photogrammetry, Remote Sensing and Spatial Information Sciences* XLII-2/W2, 107–114. doi:10.5194/isprs-archives-XLII-2-W2-107-2016

Eling, C., Wieland, M., Hess, C., Klingbeil, L., and Kuhlman, H. (2015). Development and Evaluation of a UAS Based Mapping System for Remote Sensing and Surveying Applications. *ISPRS-International Archives of the Photogrammetry, Remote Sensing and Spatial Information Sciences* XL-1/W4 30, 233–239. doi:10.5194/isprsarchives-XL-1-W4-233-2015

Fras, M.K., Kerin, A., Mesarič, M., Peterman, V., and Grigillo, D. (2016). Assessment of the Quality of Digital Terrain Model Produced from Unmanned Aerial System Imagery. *ISPRS-International Archives of the Photogrammetry, Remote Sensing and Spatial Information Sciences* XLI-B1, 893–899. doi:10.5194/isprsarchives-XLI-B1-893-2016

Georgopoulos, A., Oikonomou, C., Adamopoulos, E., and Stathopoulou, E.K. (2016). Evaluating Unmanned Aerial Platforms for Cultural Heritage Large Scale Mapping. *ISPRS-International Archives of the Photogrammetry, Remote Sensing and Spatial Information Sciences* XLI-B5, 355–362. doi:10.5194/isprsarchives-XLI-B5-355-2016

Gerke, M., and Przybilla, H.-J. (2016). Accuracy Analysis of Photogrammetric UAS Image Blocks: Influence of Onboard RTK-GNSS and Cross Flight Patterns. *Photogrammetrie-Fernerkundung-Geoinformation (PFG)*, 17–30. doi:10.1127/pfg/2016/0284

Gillan, J.K., Karl, J.W., Elaksher, A., and Duniway, M.C. (2017). Fine-Resolution Repeat Topographic Surveying of Dryland Landscapes Using UAS-Based Structure-from-Motion Photogrammetry: Assessing Accuracy and Precision against Traditional Ground-Based Erosion Measurements. *Remote Sensing* 9(5), 437. doi:10.3390/rs9050437

Harwin, S., and Lucieer, A. (2012). Assessing the Accuracy of Georeferenced Point Clouds Produced via Multi-View Stereopsis from Unmanned Aerial Vehicle (UAV) Imagery. *Remote Sensing* 4(6), 1573–1599. doi:10.3390/rs4061573

Hoffman, C., Weise, C., Koch, T., and Pauly, K. (2016). From UAS Data Acquisition to Actionable Information - How an end-to-End Solution Helps Oil Palm Plantation Operators to Perform a More Sustainable Plantation Management. *ISPRS-International Archives of the Photogrammetry, Remote Sensing and Spatial Information Sciences* XLI-B1, 1113–1120. doi:10.5194/isprs-archives-XLI-B1-1113-2016

James, M.R., Robson, S., d'Oleire-Oltmanns, S., and Niethammer, U. (2017). Optimising UAS Topographic Surveys Processed with Structure-from-Motion: Ground Control Quality, Quantity and Bundle Adjustment. *Geomorphology* 280, 51–66. doi:10.1016/j.geomorph.2016.11.021

Mesas-Carrascosa, F.J., Rumbao, I.C., Berrocal, J.A.B., and García-Ferrer Porras, A. (2014). Positional Quality Assessment of Orthophotos Obtained from Sensors Onboard Multi-Rotor UAS Platforms. *Sensors* 14, 22394–22407. doi:10.3390/s141222394

Munjy, R., and Lopez, J. (2018). Rolling Shutter Effect Important in UAV Photogrammetry. Retrieved from https://lidarnews.com/articles/rolling-shutter-effect-uav-photogrammetry/

Nasrullah, A.R. (2016). Systematic Analysis of Unmanned Aerial Vehicle (UAS) Derived Product Quality. Unpublished master's thesis, University of Twente, Enschede, The Netherlands.

Rehak, M., Mabillard, R., and Skaloud, J. (2013). A Micro-UAS with the Capability of Direct Georeferencing. *ISPRS-International Archives of the Photogrammetry, Remote Sensing and Spatial Information Sciences* XL-1/W2, 317–323. doi:10.5194/isprsarchives-XL-1-W2-317-2013

Reshetyuk, Y., and Mårtensson, S.-G. (2016). Generation of Highly Accurate Digital Elevation Models with Unmanned Aerial Vehicles. *The Photogrammetric Record* 31(154), 143–165. doi:10.1111/phor.12143

Rosnell, T., and Honkavaara, E. (2012). Point Cloud Generation from Aerial Image Data Acquired by a Quadrocopter Type Micro Unmanned Aerial Vehicle and Digital Still Camera. *Sensors* 12, 453–480. doi:10.3390/s120100453

Rumpler, M., Tscharf, A., Mostegel, C., Daftry, S., Hoppe, C., Prettenthaler, R., Fraundorfer, F., Mayer, G., and Bischof, H. (2017). Evaluations on Multi-scale Camera Networks for Precise and Geo-accurate Reconstructions from Aerial and Terrestrial Images with User Guidance. *Computer Vision and Image Understanding* 157, 255–273. doi:10.1016/j.cviu.2016.04.008

Skaloud, J., Rehak, M., and Lichti, D. (2014). Mapping with MAV: Experimental Study on the Contribution of Absolute and Relative Aerial Position Control. *ISPRS-International Archives of the Photogrammetry, Remote Sensing and Spatial Information Sciences* XL-3/W1, 123–129. doi:10.5194/isprsarchives-XL-3-W1-123-2014

Tonkin, T.N., and Midgley, N.G. (2016). Ground-Control Networks for Image Based Surface Reconstruction: An Investigation of Optimum Survey Designs Using UAS Derived Imagery and Structure-from-Motion Photogrammetry. *Remote Sensing* 8(9), 786. doi:10.3390/rs8090786

Vautherin, J., Rutishauser, S., Schneider-Zapp, K., Choi, H.F., Chovancova, V., Glass, A., and Strecha, C. (2016). Photogrammetric Accuracy and Modeling of Rolling Shutter Cameras. *ISPRS-Annals of the Photogrammetry, Remote Sensing and Spatial Information Sciences* III-2, 139–146. doi:10.5194/isprs-annals-III-3-139-2016

Yanagi, H., and Chikatsu, H. (2016). Performance Evaluation of 3D Modeling Software for UAV Photogrammetry. *The International Archives of the Photogrammetry, Remote Sensing and Spatial Information Sciences* XLI-B5, 147–152. Retrieved from https://onlinelibrary.wiley.com/toc/14779730/2011/26/136

Yusoff, A.R., Ariff, M.F.M., Idris, K.M., Majid, Z., and Chong, A.K. (2017). Camera Calibration Accuracy at Different UAS Flying Heights. *ISPRS-International Archives of the Photogrammetry, Remote Sensing and Spatial Information Sciences* XLII-2/W3, 595–600. doi:10.5194/isprs-archives-XLII-2-W3-595-2017

Zajkowski, T., and Snyder, K. (2016). Unmanned Aircraft Systems: A New Tool for DOT Inspections: Final Report. State of North Carolina Department of Transportation Research and Development. Institute for Transportation Research and Education, North Carolina State University, NCDOT RP2015-06-Final Report. Retrieved from https://connect.ncdot.gov/resources/Pages/Aviation-Division-Resources.aspx

2

Unmanned Aerial Systems (UAS) and Thematic Map Accuracy Assessment

Russell G. Congalton and Benjamin T. Fraser

CONTENTS

2.1 Introduction

The use of remotely sensed imagery has become ubiquitous and provides the leading source of land cover information for mapping and monitoring the Earth. Imagery is now readily available in many spatial, spectral, and temporal resolutions offering great adaptability and efficiency for meeting our ever-increasing information needs (Turner, 2005; Radoux et al., 2011; MacLean and Congalton, 2013; Whitehead and Hugenholtz, 2014). The process of turning imagery into information is called classification and results in a thematic map (Bolstad, 2016; Jensen, 2016). This process can occur manually through photo or image interpretation or through computer processing of digital imagery (Lillesand et al., 2015). Traditionally, computer image processing was performed on a per-pixel basis with each pixel being labeled individually. More recently, an object-based approach that combines pixels into meaningful groupings (called objects) and then labels the objects has been developed (Blaschke, 2010). Regardless of the thematic mapping method used, these maps include general land cover and land use maps, forest-type maps, agricultural crop maps, as well as maps showing change over time as a result of natural and catastrophic events.

No map, no matter how it is created, is very useful without some measure of the map quality or accuracy (Anderson et al., 1976, Congalton, 1991). Validation of any map is a necessary step in using that map, and the information gleaned from it, in any decision-making process. Thematic map accuracy assessment is performed by comparing selected sample areas on the map with the same areas from some reference data source (Congalton and Green, 2019). The reference data source is assumed to be of substantially higher accuracy than the map (i.e., correct) and can be generated from field visitation, existing data, and/ or higher spatial resolution imagery. The comparison of the map to the reference data is represented in the form of an error matrix (i.e., contingency table) allowing for the computation of accuracy metrics (Congalton, 1991). These metrics include overall accuracy, producer's accuracy, and user's accuracy (Story and Congalton, 1986). The compilation of a valid accuracy assessment is an expensive component of any mapping project and the most time-consuming step in this process is the collection of the reference data.

Therefore, any efficiencies that can be gained in the reference data collection process results in a more effective and timely assessment of the thematic map. Recent developments in the miniaturization of various components of image acquisition systems including the sensor, the Global Positioning System (GPS), and the Inertial Measurement Unit (IMU), have resulted in low-cost, lightweight, and incredibly flexible Unmanned Aerial Systems (UAS). UAS are an interrelated combination of hardware and software technologies managed by a remote pilot that can quickly, effectively, and efficiently collect very high spatial and temporal resolution imagery (Colomina and Molina, 2014; Fraser and Congalton, 2019). UAS offer the potential to collect imagery that can be used as appropriate reference data for thematic maps generated from imagery of coarser spatial resolution.

Therefore, the goal of this chapter is to provide an overview of the thematic map accuracy assessment process and then a discussion of how the use of UAS can fit into this process. A pilot study of some preliminary work done using UAS to collect reference data in New England forests is included to demonstrate the feasibility of this concept. The chapter ends with a summary and conclusions for future work.

2.2 An Overview of Thematic Map Accuracy Assessment

Map accuracy assessment can be divided into two parts: positional accuracy and thematic accuracy. This chapter is about thematic accuracy and how UAS can be used to collect the required reference data necessary for conducting a thematic accuracy assessment. However, it is still important to consider positional accuracy as well. Positional accuracy assesses whether the location on the map is in the correct place while thematic accuracy evaluates whether that location is labeled correctly. Both are important and are highly related to each other. If you are in the wrong location, but label it correctly, you have made an error. Conversely, if you are in the right location, but label it wrong, you have also made an error. Therefore, failure to account for positional error when conducting a thematic accuracy assessment results in positional errors being represented as thematic errors.

Whether assessing positional or thematic map accuracy, the process requires considering a number of factors and making appropriate decisions in order to be as efficient and effective as possible. The flow chart in Figure 2.1 shows the major steps taken to conduct a thematic map accuracy assessment. Associated with each of these steps are the considerations and decisions that must be factored into the process. These steps include some initial considerations, the reference data collection, the error matrix generation, and the metrics/analysis of the results.

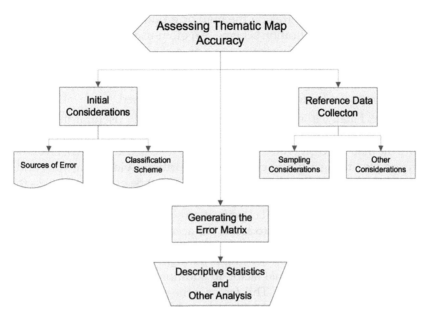

FIGURE 2.1
A flowchart of the key steps in assessing thematic map accuracy.

2.2.1 Initial Considerations

There are two initial considerations when conducting a thematic accuracy assessment. These are (1) sources of error and (2) classification scheme. The impetus for using remotely sensed imagery for making maps is that there is a very strong relationship between what is occurring on the ground and what is visible in the imagery. Any source of error reduces this correlation and therefore, reduces the accuracy of the thematic map. A number of researchers have suggested methodologies for evaluating these error sources including Lunetta et al. (1991), Congalton and Brennan (1999), and Congalton (2009). Careful consideration of the sources of error in the use of the remotely sensed imagery for mapping and thoughtful planning to minimize them is key to producing the best thematic map possible.

It is important when creating any thematic map to carefully consider the classification scheme selected; that is, the map classes to be used. The choice of the scheme determines the type of imagery needed to create the map as it dictates the required level of detail. It is also critical that the same classification scheme be used to collect the reference data as was used to make the map or else needless errors will be introduced into the process. Every classification scheme should have the following four characteristics:

(1.) *Rules/Definitions*: A scheme must have rules that explicitly distinguish each of the map classes. A simple list of the map classes is not sufficient as it promotes ambiguity. Everyone must be able to use the classification scheme rules to arrive at the exact same map class label.

(2.) *Mutually Exclusive*: The class rules/definitions must be complete and clear so as to eliminate any overlap or confusion between classes so that every area on the map or ground falls into one and only one class in the scheme.

(3.) *Totally Exhaustive*: Every area on the map or ground must be included in one of the classes in the classification scheme. An often-used way of insuring that your scheme is totally exhaustive is to have a map class called "Other." This "Other" class should only be a very small portion of the thematic map. If the "Other" class becomes a large portion of the map, then the classification scheme is missing some important information and should be revised.

(4.) *Hierarchical*: Finally, it is very common and extremely useful if the classification scheme is hierarchical (i.e., having multiple levels of detail). A hierarchical scheme allows the map to be generated and especially assessed at different levels of detail. For example, the agriculture class may be further divided into corn and wheat classes that add more detail. Sometimes, it may not be possible to assess the accuracy of the map at the most detailed level because of costs of collecting the reference data. Using a hierarchical classification scheme, it is then possible to map to one level of the hierarchy but to assess the accuracy to a lesser level of detail.

Because the choice of classification scheme is so highly influenced by the level of detail required in the thematic map, a final factor, in addition to the four discussed in the previous paragraph, must be considered. This factor is the minimum mapping unit (mmu). The mmu is the smallest area that will be identified/delineated on the thematic map. Once selected, any area smaller than the mmu is absorbed into a surrounding map class while any area the size of the mmu or larger is identified and delineated on the map. The use of an mmu is a standard component of any photo interpretation; however, the concept has unfortunately not been universally adopted for digital imagery. The idea of the smallest identifiable area becomes especially important when selecting the appropriate sample unit which will be discussed more later in this chapter.

2.2.2 Reference Data Collection

There are many factors to consider when collecting the reference data used in assessing thematic map accuracy. These factors can be divided into two general categories: other (non-sampling) considerations and sampling considerations. Each of these factors must be carefully considered for the assessment to be as efficient and as effective as possible. It is often necessary to accept trade-offs between certain factors and therefore, it is vital that a full and complete reporting of all the factors considered and decisions made be part of the accuracy assessment report.

There are four other (non-sampling) considerations when collecting thematic map reference data. These are: data independence, data source, timing of data collection, and collection consistency.

(1.) *Data Independence*: It is critical that any reference data used in the thematic map accuracy assessment be *independent* of any other data used in the mapping project. This basic concept is well understood by the geospatial analysis community. For example, using the same data to both train the classifier and then test the results has not commonly occurred since very early on in the digital imagery age. Data independence with regard to reference data collection can be achieved in one of two ways. First, the reference data collection can be performed at a different time and/or by different personnel from the training data collection. This method is effective, but inefficient as it involves either a second collection effort (e.g., field visit) or a second collection crew. An alternative approach is to collect the training and reference data simultaneously and then split the data into two sets. In this case, the reference data set used for the assessment is set aside and hidden from any project personnel until it is time to conduct the accuracy assessment.

(2.) *Data Source*: The source of thematic accuracy reference data can vary and typically depends on the project budget and the complexity of the classification scheme (number and detail of the map classes). Ideally, but rarely, some existing reference data can be used. These data would have to have a similar classification scheme (or at least be able to crosswalk between schemes), have a sufficient size sample unit, and be of similar age (not too old) to be useful. These conditions are almost never met and therefore, most reference data are collected as part of the project. The most common sources of new reference data are manual image interpretation and ground/field visits. Again, whether to use image interpretation or field visits is highly dependent on the classification scheme and the budget. Given today's high spatial resolution imagery, manual interpretation is often a feasible solution. If such a method is selected, some sub-sample of the data should be compared to the ground to confirm that the interpretation is accurate. However, for more detailed map classes, there is no substitute for field visits. Sometimes, simple observations can be made in the field while some map classes may require actual measurements to accurately generate the reference data (Congalton and Green, 2019).

(3.) *Timing of Data Collection*: The timing of the reference data collection is also highly dependent on the thematic map (i.e., classification scheme) being assessed. Agricultural crops can change rapidly, especially in areas where multiple crops are grown in a single year. Therefore, in these cases, the reference data must be collected as near as possible to the date of the imagery used to make the map to minimize temporal errors in the reference data. Other map classes do not require such timely collections and a year or two or even five may have no effect on the accuracy of the reference data. It is important to note that if some catastrophic change (e.g., fire, flood, earthquake, hurricane, etc.) has occurred in the area between the time of the mapping and the time of the reference data collection, then these reference data will not be appropriate for assessing the accuracy of that map.

(4.) *Consistency*: Finally, consistent collection of the reference data is very important. Even if a single individual is collecting all the data, it is important that the definitions/rules developed for the classification scheme be closely followed. If a group of collectors are used, consistency becomes even more of an important consideration to reduce variation between the collectors. One effective way to impose objectivity in the reference data collection is through the use of a field form. Depending on the project, this form can be a simple sheet of paper. It could also be a digital form on a laptop, tablet, or smartphone that aids the collector in gathering the information needed to label the sample unit. Whether using a simple paper or a detailed digital form, there are some common components that must be included. These include the names of the collectors, the date of the collection, the locational (GPS) information, a key to the map classes (dichotomous keys are excellent), the actual map class determined by the collection, and a place to describe any issues, special findings, or problems that occurred at the collection site. Every effort must be taken to ensure that all reference data are collected as consistently as possible. If multiple collectors are used, it is suggested that some sites be visited/interpreted by two separate collectors to evaluate consistency.

There are three sampling considerations when collecting thematic map reference data. These are: sampling scheme, sample size, and sample unit.

(1.) *Sampling Scheme*: The sampling scheme provides the method by which the reference data samples are collected. There are many choices including simple random sampling, systematic sampling, stratified (random) sampling, cluster sampling, and others (Congalton and Green, 2019). Random sampling is often suggested because of its powerful statistical properties. However, it is often impractical to implement due to the costs (in money, time, and logistics) of travel to each random sample unless one is using image interpretation to collect the reference data. In addition, simple random sampling tends to oversample the larger area map classes while leaving the rarer (but perhaps most important) map classes under-sampled. Stratifying the map by map class and then collecting a minimum number of samples in each stratum can help solve this problem. Likewise, systematic sampling ensures that samples are distributed over the entire map but has similar cost issues in obtaining samples from all locations. In some cases, cluster sampling is used to increase collection efficiency by taking a number of samples in close proximity to each other. However, any time samples are taken close to each other, spatial autocorrelation must be considered. Spatial autocorrelation is said to occur when the presence, absence, or degree of a certain characteristic affects the presence, absence, or degree of that same characteristic in neighboring units (Cliff and Ord, 1973). In the thematic map accuracy case, the characteristic of interest is error and the question to consider is whether making an error in one place has a positive or negative impact on the surrounding neighbors. Congalton (1988a) and Pugh and Congalton (2001) have shown that spatial autocorrelation is a key factor in reference data collection and therefore, it is important to maximize the separation between samples in order to minimize this impact. Therefore, care must be taken when deciding on an appropriate sampling scheme. There is always a trade-off between what can practically be achieved (costs) and statistical validity, especially when the reference data are collected in the field. It is often not possible to have a completely random sample because of costs, time, and access issues. However, it is also not valid to just drive the major highway and observe out the window. In most field data collections, samples are selected in some random fashion along viable access routes and within a certain distance of that route. Effort is made to distribute the samples by stratifying the map by class and taking some minimum number of samples in each class. In some situations, map classes that cannot be collected in this way are augmented from high-resolution imagery.

(2.) *Sample Size*: In most accuracy assessments, the biggest challenge is to collect a sufficient number of samples to have a statistically valid reference data set from which to generate an appropriate error matrix. Many researchers have published equations and guidelines for choosing the appropriate sample size (Congalton and Green, 2019). These equations either dealt with the binomial equation to determine the number of samples to determine overall accuracy or with the multinomial equation (Tortora, 1978) to determine the number of samples needed to generate an error matrix. Congalton (1988b) conducted many Monte Carlo sampling simulations varying the map complexity and the sample size which yielded a general guideline of 50 samples per map class. Any method that can be used to collect the necessary number of samples more efficiently and effectively will greatly benefit the assessment process.

(3.) *Sample Unit*: The sample unit is the area used to collect the reference data. Often, a single pixel has been used as the sample unit to assess thematic map accuracy. However, a single pixel is not an appropriate sample unit in most cases because a pixel is arbitrarily defined on the imagery, difficult to locate on the imagery and the ground (i.e., register), and is typically smaller than the minimum mapping unit for the project. It is for this reason that positional accuracy becomes an important factor in thematic accuracy assessment. Again, remember that a sample unit located in the wrong place but labeled correctly will still be considered a thematic error in the thematic accuracy assessment. Despite many improvements in Global Positioning Systems (GPS), terrain correction algorithms, and geometric registration of the imagery to the ground, all reference data sample units contain positional error. Therefore, a single pixel as a sampling unit cannot be reliably identified on the image or the ground and should not be used as the sample unit. Medium-resolution sensors such as Landsat consider positional error of a half a pixel (15 m) to be satisfactory. Higher spatial resolution sensors have smaller pixels but positional errors still in the range of 5–15 m depending if the image was acquired in nadir-view or at some angle (many high spatial resolution images are collected off-nadir). GPS units that are typically used in the field also have positional error. Combining these positional errors ensures that a single pixel should not be used as the sample unit. Instead, some grouping of pixels considered together as a single unit such as either a cluster or a polygon should be used as the sample unit. If the goal is to have the smallest sample unit possible, then a cluster of pixels together counting as a single sample should be used. For example, given the positional accuracy of Landsat imagery with 30-m pixels and the positional accuracy of GPS, a homogeneous cluster of 3×3-Landsat pixels could be used together as a single sample unit. Using the GPS to locate the center of this homogeneous cluster and considering image registration and GPS error, the reference data collection would still occur within this homogeneous 3×3 (90 m $\times$ 90 m) area and therefore correctly label the sample unit. The same logic applies when selecting a cluster of pixels for high-resolution imagery. The positional error of the image registration is combined with the GPS error to determine the number of pixels that must be combined as a square, homogeneous cluster to ensure that the sample unit is appropriately labeled. With the advent of object-based image analysis (OBIA), imagery is classified in segments or objects (polygons) instead of by individual pixels. Therefore, the appropriate sample unit to assess an OBIA classification is a polygon.

2.2.3 Generating the Error Matrix

Once the initial considerations have been deliberated and the reference data collected, it is time to generate the error matrix. An error matrix is a square array of numbers set out in rows and columns which expresses the number of sample units assigned to a particular map class (i.e., land cover) in one classification as compared to the number of sample units assigned to a particular map class (i.e., land cover) in another classification (Figure 2.2). In an accuracy assessment, one of the classifications is considered to be correct (or at least of significantly higher accuracy than the other) and is called the reference data. Unfortunately, the term "ground truth" has often been used in place of reference data. While the reference data should have high accuracy, there is no guarantee that it is 100% correct and using the

Reference Data

	F	OV	D	W
F	57	11	6	3
OV	9	51	5	8
D	0	8	53	9
W	4	7	3	60

Map

Land Cover Categories

F = Forest

OV = Other Vegetation

D = Developed

W = Water

FIGURE 2.2
An example error matrix.

term "ground truth" is inappropriate. The authors hope that the geospatial community will abandon this term for more appropriate ones such as reference data or ground data or field data or the like.

Typically, the columns (x axis) of the error matrix represent the reference data while the rows (y axis) indicate the map derived from remotely sensed data or other geospatial data set (Figure 2.2). Some researchers switch the headings for the row and columns so the analyst must take care to notice how the axes of the matrix are labeled before any further analysis is undertaken.

2.2.4 Descriptive Statistics and Other Analysis

Once the error matrix has been generated using a valid reference data collection approach as described earlier in this chapter, the matrix is then the starting point for a number of descriptive statistics and other analysis techniques. Standard descriptive statistics that should be computed for every error matrix include overall accuracy, producer's accuracy, and user's accuracy (Story and Congalton, 1986). Figure 2.3 presents the same error matrix as in Figure 2.2, but now shows the computation of these key descriptive statistics. As can be seen in this figure, the error matrix is a very effective way to represent map accuracy because the individual accuracies of each category are easily discerned along with both the errors of inclusion (commission errors) and errors of exclusion (omission errors) present in the map. Map errors have two components (omission error and commission error). A commission error can be defined as including an area into a thematic map class when it doesn't belong to that class, while an omission error is excluding that area from the thematic map class in which it truly does belong. Therefore, every error is an omission from the correct thematic map class and a commission to a wrong thematic map class.

As seen in Figure 2.3, overall accuracy is computed by summing up the major diagonal of the error matrix (i.e., the correctly mapped sample units) and dividing by the total number of sample units in the matrix. This statistic has become a standard measure of thematic map accuracy. However, reporting just this statistic is insufficient as the complete error matrix should be reported so that any additional measures of accuracy can be computed and the errors in the map can be fully investigated.

In addition to overall accuracy, the geospatial analyst may want to know the accuracy of an individual thematic map class. Computation of individual class accuracies is a little more complicated than overall accuracy and requires two measures: producer's accuracy and user's accuracy as introduced by Story and Congalton (1986).

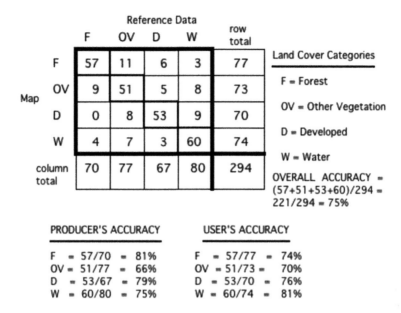

FIGURE 2.3
An error matrix demonstrating the calculations for overall, producer's, and user's accuracies.

The producer of the map may want to know how well they mapped a certain thematic map class, the producer's accuracy. This value is computed by dividing the value from the major diagonal (the agreement) for that class by the total number of samples in that map class as indicated by the sum of the reference data for that class. Looking at Figure 2.3 shows that the map producer labeled 57 areas Forest while the reference data indicates that there is a total of 70 Forest areas. Thirteen areas were omitted from the Forest and of these nine were committed to Other Vegetation and four were committed to Water. So, 57/70 samples were correctly labeled Forest for a Forest Producer's accuracy of 81%. Now consider the user's perspective of the map. Once again 57 samples were labeled Forest on the map that were actually Forest, but the map also called 11 samples Forest that were actually Other Vegetation and six samples Forest that were actually Developed and three samples Forest that were actually Water. The map, therefore, labeled 77 samples Forest, but only 57 are actually Forest. There was commission error of 20 samples into the Forest map class that were not Forest. The Forest User's accuracy is then computed by dividing the major diagonal value for the Forest class by the total number of samples mapped as Forest, 57/77 = 74%. In evaluating the accuracy of an individual map class, it is important to consider both the producer's and the user's accuracies.

While overall, producer's, and user's accuracies are now standard descriptive statistics computed for every error matrix, there are also a large number of other analysis techniques that have been suggested to further investigate the error matrix (Congalton and Green, 2019). Liu et al. (2007) performed a useful comparative analysis of 34 different measures that can be computed from the error matrix. While each of these measures provides information about some aspect of the map accuracy, it is really up to the geospatial analyst and/or the map user to determine which, if any, of these measures are useful in their situation. There has certainly been no agreement on standard techniques beyond overall, producer's, and user's accuracies.

2.3 Collecting Reference Data Using an Unmanned Aerial System (UAS)

Given what we now understand about thematic accuracy assessment from the review in this chapter, it is easy to see how the use of UAS fits effectively into the assessment process. It is clear that the most expensive, difficult, and time-consuming component of thematic map accuracy assessment is the collection of appropriate and sufficient reference data needed to perform the assessment. Therefore, the rest of this chapter will provide an overview of how UAS can be used to collect reference data concluding with a pilot study where such use has been applied in a forested environment (Fraser and Congalton, 2019).

2.3.1 Unmanned Aerial Systems (UAS)

A UAS is an interconnected system of hardware and software technologies controlled by a remote pilot in command (Colomina and Molina, 2014; Marshall et al., 2016; Fraser and Congalton, 2018). These systems offer users the ability to capture large amounts of spatial data at unprecedented scales and in a timely manner due to their flexible designs, relatively low costs, automated workflows, and minimal technical knowledge barrier. In addition, new photogrammetric modeling techniques allow processing of this high-resolution remote sensing imagery to create heretofore unavailable end-user products easily and effectively (Smith et al., 2016). The creation of these products such as 3D point clouds and planimetric models using a process called Structure from Motion (SfM) utilizes optical imagery and redundancy (i.e., overlap) between features in the imagery to efficiently produce these outputs (Westoby et al., 2012; Fonstad et al., 2013; Mancini et al., 2013). Although the basic processes for the creation of orthographic and geometrically corrected (planimetric) surfaces using photogrammetry have been applied for more than a century (McGlone, 2013), the field of computer vision has only recently linked the SfM methodology with Multi-View Stereo (MVS) and bundle adjustment algorithms to provide an automated workflow for these products (Smith et al., 2016).

Although many varieties of UAS exist today, the majority consist of the following components: (1) a launch and recovery system or flight mechanism, (2) a sensor payload, (3) a communication data link, (4) the unmanned aircraft, (5) a command and control center, and (6) most importantly the human operator (Colomina and Molina, 2014). A more common way to describe UAS may be whether it is a rotary-wing or fixed-wing aircraft. Each of these systems provide unique benefits and limitations corresponding to the structure of their design. Rotary-wing or copter-based UAS (i.e., those with horizontally rotating propellers) are especially useful for their added maneuverability and hovering capabilities and for their ability to take-off and land vertically. However, they have the disadvantage of typically shorter flight duration (Marshall et al., 2016). Fixed-wing UAS, unlike rotary-wing systems, have longer duration flights and higher altitude thresholds which allows more area to be covered in a single flight. However, these systems cannot hover over a selected area and need a large, flat, obstruction-free area in which to take-off and especially to land. The appropriate system with its corresponding advantages and disadvantages should be selected depending on the type of reference data needed to be collected.

2.3.2 UAS Reference Data Collection

Like any other project involving the use of UAS in the United States, the collection of reference data must abide by the current Federal Aviation Administration (FAA) regulations.

As of August 29, 2017, operational permission for sUAS (small UAS) use was established through the Remote Pilot in Command (RPIC) license under 14 Code of Federal Regulations (CFR) Part 107 (FAA, 2017). A sUAS is defined as being under 25 kg (55 lbs.). In addition, the UAS must fly under an altitude restriction of 121 m (400ft), must be within visible line of sight at all times, and be registered with the FAA and visibly display the registration number on the UAS. The RPIC license requires passing an aeronautical knowledge exam and must be renewed every two years.

These operational constraints dictate how and when UAS can be flown to acquire imagery used for reference data. The total weight of the sUAS may limit the potential payloads (i.e., cameras and sensors) that the vehicle may carry. Flying at or below 121 m limits the area that can be covered given the life of the onboard batteries and also impacts the visible line of sight. All these factors must be considered, and trade-offs made to determine not only the type of UAS to use, but also exactly what type of reference data can be collected. For example, if one was interested in collecting reference data for an agricultural mapping project there are a number of questions to be considered. What is the average field size? If the field size is small, almost any UAS could fly many fields using some type of transect (systematic) sample in a single flight. If the field size was large, then a more sophisticated (perhaps, fixed-wing) system would be needed. What type of reference data are needed? If agricultural crop types are to be identified, it would be possible to use an inexpensive UAS with a typical R, G, B camera to collect the imagery. If information about crop health was needed, then a more expensive UAS with a multispectral imaging system would be required. Were the crops of interest planted at the same time? If yes, it may be possible to fly once and collect the information needed. If not, regularly scheduled, multi-temporal image collection may be required. It is important to ask the right questions so that all these factors can be considered, and the best system chosen. No matter what the situation, the goal of using UAS imagery for reference data collection is to obtain accurate information about the ground more efficiently and effectively than using field-based methods.

2.3.3 Example Case Study

A pilot study was conducted that investigated and evaluated if UAS could provide thematic map accuracy assessment reference data of necessary quality and operational efficiency to use for assessing forest maps created from lower spatial resolution imagery in New Hampshire (Fraser and Congalton, 2019). This study evaluated the agreement between the UAS-collected samples and what would traditionally be collected on the ground (in the field) as reference data. In this case, the ground-based data were Continuous Forest Inventory (CFI) data for plot classifications of deciduous, mixed, and coniferous forest cover types of 377.57 ha of New Hampshire forests, over six University of New Hampshire woodland properties (Figure 2.4). This pilot study specifically investigated the collection of UAS reference data for assessing the accuracy of these forest thematic maps created from either a (1) pixel-based or (2) object-based classification approach.

To evaluate the quality of the UAS-collected reference data, a comparison was performed with existing ground reference data provided by a systematic network of CFI ground sampled plots, collected on approximately ten-year intervals, for each of the study areas. These plots are located at one plot per acre and collect information about the forest including tree species, tree diameter, and basal area (Fraser and Congalton, 2019) (Figure 2.5).

The CFI information was used to characterize the study areas into forest stands of three types: coniferous forest, mixed forest, or deciduous forest (Figure 2.6). UAS imagery was collected using an eBee Plus (SenseFly Parrot Group), fixed-wing platform, and the Sensor

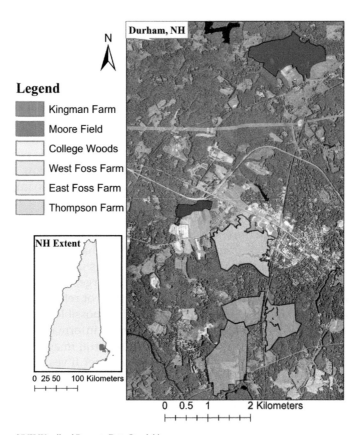

UNH Woodland Property Data Overlaid
on 2016 NAIP Imagery from NH GRANIT.

FIGURE 2.4
A map showing the six University of New Hampshire study areas used in this pilot analysis. From North to South (with total area): Kingman Farm (135.17 ha), Moore Field (47.76 ha), College Woods (101.17 ha), West Foss Farm (52.27 ha), East Foss Farm (62.32 ha), and Thompson Farm (118.17 ha).

Optimized for Drone Applications (SODA), a 20 mega-pixel, natural color, proprietary sensor designed for photogrammetric analysis. All the imagery was processed using Agisoft PhotoScan 1.3.2 for high-accuracy photo alignment, image tie point calibration, medium-dense point cloud formation, and planimetric model processing workflow. Mission planning was designed to capture both plot- and stand-level forest composition. The maximum allowable flying height of 120 m above the forest canopy with an 85% forward overlap and 75% side overlap was used for all photo missions. Additional factors such as optimal sun angle (e.g., around solar noon), perpendicular wind directions, and consistent cloud coverage were considered during each mission to obtain the best image quality and precision (Fraser and Congalton, 2019).

As previously mentioned, the goal of this pilot study was to not just collect reference data, but to specifically collect it such that both a Pixel-Based Classification (PBC) and an Object-Based Classification (OBC) approach to generating the thematic map could be assessed. A typical pixel-based assessment optimally uses a grouping of homogeneous pixels (e.g., 3×3-pixel sample unit for Landsat imagery) as the sample unit to account for positional accuracy in the sample. Object-based assessments tend to use the entire polygon as the

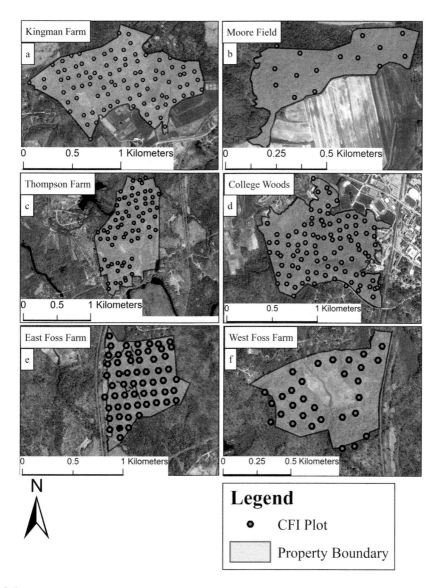

FIGURE 2.5
Maps showing the Continuous Forest Inventory (CFI) plot networks totaling 354 horizontal point sampling plots over 377.57 ha of forested land. Pictured are (top left to bottom right): (a) Kingman Farm, (b) Moore Field, (c) Thompson Farm, (d) College Woods, (e) East Foss Farm, and (f) West Foss Farm.

sample unit for the assessment. Figure 2.7 presents a graphical representation of these two strategies. Therefore, in order to evaluate the effectiveness of the UAS-derived reference data, six different approaches were used; two for the PBC and four for the OBC. The first PBC (PBC-1) reference data collection method placed a single 3×3-pixel (90×90-m) sample unit at the center of each forest stand and that sample unit was labeled by interpreting the forest type (coniferous, mixed, or deciduous) at that location. The second PBC (PBC-2) reference data collection method was similar but rather than locating the single sample unit at the center of the stand, the sample unit was located directly over one of the CFI plots in the stand. The sample plot was again interpreted to be one of the three forest types.

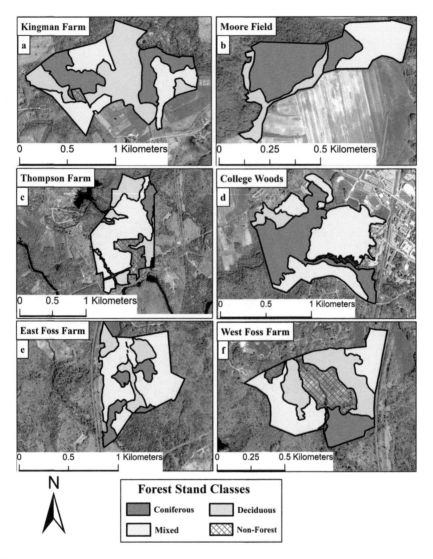

FIGURE 2.6
Ground-based forest stands as identified from the CFI plots. Pictured are (top left to bottom right): (a) Kingman Farm, (b) Moore Field, (c) Thompson Farm, (d) College Woods, (e) East Foss Farm, and (f) West Foss Farm.

There were four reference data collection methods evaluated for the OBC; two evaluate the entire polygon while the other two evaluate multiple samples within the polygon. The first OBC (OBC-1) reference data collection method used a stratified random sampling approach to place sample units within each forest stand. These samples were spatially independent with a minimum of two samples per polygon. Each sample unit in each polygon was interpreted and labeled as one of the three forest types. The second OBC (OBC-2) reference data collection method used all of the sample units with in each polygon to determine a label for the entire polygon using the majority rule of all the sample units. The third and fourth OBC (OBC-3 and OBC-4) mimic what was done in OBC-1 and OBC-2 but rather than randomly locating the sample units, they were aligned directly over each of the CFI plots in the stand. OBC-3 evaluated each sample unit in the stand while OBC-4 used all the samples in the stand and the majority rule to label the entire polygon.

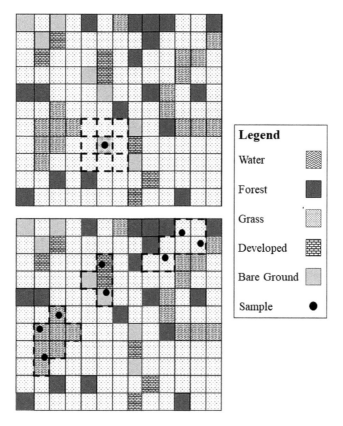

FIGURE 2.7
Sample units for reference data collection methodologies. The top graphic shows Pixel-Based Classification (PBC) with a 3×3-pixel sample unit and the bottom graphic shows Object-Based Classification (OBC) with multiple sample units within each polygon.

As described in the Generating the Error Matrix section (Section 2.2.3), error matrices were produced to quantify the agreement between the interpreted, sample unit labels generated from the UAS orthomosaics and the CFI ground-based samples. Each matrix included computing overall, producer's, and user's accuracies for each of the study sites (Fraser and Congalton, 2019). Table 2.1 summarizes the overall accuracies for each of the six evaluated reference data collection methods (two for PBC and four for OBC).

A key component in evaluating the utility of collecting reference data from UAS is to investigate the intrinsic uncertainty in this approach. One component contributing to this uncertainty is the difficulty in labeling the mixed forest sample units. Given the spatial complexity of New England forests, it is understandable that there is difficulty separating a mixed forest from a conifer forest given the ease in recognizing the conifers and the tendency for the deciduous to blend in more. The confusion caused by this heterogeneity led to larger amounts of producer's and user's errors in the mixed class than in the other two classes. In addition, while the CFI plots provide a ready source of comparison with the UAS results, the CFI plots are actually various radius plots based on tree basal area while the UAS sample plots were fixed in size. Finally, there were other potential issues with using the CFI plots including age of the collection (some study areas were collected up to ten years ago) and GPS positional error caused by being under dense forest canopies.

TABLE 2.1

Overall Accuracies for the Six Reference Data Collection Methods

Reference Data Classification Method	Overall Accuracy
PBC-1: sample unit in Forest stand Center	69%
PBC-2: Sample unit Over CFI Plot	74%
OBC-1: Random Sample Units In stand	64%
OBC-2: Majority Label from Random Samples	71%
OBC-3: Samples aligned with CFI in Stand	62%
OBC-4: Majority Label from CFI aligned	86%

All these issues together tended to reduce the agreement (accuracy) between the ground CFI plots and the reference data derived from the UAS imagery.

However, this was only a first pilot study to begin to evaluate the potential for UAS as a source of effective and efficient reference data. It should be noted that in just a few weeks of work, this study was able to collect imagery and corresponding reference data for almost 400 ha of forest land at a promising level of accuracy. Given this experience, work is continuing in this area to improve our ability to effectively and efficiently collect forest reference data from UAS.

2.4 Conclusions

This chapter has provided an overview of the considerations and techniques used to assess the accuracy of a thematic map. It is very clear from this overview that the collection of the reference data required to perform the assessment is the most expensive and time-consuming component of the entire process. Any efficiencies that can be gained in collecting these data are invaluable in improving the effectiveness of the entire assessment. The development and wide-spread application of UAS technology over the last few years provides an incredible opportunity for the efficient and effective collection of reference data in a way that was not previously possible. UAS offer the advantages of timely and relatively inexpensive collection of imagery and the processing of this imagery into useful products such as orthomosaics and point clouds for many applications including thematic accuracy reference data.

Recent rulings by the FAA in the United States (Remote Pilot in Command licensing) has prompted an explosion of research and applications in the use of UAS. The potential of such a system to collect imagery for a myriad of applications seems limitless. The challenge now is to apply and evaluate many of these possibilities. This chapter has suggested how UAS could be used to collect a variety of reference data types for thematic map accuracy. It has also provided a summary of a pilot study performed to evaluate the UAS collection of reference data for complex forests in New Hampshire (Fraser and Congalton, 2019). This study experienced significant challenges in collecting such reference data for the first time. However, the results are promising and show great potential for even more success in the future. Clearly, there is still a lot of work to be done. Yet, the potential of UAS is obvious and represents a paradigm shift in the way that imagery can be acquired and therefore, the applications of this imagery.

Literature Cited

Agisoft PhotoScan Professional Edition. Version 1.3.2 Software. Available Online: www.agisoft.com/downloads/installer/ (accessed July 2017).

Anderson, J.R., E.E. Hardy, J.T. Roach, and R.E. Witmer. 1976. A land use and land cover classification system for use with remote sensor data. *Geological Survey Professional Paper, 964.* doi:10.3133/pp964

Blaschke, T. 2010. Object based image analysis for remote sensing. *ISPRS Journal of Photogrammetry and Remote Sensing, 65,* 2–16. doi:10.1016/j.isprsjprs.2009.06.004

Bolstad, P. 2016. *GIS Fundamentals: A First Text on Geographic Information Systems* (5th edn, 769 p.). White Bear Lake, MN, USA: Eider Press.

Cliff, A.D., and J.K. Ord. 1973. *Spatial Autocorrelation* (178 pp.). London, England: Pion Limited.

Colomina, I., and P. Molina. 2014. Unmanned Aerial Systems for photogrammetry and remote sensing: A review. *ISPRS Journal of Photogrammetry and Remote Sensing, 92,* 79–97. doi:10.1016/j.isprsjprs.2014.02.013

Congalton, R.G. 1988a. Using spatial autocorrelation analysis to explore errors in maps generated from remotely sensed data. *Photogrammetric Engineering and Remote Sensing, 54*(5), 587–592.

Congalton, R. G. 1988b. A comparison of sampling schemes used in generating error matrices for assessing the accuracy of maps generated from remotely sensed data. *Photogrammetric Engineering and Remote Sensing, 54*(5), 593–600.

Congalton, R. G. 1991. A review of assessing the accuracy of classifications of remotely sensed data. *Remote Sensing of Environment, 37,* 35–46. doi:10.1016/0034-4257(91)90048-B

Congalton, R. 2009. Accuracy and error analysis of global and local maps: Lessons learned and future considerations. In P. Thenkabail, J. Lyon, H. Turral, and C. Biradar (Eds.), *Remote Sensing of Global Croplands for Food Security* (pp. 441–458). Boca Raton, FL, USA: CRC/Taylor & Francis.

Congalton, R., and M. Brennan. 1999. Error in remotely sensed data analysis: Evaluation and reduction. Proceedings of the Sixty Fifth Annual Meeting of the American Society of Photogrammetry and Remote Sensing, Portland, OR, USA, pp. 729–732 (CD-ROM).

Congalton, R., and K. Green. 2019. *Assessing the Accuracy of Remotely Sensed Data: Principles and Practices* (3rd edn, 328 p.). Boca Raton, FL, USA: CRC/Taylor & Francis.

FAA (Federal Flight Administration). 2017. Fact sheet – Small unmanned aircraft regulations (Part 107). U.S. Department of Transportation Federal Aviation Administration 800 Independence Avenue, SW Washington, DC, USA, 20591. Retrieved from www.faa.gov/news/fact_sheets/news_story.cfm?newsId=20516 (accessed August 2019)

Fonstad, M.A., J.T. Dietrich, B.C. Courville, J.J. Jensen, and P.E. Carbonneau. 2013. Topographic structure from motion: A new development in photogrammetric measurement. *Earth Surface Process and Landforms, 38,* 421–430.

Fraser, B.T., and R.G. Congalton. 2018. Issues in Unmanned Aerial Systems (UAS) data collection of complex forest environments. *Remote Sensing, 10,* 908. doi:10.3390/rs10060908

Fraser, B., and R.G. Congalton. 2019. Evaluating the effectiveness of Unmanned Aerial Systems (UAS) for collecting thematic map accuracy assessment reference data in New England Forests. *Forests, 10,* 24. doi:10.3390/f10010024

Jensen, J. 2016. *Introductory Digital Image Processing: A Remote Sensing Perspective* (4th edn, 623 p.). Upper Saddle River, NJ, USA: Prentice Hall.

Lillesand, T., R. Kiefer, and J. Chipman. 2015. *Remote Sensing and Image Interpretation* (7th edn, 720 p.). New York, USA: John Wiley & Sons.

Liu, C., P. Frazier, and L. Kumar. 2007. Comparative assessment of the measures of thematic classification accuracy. *Remote Sensing of Environment, 107,* 606–616.

Lunetta, R., R. Congalton, L. Fenstermaker, J. Jensen, K. McGwire, and L. Tinney. 1991. Remote sensing and geographic information system data integration: Error sources and research issues. *Photogrammetric Engineering and Remote Sensing, 57*(6), 677–687.

MacLean, M., and R. Congalton. 2013. Applicability of multi-date land cover mapping using Landsat 5 TM imagery in the Northeastern US. *Photogrammetric Engineering and Remote Sensing*, *79*(4), 359–368.

Mancini, F., M. Dubbini, M. Gattelli, F. Steechi, S. Fabbri, and G. Gabbianelli. 2013. Using unmanned aerial vehicles (UAV) for high-resolution reconstruction of topography: The structure from motion approach on coastal environments. *Remote Sensing*, *5*, 6880–6898.

Marshall, D.M., R.K. Barnhart, E. Shappee, and M. Most. 2016. *Introduction to Unmanned Aircraft Systems* (2nd edn, 233 p.). Boca Raton, FL, USA: CRC Press. ISBN:978-1482263930.

McGlone, J.C. (Ed.). 2013. *Manual of Photogrammetry* (6th edn, 1318 p.). Bethesda, MD, USA: American Society for Photogrammetry and Remote Sensing.

Pugh, S., and R. Congalton. 2001. Applying spatial autocorrelation analysis to evaluate error in New England forest cover type maps derived from Landsat Thematic Mapper Data. *Photogrammetric Engineering and Remote Sensing*, *67*(5), 613–620.

Radoux, J., P. Bogaert, D. Fasbender, and P. Defourny. 2011. Thematic accuracy assessment of geographic object-based image classification. *International Journal of Geographical Information Science*, *25*, 895–911.

Smith, M.W., J.L. Carrivick, and D.J. Quincey. 2016. Structure from motion photogrammetry in physical geography. *Progress in Physical Geography*, *40*(2), 247–275.

Story, M., and R.G. Congalton. 1986. Accuracy assessment: A user's perspective. *Photogrammetric Engineering and Remote Sensing*, *52*, 397–399.

Tortora, R. 1978. A note on sample size estimation for multinomial populations. *The American Statistician*, *32*(3), 100–102.

Turner, M.G. 2005. Landscape ecology: What is the state of the science? *Annual Review of Ecology, Evolution, and Systematics*, *36*, 319–344. doi:10.1146/annurev.ecolsys.36.102003.152614

Westoby, M.J., J. Brasington, N.F. Glasser, M.J. Hambrey, and J.M. Reynolds. 2012. 'Structure-from-motion' photogrammetry: A low-cost, effective tool for geoscience applications. *Geomorphology*, *179*, 300–314.

Whitehead, K., and C. Hugenholtz. 2014. Remote sensing of the environment with small unmanned aircraft systems (UASs), part 1: A review of progress and challenges. *Journal of Unmanned Vehicle Systems*, *02*, 69–85.

3

Multiuser Concepts and Workflow Replicability in sUAS Applications

Jason A. Tullis, Katie Corcoran, Richard Ham, Bandana Kar, and Malcolm Williamson

CONTENTS

3.1 Introduction

Accelerated developments in sUAS and other "unstaffed systems" (Reichardt and Simmons 2019) are poised to influence location-based services and the economies of the future. As new applications of these systems are rapidly being discovered, educational and research programs are developing new "data science" initiatives to respond to unprecedented shifts in associated cyberinfrastructure, artificial intelligence (AI), and deep learning (DL). Geospatial sUAS is described as a "disruptive technology" because it challenges the operational capacity of traditional cyber and personnel investments by introducing rapid innovation, adoption of which is dependent upon new skill, computation, and collaboration. Understanding longstanding traditions as well as current sUAS forces at work is critical for leveraging the positive potential of sUAS across many applications.

One tradition that preceded recent widespread geospatial sUAS acceleration was that highly experienced and reputable agencies, corporations, institutes, and teams were both trusted and tasked with remote sensing data collection (Jensen 2007). Much like the advent of the personal computer, "personal remote sensing" (Jensen 2017) increases the power of the individual. As of mid-2019, in the United States, there are over 100,000 remote pilots certified by the Federal Aviation Administration (FAA) who can now legally oversee commercial remote sensing operations. While few of these individuals are trained in remote sensing (and even fewer in photogrammetry), they along with many affiliated organizations now have access to smaller, more sophisticated, and affordable sUAS vehicles, gimbals, sensors, inertial measurement units (IMUs), global navigation satellite system (GNSS) receivers, etc. Integration and automated technologies aside, evidently, the geospatial knowledge, skills, and roles of *people* involved in remote sensing data collection are shifting, and new traditions are emerging.

Among various significant sUAS-related forces currently at work is geographic information science (GIScience) and technology developed over the last 50 years, and (less understood) its capacity for multiuser collaboration and replicability (Tullis et al. 2015; formally defined below). Most workstation-based GIS, photogrammetry, and remote sensing environments are designed for only one user at a time such that multiuser collaboration and interoperability are inherently frustrated and often largely ignored. In an international workshop on replicability in GIScience, Goodchild et al. (2019) identified extensive replicability failures in traditional geospatial software design and the need to harness the capabilities of information and communication technologies and data science to develop multiuser cyberinfrastructure to solve this problem. Recent disruptive innovations in sUAS provide both opportunity and motive to help overcome such challenges within the geospatial data sciences community and beyond.

After introducing selected working definitions, this chapter examines the historical context of geospatial interoperability and common points of replicability failure in sUAS. Related interoperability requirements in selected applications, such as emergency response and food security, are then discussed. Finally, recommendations to enable multiuser sUAS workflow replicability are explored within the broader goals of geospatial interoperability and collaborative real-world problem solving.

3.1.1 Working Definitions

With advancements in sUAS innovation and geospatial data science, consensus on terminology is both important and difficult to achieve. While it is the authors' belief that definitional controversy is less important than sUAS-based real-world problem solving, the latter may sometimes be highly dependent upon clear definitions. Terms presented here are matched with the definitional school or camp that seems most in harmony with the emerging geospatial data science, though it is acknowledged that preferred definitions will vary among readers and are subject to change.

The Association for Computing Machinery (ACM) carries significant weight in the computational sciences and engineering, and therefore influences terminologies associated with sUAS-based GIScience, remote sensing, and photogrammetry. In ACM's (2018) artifact review and badging policy, an *artifact* is defined broadly to include study-specific digital objects such as raw input, intermediate, and output datasets, scripts, and even entire software systems that influence computation. They further define *repeatability, replicability,* and *reproducibility* in a computational context (Table 3.1) that is adopted by the authors for geospatial sUAS applications and more broadly for discussion of the GIScience context. Significant confusion and sometimes heated controversy surrounding definitions of replicability and reproducibility, while beyond the scope of this chapter, are addressed by Plesser (2018) and by National Academies of Sciences, Engineering, and Medicine (2019).

TABLE 3.1

Definitions from ACM's (2018) Artifact Review and Badging Policy and Corresponding Geospatial sUAS Scenarios; Consensus on Computational Repeatability, Replicability, and Reproducibility Is a Precursor to Multiuser sUAS Interoperability and More Broadly to Its GIScience Context; Given the Complexities Involved (e.g., Continual Software Versioning), It Is Unlikely That a Given Scenario Will Perfectly Match the ACM Definitions

Term	ACM Summary	ACM Definition	Geospatial sUAS Scenario
Repeatability	Same team, same experimental setup	The measurement can be obtained with stated precision by the same team using the same measurement procedure, the same measuring system, under the same operating conditions, in the same location on multiple trials. For computational experiments, this means that a researcher can reliably repeat her own computation.	A South Carolina, USA forestry consultant demonstrates an sUAS hexacopter LiDAR-derived pine timber stump wood price workflow to a company trainee using the original workstation and LiDAR data, as well as original versions of Windows, in house Python code, LAStools, and fixed area plot estimates based on a specific USDA Forest Service (2019) National Biomass Estimator Library function
Replicability	Different team, same experimental setup	The measurement can be obtained with stated precision by a different team using the same measurement procedure, the same measuring system, under the same operating conditions, in the same or a different location on multiple trials. For computational experiments, this means that an independent group can obtain the same result using the author's own artifacts.	Three forestry companies in Tennessee and Arkansas, USA test the above South Carolina workflow using the same LiDAR data, Python code, and LAStools version running on functionally similar builds of Windows; they each obtain the exact same estimation for dollars per ha stump wood
Reproducibility	Different team, different experimental setup	The measurement can be obtained with stated precision by a different team, a different measuring system, in a different location on multiple trials. For computational experiments, this means that an independent group can obtain the same result using artifacts which they develop completely independently.	After replicating and carefully reviewing the above workflow, a New South Wales, Australia-based forestry company obtains similar precision in price per ha pine stump wood using a fixed wing drone, a color aerial camera, photogrammetry software, an automated tree count algorithm, and fixed area plot data in homogeneous plantation stands

Given the capacity for sUAS-based geospatial replicability, it seems reasonable that repeatability should be most easily achieved. In contrast, while geospatial replicability provides no guarantee that corresponding reproducibility is achievable, it may provide valuable context. In lieu of continual references to all three terms throughout the chapter, the authors focus on replicability as an important starting point that is sufficiently challenged in sUAS applications to warrant the chapter's full attention.

GIScience or computational definitions of *geoprocessing, workflow*, and *provenance*, are critical to replicability and interoperability. As the root of geoprocessing, *process* implies an instance (execution) of a computer program. Just having the exact same computer program is insufficient for repeatability or replicability because geoprocessing may further be subject to software environment settings such as "snap raster" in Esri's ArcGIS platform, or the hardware environment such as graphical processing unit (GPU) resources required by Pix4D's desktop-based Pix4Dmapper. Hardware and software environments are thus critical when sUAS-based geospatial processes are chained together into a repeatable or replicable *workflow* (e.g., Python scripting or ModelBuilder graphic block programming within Esri's ArcGIS Pro platform). *Provenance* is perhaps the least frequently understood and yet one of the most important terms in the context of sUAS-based replicability and interoperability.

Tracing back more than 700 years to the Old French derivative of the Latin *provenire*, definitions of provenance vary between a concept (e.g., valuable ownership history of a painting) and a record of such provenance (Moreau 2010). The Worldwide Web Consortium (W3C) Provenance Incubator Group's working definition states that "[provenance] of a resource is a record that describes entities and processes involved in producing and delivering or otherwise influencing that resource," and that such information "provides a critical foundation for assessing authenticity, enabling trust, and allowing reproducibility" (Gil et al. 2010). A detailed treatment of geoprocessing, workflows, and provenance in a computational/GIScience context is found in Tullis et al. (2015). However, the recent proliferation in sUAS applications may create new practical complexities for provenance as a record. For example, such provenance may be scattered across variable formats such as XML processing history, binary drone log files in a commercial format, remote pilot in command (RPIC) preflight checklists and flight logbooks, previously executed scripts and graphic block programming repositories, and in their full detail the artifacts (ACM 2018) associated with an sUAS-based product. As Gil et al. (2010) continue, "[provenance] assertions are a form of contextual metadata and can themselves become important records with their own provenance." If such records are indeed considered important in sUAS and other geospatial applications, then prospects for replicability seem more likely. Other possible benefits of provenance extend beyond replicability, such as digitally assisted comparison of dissimilar workflows in order to understand potential advantages in quality or accuracy, concepts that are sometimes difficult to judge in remote sensing derivatives (Congalton 2010).

3.2 Historical Context

One may ask why interoperability, multiuser concepts, and replicability are typically not "baked in" to common approaches to sUAS software and related artifacts and workflows. The answer is at least partially dependent upon major forces in computational history.

GIS-class workstation and operating system (OS) technology is *modus operandi* for the vast majority of sUAS-based geoprocessing, and these are generally designed for single user (one at a time) login and geoprocessing. Constant OS improvements or patches, often driven by user feedback or identification of security vulnerabilities, introduce potentially disruptive versioning. Fortunately, in computer engineering, machine-managed provenance in software version control systems resists such chaos (Buneman 2013). Understanding these and other historical connections within GIScience and sUAS and between variable disciplines helps set the stage for best practices, recommendations, and potential improvements in sUAS and related applications.

3.2.1 Data Quality and Geospatial Provenance

In 1988 the National Committee for Digital Cartographic Data Standards (NCDCDS), established in 1982 by the American Congress of Surveying and Mapping (Bossler et al. 2010), proposed a structure for a geospatial data quality report to include 1) lineage, 2) positional accuracy, 3) attribute accuracy, 4) logical consistency, and 5) completeness (Moellering et al. 1988). *Lineage* (comparable to provenance) was the *first* quality component. David Lanter's (1992) *Geolineus* project became the first provenance-focused geospatial workflow system and leveraged the "lineage diagram" (Essinger and Lanter 1992) as a visual representation of provenance. Geolineus' graphical user interface (GUI) enabled maintenance and propagation of metadata, workflow execution, and replicability, and augmented workflow tools such as "merge" and "condense" to simplify and combine overlapping workflows (Figure 3.1a). The value of graphic-block programming with these and other capabilities cannot be overstated, because it provides a visual representation of provenance that is directly connected to workflow execution. For example, shadows in Esri's ModelBuilder GUI (Figure 3.1b) indicate retrospective provenance (the workflow was previously executed), and lack of shadows in a validated workflow indicates prospective provenance (the potential for successful execution). Del Rio and da Silva (2007) found significant gains in experienced user and more so in novice user understanding of data quality as a result of visual representations of provenance.

As Congalton (2010) points out, assessment of accuracy cannot be reduced to following specific steps or a "cookbook" approach. Instead, a number of considerations and factors influence interpretation of accuracy. Multiuser access to a visual representation of the workflow-level provenance, or to a "lineage diagram" as proposed by Essinger and Lanter (1992) can allow sUAS researchers to understand many of the nuances that may impinge upon final accuracy determinations and judgments. As Chrisman (1986) suggested, while users are responsible for determining fitness for use, they need more information than they often receive to make such a determination. An important question that remains to this day in sUAS and other GIScience applications is how to convey the provenance information so that it isn't scattered across various files (e.g., personal logbooks versus binary drone log files in a commercial format), and how provenance may influence data quality and accuracy of outputs as well as augment trust, capacity building, technology transfer, and other research and educational endeavors.

3.2.2 International Specifications and Standards

In recent decades international metadata interchange standards have been established with relation to geospatial data collection and processing relevant to sUAS applications. For example, in the United States, the Federal Geographic Data Committee (FGDC) included

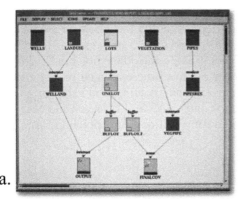

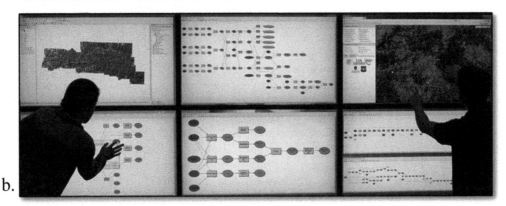

FIGURE 3.1
Historical and recent graphic block programming examples including GUI representations of a) "merge" and "condense" commands associated with or derived from Lanter's (1992) Geolineus project; and b) example of two users simultaneously working with ArcGIS ModelBuilder using a single high-end workstation with a tiled display (Tullis et al. 2012). (Reprinted with permission from Prasad Thenkabail, *Remote Sensing Handbook*, Boca Raton: CRC Press, 2015, 1:409–410.)

processing history in its 1994 metadata standards. ISO's Technical Committee (TC) 211 later published ISO 19115, derived in part from the work by FGDC, including ISO "extensions for imagery and gridded data" (ISO 2009) together with its XML implementation. This standard, endorsed by FGDC, was revised in January 2019 as ISO "extensions for acquisition and processing" (ISO 2019). Clearly there is an emphasis on imagery and more recently additional focus on acquisition and processing, all concepts highly relevant to sUAS applications. Authoritative metadata recognition notwithstanding, there is an important discrepancy in the concept of provenance as "contextual metadata" (Gil et al. 2010) and provenance as ISO metadata (Tullis et al. 2015). The concept of contextual metadata suggests at least the potential for a many-to-many cardinality (e.g., a provenance record shows that a specific FLIR Tau 2 sensor is used in a series of collections that support a variety of thermal derivatives). In contrast, provenance as ISO metadata seems to restrict the record into a one-to-one relationship (e.g., a metadata record associates one FLIR Tau 2 with one image stream).

In addition to metadata options for sUAS and other geospatial applications, W3C and ISO have also developed provenance-specific (non-metadata) interchange standards. For example, ISO 8000-120:2016 specifically addresses "requirements for capture and exchange of data provenance information," with a focus on master data (e.g., raw sUAS LiDAR).

In a broader and more exhaustive effort, W3C produced a series of documents and recommendations (each beginning with the word PROV) for interchanging provenance information online. The PROV data model (PROV-DM; Moreau and Missier 2013; Figure 3.2) enables machine readable "representations of the entities, people and processes involved in producing a piece of data or thing in the world" (Gil and Miles 2013). To date only a very small number of published studies (e.g., Masó et al. 2014) have successfully implemented PROV-DM in a geospatial or sUAS environment.

In a hypothetical implementation of PROV-DM, sUAS LiDAR is collected and processed to a high-accuracy point cloud (Figure 3.2). As part of this workflow, postprocessing incorporates raw Phoenix LiDAR data, real-time kinematic (RTK), global navigation satellite system (GNSS), data collected with a local base station, smoothed best estimate trajectory (SBET) data, the online positioning user service (OPUS) from National Geodetic Survey (NGS), and LiDARMill (a cloud-based processing tool from Phoenix LiDAR Systems). While details of arrows (relationships) are not shown (Figure 3.2 diagram versus legend), they point from future to past in terms of processing history.

In addition to international standards developed by W3C, the Open Geospatial Consortium (OGC) has created specifications and standards for interoperability more particularly focused within the geospatial community. While keyword searches for replicability and reproducibility yield limited results within OGC's website (www.opengeo-spatial.org), the word provenance appears in a far more extensive list of activities. For example, a recently formed "Workflow DWG" (domain working group; OGC 2019) enables visitors to join an email list dedicated to geospatial workflow issues such as optimization, provenance, security, interfaces, and web service harmonization. Tullis et al. (2015) review additional OGC activities related to provenance.

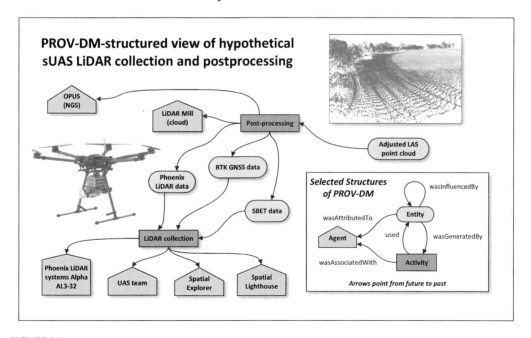

FIGURE 3.2
Selected provenance diagram of a hypothetical sUAS LiDAR collection and associated postprocessing according to W3C's PROV data model (PROV-DM; Gil and Miles 2013; Moreau and Missier 2013). (LiDAR drone image adapted from Phoenix LiDAR Systems 2018, "Alpha Series: AL3-32"; point cloud derived from Phoenix AL3-16 data collected by ArkUAV in Jun 2018.)

3.2.3 sUAS Proliferation and Practical Limitations

In an age of rapid sUAS proliferation, there are some practical limitations to multiuser collaboration and replicability as defined in the previous section. These include safety, socioeconomic, political, and technical factors that are mostly beyond the scope of this chapter (such as legal authority for airspace). Advanced skillsets are required to work in teams with sUAS hardware and software, and after huge time investments are made, new innovations may quickly become outdated. It may not seem worthwhile to be able to replicate a workflow, for example, if the sUAS hardware is no longer available in the market. As sUAS experts make career or position changes (or retire), their continued contribution to interoperability and replicability within an sUAS team is necessarily limited. Market interest or motivation is a highly important factor to replicability, as almost all sUAS applications are supported by industry partners who depend upon trade secrets, intellectual property, patents, non-disclosure agreements, etc., to maintain market viability.

In a market-related example, sUAS-based hyperspectral sensors generally require real-time kinematic (RTK) or post processing kinematic (PPK) global navigation satellite system (GNSS) capacity that implies a special GNSS antenna on the drone as well as a local base station. A relatively high quality IMU is also generally required. However, as noted by Laser Focus World Editors (2016), BaySpec has developed sUAS-based hyperspectral systems that do not depend upon GNSS or IMU data and are unlike traditional hyperspectral arrays. For example, some of these systems capture approximately 50 frames per second of raw data while in motion, and associated software subsequently enables production of geometrically corrected hyperspectral data cubes. The ability to avoid expensive and complicated GNSS/IMU hardware and software services is potentially a remarkable innovation, especially for GNSS-denied study areas (e.g., a deep rock quarry with insufficient satellite coverage). As with many other systems on the market, the sensor design and software processing for these hyperspectral BaySpec systems are commercially proprietary.

While it seems clear that practical considerations such as economics, intellectual property, safety, human lifecycles, etc. must necessarily bracket our discussion of failures to replicate and recommended practices (to be discussed), the precise boundary of what is truly a limitation is dynamic. It is the authors' position that the greater the understanding within the sUAS and geospatial communities regarding such concepts as geospatial workflows and provenance, the greater the opportunity to positively influence traditional boundaries.

3.2.4 Common Points of Failure to Replicate in sUAS Context

3.2.4.1 sUAS Platform and Sensor Hardware

The sheer variety of sensors, optics, gimbals, anti-vibration dampeners, GNSS receivers, etc., and the fact that many platforms (e.g., ultra-professional drones such as the DJI M600 Pro) can be configured in various ways depending upon need is inherently counterproductive for replicability. Also, while sUAS platform and sensor hardware transfers between sUAS laboratories are not impossible (though some equipment such as certain sUAS batteries is restricted from air travel), such transfers for the sole purpose of replicability seems impractical.

In a regulatory context, while commercial sUAS operational authority may require maintenance/modification logs, such records may be only loosely tied to specific platforms and hardware configurations. In the case of FAA 14 CFR Part 107 in the United States, the remote pilot in command (RPIC) is responsible for the maintenance of such logs, and RPICs may vary for a given (especially ultra-professional) platform. There is a natural entropy associated with the cardinalities of RPICs, hardware, and hardware configuration history and while pilot due diligence may be met by following the letter of the law, replicability in such a complex web of manually verified records and hardware is an unlikely beneficiary.

As sUAS experts well know, wireless transmission of imagery or other geodata from the sensor/platform to the ground station or field computer is constrained by limited bandwidths. For example, the Phoenix Alpha AL3-16 LiDAR sensor incorporates 16 lasers all used during data collection, with the raw data stored on a solid-state hard drive onboard the aircraft. Depending on the WiFi or 4G LTE wireless bandwidth employed for the data link, the system is typically configured to only transmit data from a fraction of the lasers to the Spatial Explorer software running on the field computer (e.g., 2/16 lasers or 1/8th of the full data). Some video streams such as the hyperspectral data collected using BaySpec OCI-F sensors may consume many GB storage per min of flight. As a result of high volumes and limited wireless bandwidths of the sUAS hardware infrastructure, sensor data is typically stored onboard the aircraft and must be manually transferred post flight (e.g., via micro-SD cards up to 1 TB in size, via Ethernet, USB etc.). Depending on the manual activities or idiosyncrasies associated with moving sUAS data to a workable environment, the data may become inaccessible to team members interested in replicability.

3.2.4.2 Mission Planning

A significant impediment to replicability is found in mission planning and associated common workflows and delivered systems used in the low- to mid-range of the market. While a detailed discussion is beyond the scope of this chapter, several common points of failure are identified in this section with a focus on popular platforms used for photogrammetry that often employ red-green-blue (RGB) cameras as well as some of the basic multispectral systems emphasizing near-infrared and red-edge bands.

When comparing RGB versus multispectral platforms (e.g., for extraction of agricultural reflectance indices), the latter clearly has more robust demands for replicability. For a static comparison of relative plant health/vigor in a single field at a single point in time, replicability may not be a high priority. However, multispectral passive sensors in the visible to short wavelength infrared (SWIR) range are based on reflected EMR and are thus greatly influenced by changes in scene illumination. The time of day, time of year, cloud cover, humidity, dust, haze, and other factors can all alter the ambient spectral distribution of light (Jensen 2007). In harmony with requirements for handheld spectroradiometers, sUAS-based multispectral sensors must be carefully calibrated to produce standardized data, necessary for comparison between acquisitions across space and time.

As big as the market is for multispectral sUAS data, an even larger market is found in RGB orthoimagery, usually for survey, mapping, or change detection. While direct georeferencing (e.g., from DJI Mavic Pro) may satisfy some applications requiring visual interpretation of single epochs, a major replicability shortfall in change detection and related applications is ground control, or lack of it. If such *in situ* data is properly collected it requires a setting that is amenable to high accuracy GNSS (dependent upon satisfactory terrain and overhead vegetation cover, etc.), and even in those settings there is a need for participation of skilled field technicians to help with data curation and collection of ground control points, etc. Another pitfall is in the use of mission flight trajectories that may be readily available within mission planning software, but which do not correspond to final product requirements. For example, since a mix of oblique aerial images with the typical nadir images significantly improves vertical accuracy and thus replicability of orthoimage geometric and positional accuracy, specialized flight planning and hardware may be required.

3.2.4.3 *Data Curation and Warehousing Considerations*

Principal challenges in data curation and warehousing include storage and multiuser access considerations, collaborative decision making throughout the full data lifecycle, and allocation of requisite IT resources that may only be partly funded for sUAS applications. Breakdowns in replicability can occur at any of these levels and may involve complex linkages to other points of failure. One of the most perplexing challenges in a collaborative curation venture is determining "who" is responsible for curation, metadata verification, maintaining an ongoing relationship with IT staff, etc.

In the simple case of a local area network (LAN), the administrator may easily grant high-speed access to an sUAS repository for individuals within the associated enterprise directory, but special administrative approvals may be required to extend access to members of an sUAS team who are credentialed in different administrative organizations. Also, the organizational separation between sUAS activities and IT investment may further complicate the use of customized LAN approaches which may be required for working with large data volumes (e.g., multiple GB per min of flight from BaySpec OCI-F hyperspectral sensor). While the "cloud" is often touted as a solution for large storage volumes, it is rarely appropriate for large data upload. Ongoing sUAS "storage wars" (Peters 2014) do not end simply because storage costs are lowered; access and curation over time are paramount. An additional storage consideration is compression; obviously lossless methods are ideal but even these may present an interoperability challenge in the face of persistent versioning of sUAS data processing software.

3.2.4.4 *Commercial and Open-Source Software*

An expanding variety of commercial and open source GIS, photogrammetry, and other sUAS workflow-related software is now available. As in other domains the variety and frequent versioning of the software is inherently inhospitable to replicability. While open-source solutions (e.g., OpenDroneMap) usually offer deprecated versions via the organization's website or a participating community repository for versioned code (e.g., GitHub), versions of commercial projects (e.g., Pix4Dmapper, Metashape, RealityCapture, LAStools, ArcGIS, etc.) are constrained by license negotiations. (Of course, there are very good reasons why commercial software is preferred, including high-quality documentation and commercial investment in expensive innovation). To further complicate replicability, the ecosystem of operating systems (e.g., Windows, Mac, Linux, 32/64-bit, etc.) requires careful planning for replicability (e.g., the use of software containerization such as Docker) which as of 2019 is atypical.

While some software vendors approach end-to-end solutions, many cutting-edge remote sensing activities involve special calibration procedures for the sensor that may or may not be recognizable by data processing software. While some major software vendors attempt to keep up with a great variety of sensor options, organizational disconnects between sensor manufacturers and engineers and the software used to process associated geodata also complicate the issue of replicability. Customer education on the nuances of any discrepancies may be difficult to obtain, which ultimately could lead to increased uncertainty and reduced usability of the data. Many other common points of sUAS software failure not identified here could be equally if not more significant to multiuser concepts and replicability.

3.2.4.5 *sUAS Finance and Technology Transfer*

Faculty, staff, and student employees at educational and commercial organizations are often financially compensated for activities that fall within generalized categories such as research, teaching, service, sales, development, etc. and sUAS hardware and software investments often involve multiple contributors that may span more than one organization. Following the money trails inevitably leads to agreements and understandings that may impede sUAS technology transfer as a prerequisite for replicability. For example, if a sUAS laboratory is funded by a government agency not authorized to conduct international activities, sUAS replicability may be constrained by national borders. In a commercial context, while technology transfer between competing interests is carefully controlled, there nonetheless exists a proliferation of non-disclosure agreements between educational and commercial entities seeking collaboration that may severely limit authorization of commercially beneficial workflow replicability.

3.3 Selected Interoperability Considerations in sUAS Applications

3.3.1 Emergency Response

Emergency management measures are generally taken to mitigate adverse impacts of hazards on social and physical environments, thereby preventing them from becoming disasters that far exceed impacted communities' coping capacities (UN/ISDR 2004). High quality, up-to-date, accurate, timely, and complete data and information is needed to effectively mitigate, respond to, and recover from disasters (Poser and Dransch 2010; Gao et al. 2011; Ostermann and Spinsanti 2011; Erskine and Gregg 2012). Geospatial data and tools (e.g., remote sensing, GIS) are extensively used in emergency management activities (Cova et al. 1999; National Research Council 2007; Kevany 2008) as they may rapidly provide valuable spatiotemporal information. Such data are not always available at appropriately high spatial and temporal resolutions, may contain erroneous information due to multiple known or unknown factors, and may be expensive to procure for near real-time use in emergency management efforts (Gao et al. 2011; Pu and Kitsuregawa 2014). Remote sensing data often needs to be obtained within the first 24–72 hours following a disaster and requires trained professionals with specialized knowledge to help with processing and interpretation (Hodgson et al. 2010). A remarkable example of cooperative space-based remote sensing for emergency response is the International Charter Space and Major Disasters (International Charter 2019), where member nations rapidly contribute space-borne observations and expert volunteers to support emergency response efforts during a charter activation.

A definite rise in crisis mapping efforts followed the 2010 Haiti earthquake and enabled the creation of the United Nations Office for Outer Space Affairs' (UNOOSA) platform for space-based disaster management and emergency response, the Satellite Sentinel Project, Global Earth Observation Catastrophe Assessment Network (GEO-CAN), and the Copernicus Emergency Management Service (Shanley et al. 2013; Havas et al. 2017; Zollner 2018). GEO-CAN, for example, has operated as a do-it-yourself grassroots group that collects images and aerial photos of disaster-impacted areas to aid in near real-time disaster-damage assessment.

With the rise of sUAS, similarly interested crowdsourcing groups such as SkyEye in the Phillipines, CartONG in Haiti, and Humanitarian UAV Networks (UAViators) have started using sUAS to collect aerial imagery at higher spatial resolution, in near real-time and at much lower cost to meet the demands of emergency response. However, when these disaster-related sUAS data streams are combined with traditional geospatial data during a disaster, decision support products often suffer from conflated precisions and provenance (Sui and Goodchild 2011; CRICIS Workshop Steering Committee 2012). This is at least in part due to a lack of interoperability standards for multiuser collection and processing, and the regulatory framework (e.g., airspace) that governs sUAS activities, especially those in an emergency response context.

Quality and interoperability of emergency response related sUAS data are strongly influenced by spatial, spectral, and temporal resolutions, and these in turn are constrained by flying altitude time of day or night (illumination), atmospheric conditions (smoke, haze), and processing workflows. For example, within authorized airspace (e.g., the FAA in the United States generally requires sUAS to fly below 400 feet), the flying altitude is determined by designated remote pilot(s) and their sponsors, thereby fundamentally generating data at varying spatial resolutions. The variable software and photogrammetric techniques to create 3D point clouds and topographic models from overhead image streams (e.g., using Structure from Motion or SfM) are also not consistent and depend upon agency decisions and investments made before, during, and after flight operations.

When sUAS data are collected by private institutions to help with emergency response activities, the institutions are generally not required to provide continuing data access after an event. Furthermore, although airspace controllers may waive certain protocols to ensure legitimate authorized sUAS operations during a disaster (e.g., as did the FAA during the 2017 and 2018 hurricane seasons in the United States), there is no universal protocol in place about the flight planning process, acquisition altitudes, the format of the output imagery or other geodata, and how sUAS data should be processed to enable effective access by on-the-ground emergency responders. The inherent variability in data quality and accessibility is often a drawback to the successful use of drone imagery for emergency management activity.

As mentioned previously, during and following a disaster event there is a need for real-time response, which necessitates real-time access to drone data. Very often private organizations may have relatively limited mechanisms in place for network or cloud upload to accelerate sUAS data access. The private institutions collecting imagery may also not be responsible for the processing and integration of imagery with other conventional geodata required for emergency responders. While interoperability standards exist for emergency communication (e.g., the U.S. Department of Homeland Security is working on the development of an interoperability standard for spatial data for disasters), currently no such standard exists for sUAS data collection, storage, access, and processing. The lack of such standards also can lead to location privacy violations that further increase failures to generate replicable outputs and methods for responding to emergency events.

3.3.2 Law Enforcement

Many of the same sUAS interoperability considerations related to emergency response also pertain to (or even overlap with) time sensitive sUAS applications in local and national law enforcement. One provenance-related consideration is privacy legislation governing the lifecycle of and access to data collected for law enforcement purposes. For example, the Illinois General Assembly's (2014) Freedom from Drone Surveillance Act generally restricts drone usage by the police except for specific purposes (e.g., aerial videography of an accident scene, a very common usage of sUAS in law enforcement). It further prevents "all information gathered by the drone" from being disclosed and mandates that this information be destroyed within 30 days barring criminal suspicion or investigative procedures (Lippmann 2015). A critical interoperability requirement is thus the ability to share lessons learned (or methodologies) for purposes of law enforcement training while at the same time complying with national and local privacy laws (e.g., that may disallow the police from using their own sUAS data for training).

3.3.3 Food Security and Agriculture

sUAS applications to improve food security and agriculture increasingly garner international interest. While these applications are generally less concerned with privacy and data sharing restrictions (than some aspects of emergency response or law enforcement), data volumes and variability in sensor/flight configurations are overwhelming. In a recent podcast sponsored by the Food and Agricultural Organization of the United Nations (FAO 2016), several sUAS dynamics are discussed. These dynamics include a) the need for international sUAS capacity building to determining farmland disaster risk and for aiding in the recovery process (e.g., from tropical storms), b) crop-specific threats from a variety of pests and invasive species worldwide, c) massive data volumes that are achieved with agriculture observation due to large areas, low flight altitudes, and high spatial and spectral resolutions, and d) innovative and sophisticated sensors (e.g., hyperspectral cameras) that are being designed by biological and agricultural engineers.

Farmers, farm agencies, and governments need interoperable means for distributed processing of large volume agricultural data. These stakeholders also need the distribution of trusted platform-, sensor-, flight configuration-specific workflows to enable effective sharing of sUAS capacity in areas most vulnerable to food insecurity. Such workflows should include robust methods, such as sensor calibration (e.g., perhaps through use of downwelling "sunlight sensors" to compensate when ambient light conditions change during a 45-min flight). Simply providing a GUI that reports results from the data is insufficient; rather, agricultural decision makers (data end users) require enough understanding of the detailed sUAS/geospatial workflow to be able to trust numbers they are presented with (crop yield, aggregate quantity of necessary fertilizer, etc.).

3.4 Recommendations for Multiuser sUAS Workflow Replicability

What follows is a concise attempt to provide, within the historical context and practical limitations, recommendations for maximizing multiuser sUAS replicability. These recommendations are by no means exhaustive (the variety of sUAS applications is simply too complex), and our purpose is to emphasize high priority opportunities for individuals and organizations participating in sUAS capacity building. References made to "team" in this context implies (to the extent allowed by policy) users, working groups, supervisors, resource allocators, participants, technicians, decision makers, etc., and membership may change over time. These recommendations further assume good faith intentions of all participants to share sUAS capacity building knowledge with other members of the team.

Complex decisions may surround the many options for effective use of cloud-based or locally hosted logbooks, open source and commercial off-the-shelf (COTS) data processing, and other sUAS technologies that are constantly undergoing changes or upgrades. Therefore, it is recommended that documents, spreadsheets, reports, databases, forums, etc. be curated and made available to all members of the team through a high-quality cloud-based collaboration portal. The UAS working group at the University of Arkansas, for example, utilizes Microsoft Teams in part because of its campus-wide availability and organizational support for the associated Office 365 Education platform over time. Microsoft Teams enables collaborative access to storage, notes, office documents, discussion boards, video chat, etc. and is being leveraged to provide a longer term view of and instructions for more specific retrospective and prospective workflows. Templates are available for a "professional learning community" to set and achieve goals (in this case increase sUAS research, teaching, and service capacity within the enterprise). An sUAS portal is a good place for the team to agree upon valuable record-keeping workflows (e.g., where and how to curate sUAS logfiles).

3.4.1 sUAS Hardware and Software Inventory and Maintenance

An important and sometimes difficult question to answer is: what sUAS resources are available to the team and what is the status of their maintenance? While it may seem old fashioned to keep a platform-centric *hardcopy* logbook for data entry in a variety of field or laboratory settings, this practice is highly recommended as a backup (or perhaps a precursor) to the use of cloud-based logbooks. Such an approach allows rapid and flexible data entry in cellular network–denied locations and provides a quick workaround for any portable device or software issues (e.g., screen glare, battery life, mobile operating system, limitations in data dictionary, etc.). Hardcopy logs should of course be scanned and curated through the team portal, and digital logbooks should subsequently be updated accordingly to minimize access costs associated with replicability.

The digital logbook should facilitate manufacturer-recommended maintenance, and maintenance scheduled based on the team's requirements, and while it may take many forms it should not be constrained by a platform-specific view but should preferably provide multiuser and multi-application support. For example, team members may need to be able to query the logbook in order to ascertain a specific sensor's interchangeable usage history and maintenance history within a fleet of drones. This is critical to determine whether any modifications (e.g., installation of upgraded optics in an aerial camera) may have taken place between flights whose workflows are being compared.

An inventory process or check-out/check-in system is vital to housing and maintaining sUAS resources. While there are several COTS options available, price may be a discouraging aspect given that these systems are often designed for large commercial inventories. One consideration is how often such systems are upgraded and the interoperability of the inventory data (e.g., for linkage to or comparison with digital logbooks, etc.). There must be a value-added proposition from a user community perspective to go with a COTS as opposed to an open source option (e.g., Grokability's 2019 Snipe-IT). Regardless, it is recommended that any solution provide full access to the "raw" data in industry standard formats so that the valuable inventory history can be curated beyond the lifecycle of the software or cloud-based service.

3.4.2 Airspace Authority and Operational Qualifications

A major consideration in replicability is the skill of the RPIC and whether the flight team is prepared for airspace authority, safety, risk management, equipment/sensor training, and other operational requirements. The Association for Unmanned Vehicle Systems International (AUVSI 2019) has begun to offer a "Trusted Operator Program" of "TOP

Operator" certification consisting of three tiered levels of increasing skill. The first level is closely associated with the FAA's 14 CFR Part 107 certification, while the third level enables skill for pilots in extreme environments. A trusted certification process like TOP Operator can help identify workflow requirements (skill level) that may be required for operational replicability in certain applications such as observation of telecommunication infrastructure.

3.4.3 Balancing Computational Resources and sUAS Requirements

In some ways, cloud-based services are rapidly eclipsing the computational facility and throughput of local resources and GIS-class workstations for sUAS workflows. For example, Phoenix LiDAR System's (2019) LiDARMill platform enables online processing of raw LiDAR together with all available GNSS/IMU data from the mission. It automatically incorporates NGA's Online Positioning User Service (OPUS) and generates a smoothed best estimate of trajectory (SBET) with an adjusted LiDAR point cloud in an interoperable open source format (Figure 3.3).

Despite the advantages of many cloud-based platforms, many applications (e.g., large area agricultural monitoring, hyperspectral video capture, etc.) simply generate too much data (e.g., multiple GB per min from BaySpec's OCI-F hyperspectral sensors) for cloud-based processing to be practical. In contrast, a single user-oriented GIS-class workstation approach hampers replicability by restricting the number of users who can access a given workflow. Even if data can be uploaded to the cloud, it is recommended that, where possible and within accepted IT policy, sUAS teams leverage local area network (LAN; see 3.4.4) approaches for curation, storage, and improved transparency of workstation-based geoprocessing.

FIGURE 3.3
Phoenix Cloud Viewer providing multiuser access to Phoenix LiDAR Systems AL3-16 data collected 24 May 2019 in Fayetteville, Arkansas. The underlying cloud-based LiDARMill platform incorporates raw LiDAR plus all available GNSS/IMU data and generates a smoothed best estimate of trajectory (SBET) with an adjusted LiDAR point cloud in an interoperable open source format. (Adapted from Phoenix LiDAR Systems 2019, "Phoenix Cloud Viewer.")

3.4.4 Development of a Multiuser sUAS Data Processing Pipeline

Extensive resources are now available for multiuser collaboration in terms of cloud-based storage (e.g., Microsoft OneDrive, Google Drive, DropBox, etc.), office communication (e.g., Microsoft Office 365, Google Docs, etc.), geospatial data collection (e.g., Esri ArcGIS Collector, Survey 123, etc.), GIS (ArcGIS Enterprise, Google Earth Engine, EOS Platform), and software programming (e.g., Jupyter Notebook, Bitbucket, etc.), and containerization (Docker, VMware ESXi, etc.). sUAS laboratories and working groups should seek multiuser capacities and limit the encroachment of traditional single-user environments, though these will still have a role given ongoing traditions. Even where personal computing is the *modus operandi* and suitable multiuser options are not available, steps can still be taken to enable multiuser capacity as illustrated in the following paragraphs.

Gigawatt (moniker for geoprocessing and workflows or GW) is a Python-based demonstration of GIS/UAS multiuser workflow interchange and replicability recently prototyped at the University of Arkansas' Center for Advanced Spatial Technologies (CAST; Figure 3.4). At its core, GW facilitates LAN-based collaboration, with optional server-based PostgreSQL and client-based Psycopg2 workflow interchange using Esri's ArcGIS Enterprise. GW allocates file server resources for necessary read/write access to input (e.g., UAS image streams), intermediate, and output geodata storage applicable to a variety of

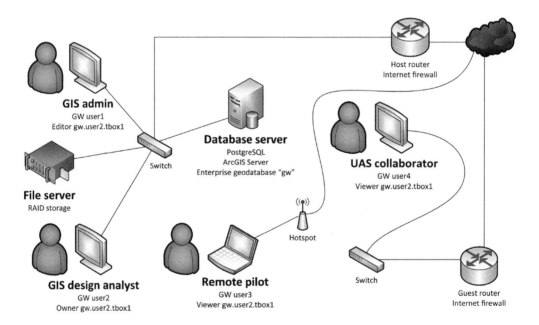

FIGURE 3.4

Hypothetical users of Gigawatt (GW) representing host and guest organizations. Both database and file servers are required for GW function. A GIS coordinator (user1) with administrative access creates and manages user accounts. A GIS analyst (user2) at the host organization creates an ArcGIS ModelBuilder toolbox (gw.user2. tbox1) for agricultural monitoring and populates it with a series of sUAS LiDAR processing workflows based on LAStools (Isenburg 2018). After granting view access to a field technician (user3) at the host organization's agricultural extension, the latter examines the detailed retrospective workflow provenance by executing (as a viewer) the same workflows *in situ*. As a toolbox editor, the GIS manager (user1) modifies one of the workflows during a phone consultation with user3. In the meantime, user2 (the toolbox owner) grants view access to a GIS analyst at a collaborating agricultural research institute, who can immediately begin to replicate all workflows in the toolbox.

local software deployed within or authenticated to the LAN (e.g., Google Earth Engine, ERDAS Imagine, Agisoft PhotoScan Professional, Pix4D Desktop, etc.). By using absolute (as opposed to relative) paths, GW resists path changes that can introduce ambiguity during replicability attempts.

When educational or research collaborators are authenticated in GW, they can immediately access original shared workflows through the familiar ModelBuilder graphic block programming environment. These user-developed workflows as well as traditional assets are stored in an enterprise geodatabase for centralized access. Users may quickly share functional toolboxes (containers for multiple workflows) with any number of GW participants through configuration of PostgreSQL table permissions such as select and update. GW's most powerful feature is that once toolboxes are shared (and become a form of prospective provenance), the recipient(s) can immediately execute the same workflow within a few seconds (as opposed to traditional delays of hours or minutes required to manage data access and file paths). This provides users with a replica of the full retrospective provenance (all the intermediate files, etc.) which is invaluable for GIS and UAS-based laboratory discussion, examination, and discovery.

3.4.5 Role of Metadata and Provenance Interchange Standards

When planning for sUAS workflow replicability, it is important to conceptualize the differences between metadata and provenance. As noted previously, the former has a one-to-one cardinality with the data, while the latter ("a form of contextual metadata"; Gil et al. 2010) is characterized by a many-to-many cardinality. Therefore, while it is recommended that ISO 19115-2:2019 (ISO 2019) be adhered to for *metadata* and where possible "lineage" or processing history, this standard is not an efficient means for provenance interchange. Until such time as robust provenance standards have been demonstrated in GIScience, preferably with effective AI-based maintenance, there is ample opportunity for developing new applications in this area. One option is to build sUAS applications that begin to leverage PROV-DM (Moreau and Missier 2013) as well as OGC (2019) standards and specifications, and to request support for some provenance-related aspects of these in sUAS/GIS workstation and cloud-based software.

3.4.6 Research/Educational Synergy and Intellectual Property

Research institutions should focus on creating pipelines that cover the entire process of planning to production so that users can better understand the different areas where problems can occur that will prohibit collaboration/replication. Fortunately, with publications like the National Academies of Science, Engineering, and Medicine (2019) the importance of replicability in software generally is becoming better understood. sUAS working groups at educational institutions should ensure that replicability is given priority if for nothing else than to enable research and educational synergy. For example, if a graduate student develops an innovative sUAS LiDAR processing workflow for her doctoral dissertation, she can share this with students, faculty, and staff on her campus as well as others in the user community to augment their knowledge of sUAS LiDAR processing. (They will learn the most about a workflow if they can replicate it.) In cases where intellectual property must be protected or non-disclosure agreements made (e.g., when private entities protect their proprietary software), a coarser grained provenance (record) can be developed to communicate replicability (including the need to work through authorized intellectual property channels).

3.5 Conclusions and Research Implications

As sUAS applications and associated technologies proliferate, their digital and socioeconomic impact, disruptive influence within IT, and sheer complexity all seem to be expanding. The problem of replicability and reproducibility in the sciences is gaining national and international attention, and arriving at a consensus on what these words mean for GIScience and sUAS is important. It is critical that the sUAS workflow be considered from preflight planning all the way to geodata processing and distribution of final products to spatial decision makers. Common points of failure can be identified in major sUAS application areas such as emergency response, law enforcement, and food security and agriculture. By seeking solutions and weighing possible recommendations to address these challenges, multiuser access and replicability can be improved and can benefit data quality, technology transfer, transparency, and trust in sUAS applications.

The GIScience, sUAS, and related research communities should be involved in the provenance/replicability standards development process in order to ensure the right kinds of standards (and associated functionality) can be incorporated into common sUAS collection and processing pipelines. These communities will also need to address the use of such standards, if possible, to both protect intellectual property and to maximize clarity of workflow knowledge that can be shared. Success in this endeavor will allow public, private, and international agencies to develop sUAS resources that are less impacted by administrative boundaries, and are given a wider range of relevance in current and emerging application areas that impact multiple levels of national and economic security, the environment, and quality of life.

Acknowledgments

Dr. Bandana Kar has participated in this project in her own independent capacity and not on behalf of UT-Battelle, LLC, or its affiliates or successors. The views and conclusions expressed in this article are those of the authors and do not reflect the policies or opinions of the funding agency, Oak Ridge National Laboratory, UT-Battelle, the Department of Energy, or the US Government.

References

Association for Computing Machinery. 2018. "Artifact Review and Badging." www.acm.org/publications/policies/artifact-review-badging

AUVSI. 2019. "TopOperator." Association for Unmanned Vehicle Systems International. www.auvsi.org/topoperator

Bossler, John D., James B. Campbell, Robert B. McMaster, and Chris Rizos, eds. 2010. *Manual of Geospatial Science and Technology*, 2nd edn. Boca Raton, FL, USA: CRC Press.

Buneman, Peter. 2013. "The Providence of Provenance." In *Big Data*, edited by Georg Gottlob, Giovanni Grasso, Dan Olteanu, and Christian Schallhart, Vol. 7968, pp. 7–12. Berlin, Germany: Springer. doi:10.1007/978-3-642-39467-6_3

Congalton, Russell. 2010. "Remote Sensing: An Overview." *GIScience & Remote Sensing*, 47(4): 443–59.

Cova, T. J. 1999. "GIS in Emergency Management." In *Geographical Information Systems: Principles, Techniques, Management, and Applications*, edited by Paul A. Longley, Michael F. Goodchild, David J. Maguire, and David W. Rhind, pp. 845–58. Hoboken, NJ, USA: Wiley & Sons.

CRICIS Workshop Steering Committee. 2012. "CRICIS: Critical Real-Time Computing and Information Systems: A Report from a Community Workshop." Computing Community Consortium. https://cra.org/ccc/wp-content/uploads/sites/2/2012/06/CCC_CRICIS_REPORT.pdf

Del Rio, Nicholas, and Paulo Pinheiro da Silva. 2007. "Probe-It! Visualization Support for Provenance." In *Advances in Visual Computing*, pp. 732–741. Lake Tahoe, NV, USA: Springer. http://link.springer.com/chapter/10.1007/978-3-540-76856-2_72

Erskine, Michael A., and Dawn G. Gregg. 2012. "Utilizing Volunteered Geographic Information to Develop a Real-Time Disaster Mapping Tool: A Prototype and Research Framework." In *CONF-IRM 2012 Proceedings*, 14. Association for Information Systems AIS Electronic Library (AISeL).

Essinger, Rupert, and David P. Lanter. 1992. "User-Centered Software Design in GIS: Designing an Icon-Based Flowchart That Reveals the Structure of ARC/INFO Data Graphically." In *Proceedings of the Twelfth Annual ESRI User Conference*. Palm Springs, California, USA.

FAO. 2016. "Drones, Data, Food Security: How UAVs Offer New Perspectives for Agriculture." Podcast Target Zero Hunger. May 20, 2016. http://www.fao.org/news/podcast/drones/en/.

Gao, Huiji, Xufei Wang, Geoffrey Barbier, and Huan Liu. 2011. "Promoting Coordination for Disaster Relief – From Crowdsourcing to Coordination." In *Social Computing, Behavioral-Cultural Modeling and Prediction*, edited by John Salerno, Shanchieh Jay Yang, Dana Nau, and Sun-Ki Chai, pp. 197–204. Lecture Notes in Computer Science. Berlin Heidelberg, Germany: Springer.

Gil, Yolanda, James Cheney, Paul Groth, Olaf Hartig, Simon Miles, Luc Moreau, and Paulo Pinheiro da Silva, eds. 2010. "The Foundations for Provenance on the Web." *Foundations and Trends in Web Science* 2(2–3): 99–241. doi:10.1561/1800000010

Gil, Yolanda, and Simon Miles, eds. 2013. "PROV Model Primer." W3C. www.w3.org/TR/prov-primer/

Goodchild Mike, Stewart Fotheringham, Wenwen Li, and Peter, Kedron. 2019. "Replicability and Reproducibility in Geospatial Research: A SPARC Workshop." Arizona State University. 2019. https://sgsup.asu.edu/sparc/RRWorkshop.

Grokability. 2019. "Snipe-IT Open Source IT Asset Management." https://snipeitapp.com

Havas, Clemens, Bernd Resch, Chiara Francalanci, Barbara Pernici, Gabriele Scalia, Jose Fernandez-Marquez, Tim Van Achte, et al. 2017. "E2mC: Improving Emergency Management Service Practice through Social Media and Crowdsourcing Analysis in Near Real Time." *Sensors* 17(12): 2766. doi:10.3390/s17122766

Hodgson, Michael E., Bruce A. Davis, and Jitka Kotelenska. 2010. "Remote Sensing and GIS Data/Information in the Emergency Response/Recovery Phase." In *Geospatial Techniques in Urban Hazard and Disaster Analysis*, edited by Pamela S. Showalter and Yongmei Lu, pp. 237–354. Geotechnologies and the Environment. Dordrecht: Springer Netherlands. http://springer.come/gp/book/9789048122370.

Illinois General Assembly. 2014. *Freedom from Drone Surveillance Act*. www.ilga.gov/legislation/ilcs/ilcs3.asp?ActID=3520&ChapterID=54

International Charter. 2019. "The International Charter Space and Major Disasters." 2019. https://disasterscharter.org/.

Isenburg, Martin. 2018. "LAStools: Converting, Filtering, Viewing, Processing, and Compressing LIDAR Data in LAS Format." www.cs.unc.edu/~isenburg/lastools/

ISO. 2009. "ISO 19115-2:2009 Geographic Information – Metadata – Part 2: Extensions for Imagery and Gridded Data." International Organization for Standardization.

ISO. 2016. "24ISO 8000-120:2016 Data Quality – Part 120: Master Data: Exchange of Characteristic Data: Provenance." ISO. www.iso.org/cms/render/live/en/sites/isoorg/contents/data/standard/06/23/62393.html

ISO. 2019. "25ISO 19115-2:2019 Geographic Information - Metadata - Part 2: Extensions for Acquisition and Processing." ISO. www.iso.org/cms/render/live/en/sites/isoorg/contents/data/standard/06/70/67039.html

Jensen, John R. 2007. *Remote Sensing of the Environment: An Earth Resource Perspective*, 2nd edn. Prentice Hall Series in Geographic Information Science. Upper Saddle River, NJ, USA: Prentice Hall.

Jensen, John R. 2017. *Drone Aerial Photography and Videography: Data Collection and Image Interpretation*. 1.2. Apple. https://itunes.apple.com/us/book/drone-aerial-photography-and-videography/id1283582147?mt=11

Kevany, M.J. 2008. "Improving Geospatial Information in Disaster Management through Action on Lessons Learned from Major Events." In *Geospatial Information Technology for Emergency Response*, edited by Sisi Zlatanova and Jonathan Li, pp. 3–19. International Society for Photogrammetry and Remote Sensing (ISPRS) Book Series. AK Leiden, The Netherlands: Taylor & Francis/Balkema.

Lanter, David P. 1992. "GEOLINEUS: Data Management and Flowcharting for ARC/INFO." 92-2. Santa Barbara, CA, USA: National Center for Geographic Information & Analysis. www.ncgia.ucsb.edu/Publications/tech-reports/91/91-6.pdf

Laser Focus World Editors. 2016. "Hyperspectral Imager from BaySpec Moves and Scans at Speeds up to 50 Frames/s." March 1, 2016. www.laserfocusworld.com/articles/2016/03/hyperspectral-imager-from-bayspec-moves-and-scans-at-speeds-up-to-50-frames-s.html

Lippmann, Rachel. 2015. "Illinois State Police Say Drones Are a Success." December 28, 2015. https://news.stlpublicradio.org/post/illinois-state-police-say-drones-are-success

Masó, Joan, Guillem Closa, and Yolanda Gil. 2014. "Applying W3C PROV to Express Geospatial Provenance at Feature and Attribute Level." In *Provenance and Annotation of Data and Processes*, edited by Bertram Ludäscher and Beth Plale, pp. 271–74. Lecture Notes in Computer Science 8628. Springer International Publishing. doi:10.1007/978-3-319-16462-5_31

Moellering, Harold, Lawrence Fritz, Dennis Franklin, Robert W. Marx, Jerome E. Dobson, Dean Edson, Jack Dangermond, et al. 1988. "The Proposed Standard for Digital Cartographic Data." *The American Cartographer* 15(1): 9–140.

Moreau, Luc. 2010. "The Foundations for Provenance on the Web." *Foundations and Trends in Web Science* 2(2–3): 99–241. doi:10.1561/1800000010

Moreau, Luc, and Paolo Missier, eds. 2013. "PROV-DM: The PROV Data Model." www.w3.org/TR/2013/REC-prov-dm-20130430/#section-example-two

National Academies of Sciences, Engineering, and Medicine. 2019. *Reproducibility and Replicability in Science*. Washington, DC, USA: National Academies Press. doi:10.17226/25303

National Research Council. 2007. *Successful Response Starts with a Map: Improving Geospatial Support for Disaster Management*. Washington, DC, USA: National Academies Press. doi:10.17226/11793

OGC. 2019. "Workflow DWG." 2019. http://www.opengeospatial.org/projects/groups/workflowdwg.

Ostermann, F.O., and L. Spinsanti. 2011. "A Conceptual Workflow for Automatically Assessing the Quality of Volunteered Geographic Information for Crisis Management." In *The 14th AGILE International Conference on Geographic Information Science: Advancing Geoinformation Science for a Changing World*. Association of Geographic Information Laboratories for Europe (AGILE). https://agile-online.org/index.php/conference/proceedings/proceedings-2011.

Peters, Chuck. 2014. "5 Reasons You Might Want to Wait on 4K Ultra HD Video Production (For Now)." *Videomaker* (blog). February 15, 2014. www.videomaker.com/article/f6/17189-5-reasons-you-might-want-to-wait-on-4k-ultra-hd-video-production-for-now

Phoenix LiDAR Systems. 2018. "Alpha Series: AL3-32." *Phoenix LiDAR Systems* (blog). www.phoenixlidar.com/al3-32/

Phoenix LiDAR Systems. 2019. "Phoenix Cloud Viewer." https://api.lidarmill.com/data_directories/1b364e13-d62e-45df-bc56-9fee50ebfa76/cloudviewer

Plesser, Hans E. 2018. "Reproducibility vs. Replicability: A Brief History of a Confused Terminology." *Frontiers in Neuroinformatics* 11 (January). doi:10.3389/fninf.2017.00076

Poser, Kathrin, and Doris Dransch. 2010. "Volunteered Geographic Information for Disaster Management with Application to Rapid Flood Damage Estimation." *Geomatica* 64(1): 89–98.

Reichardt, Mark, and Scott Simmons. 2019. "Remote Sensing, Drones, Unstaffed Systems (UxS), Sensors, and Interoperability at Open Geospatial Consortium (OGC)." Text. March 2019. doi:10.14358/PERS.85.3.157

Shanley, Lea A., Ryan Burns, Zachary Bastian, and Edward S. Robson. 2013. "Tweeting Up a Storm: The Promise and Perils of Crisis Mapping." *Photogrammetic Engineering and Remote Sensing* 79(10): 865–79.

Sui, Daniel, and Michael Goodchild. 2011. "The Convergence of GIS and Social Media: Challenges for GIScience." *International Journal of Geographical Information Science* 25(11): 1737–48. https://doi.org/10.1080/13658816.2011.604636

Tullis, J.A., J.D. Cothren, D.P. Lanter, X. Shi, W.F. Limp, R.F. Linck, S.G. Young, and T. Alsumaiti. 2015. "Geoprocessing, Workflows, and Provenance." In *Remotely Sensed Data Characterization, Classification, and Accuracies*, edited by P. Thenkabail, Vol. 1, pp. 401–21. Remote Sensing Handbook. Boca Raton, FL, USA: CRC Press.

Tullis, Jason A., Fred M. Stephen, James M. Guldin, Joshua S. Jones, John Wilson, Peter D. Smith, Tim Sexton, et al. 2012. "Applied Silvicultural Assessment (ASA) Hazard Map." University of Arkansas Forest Entomology's Applied Silvicultural Assessment. August 10, 2012. http://asa.cast.uark.edu/hazmap/

UN/ISDR. 2004. *Living with Risk: A Global Review of Disaster Reduction Initiatives*, 2 Vols. Geneva, Switzerland: United Nations.

USDA Forest Service. 2019. "Biomass Volume Estimation." www.fs.fed.us/forestmanagement/products/measurement/biomass/index.php

Zollner, Katharina. 2018. "United Nations Platform for Space-Based Information for Disaster Management and Emergency Response (UN-SPIDER)." In *Satellite-Based Earth Observation: Trends and Challenges for Economy and Society*, edited by Christian Brünner, Georg Königsberger, Hannes Mayer, and Anita Rinner, pp. 235–41. Cham, UK: Springer International Publishing.

4

The sUAS Educational Frontier: Mapping an Educational Pathway for the Future Workforce

John A. McGee III and James B. Campbell

CONTENTS

4.1 From Concept to Airborne

Small Unmanned Aircraft Systems (sUAS) represent a transformative technology that is changing the way that local governments, state and federal agencies, other non-profits, and the private sector acquire data to support their day-to-day business operations. Application areas that can utilize sUAS for data collection cut across varied disciplines and industry sectors, which include, but are not limited to: agricultural production, conservation management, forestry, public safety, disaster planning and mitigation, infrastructure inspection, construction management, surveying, and natural resources management (Preble 2015; Freeman and Freeland 2015; Lovelace and Wells 2018; Colomina and Molina 2014; Eninger 2015; Gallik and Bolesova 2016; Gibbens 2014; Tang and Shao 2015).

What has led to the proliferation of sUAS civilian applications? While military services have utilized 'drones' for some time, civil applications of sUAS have their roots in the portable electronics industry (Preble 2015). The development of smartphone devices and other compact electronics fueled technical innovations in batteries, cameras, processors, and navigation components which fueled the development of drones (Floreano and Wood 2015). Furthermore, these devices have integrated system components so that they function seamlessly. Thus, the evolution from smartphone to 'smart drone' was not a huge leap.

Rapid development and commercialization of sUAS caught many individuals, organizations, and regulatory agencies off-guard, resulting in development of new rules and regulations (Christie et al. 2016; Pasztor and Wall 2016). In February 2015, the Federal Aviation Administration (FAA) released its *Notice of Proposed Rulemaking for sUAS* (Federal Aviation Administration 2015). These provisions and rules significantly impacted sUAS operations in the United States, by reducing requirements for sUAS operators by clearly outlining their responsibilities. For example, provisions permitted the operation of an sUAS without a 'manned' pilot's license. This notice served to further improve the regulatory environment, and enhanced opportunities both for commercial and educational operation of sUAS.

The FAA subsequently developed a mandatory sUAS registration system for drones in 2016. sUAS registration numbers provide evidence of the rapid adoption of this technology. Within the first nine months, 550,000 small unmanned aircraft were registered in the United States (Kuhn 2016). Rates of registrations are anticipated to continue surging. For example, the FAA estimated that the compound annual growth rate of non-commercial sUAS from 2016 to 2017 would be approximately 40% (Federal Aviation Administration 2018b), while the commercial sUAS sector is projected to have cumulative average growth rates between 33% (lower end) and 44% (higher end) between 2017 and 2022. The FAA has further projected commercial and non-commercial sUAS fleets at between 2 million and 3.8 million by 2022 in the United States alone (Federal Aviation Administration 2018b). From a geographic perspective, sUAS registrations are non-discriminatory, as registrations are occurring across the United States, and mirrors U.S. population patterns (Federal Aviation Administration 2018b). Figure 4.1 shows the distribution of drone registrations in the United States.

4.2 Mapping with Drones

Given the continued growth of sUAS registrations, and the rapid adoption of this technology, one might ask, what is all this excitement really about? The excitement is not associated with a novel new hobbyist toy that is interesting to watch and fun to operate. sUAS represents a transformative and disruptive technology, as a major benefit of commercial sUAS operations is the ability of businesses to capture 'data on demand' (Barden, Valentine, and Culliver 2017). These systems are empowering – they provide an ability to collect and deliver data efficiently by streamlining many of the steps required for decision making. These steps include: data capture, expedited image processing, and the delivery of analytical products to the end client or decision maker. The integration of sUAS

FIGURE 4.1
Map of drone registrations in the United States by zip code (Federal Aviation Administration, 2018, acquired through FOIA request #2019-001507).

to expedite data collection also ensures that the data captured can be tailored to support specific application needs (McGee 2017). For example, previously, there were two main airborne image acquisition options available to project managers including: 1.) using imagery that was archived and already available; and 2.) contracting with a company to acquire new imagery.

4.2.1 Archived Imagery

Archived data were typically collected to support an initial program requirement. These data may not be appropriate for other ensuing applications. Therefore, archived and downloaded imagery were utilized to support an additional application, not because the imagery met needs and specifications associated with the project, but because the archived data represented 'the best imagery available at the time'. For example, the spatial, spectral, or temporal image resolutions might have been appropriate for an initial application (local government infrastructure mapping, most often collected during 'leaf off' in late winter or early spring), but might not be appropriate for the ensuing project (i.e., urban tree canopy assessments, which may be more effectively conducted at different spatial, temporal, and/or radiometric resolutions). But due to resource constraints, project managers were often required to compromise and use archived data collected to meet other priorities, even though these data may not conform to application needs of their project (Anderson and Gaston 2013). In summary, managers often sacrificed their preferred image specifications by utilizing accessible archived imagery.

4.2.2 Acquiring New Imagery

Another image acquisition option available to project managers is to contract image acquisition with a private company, employing either satellites or manned aircraft.

Such contractual options may often require lengthy negotiations (involving price, licensing issues, bids, scheduling, etc.), which result in additional lead times. Prices associated with image acquisition services can be associated with economies of scale. Therefore, this approach often demands significant financial resources, even for relatively small areas.

Furthermore, even after contractual agreements have been reached, and aircraft (or satellites) are positioned (either on a nearby runway or an appropriate flight path) to collect the imagery, data collection could be further delayed by unanticipated weather events, or personnel or maintenance issues. Due to relatively long contract lead times, it was often extremely difficult to acquire data, given optimal conditions on the ground (just prior or after a rain event, hours prior to/after a prescribed burn, etc.). After collection, data still had to be processed (and this part of the project might be subcontracted to another firm), which can potentially lead to further delays in product delivery.

sUAS represents a transformative technology. Utilizing sUAS, managers can capture imagery, especially for areas consisting of several hundred acres or smaller at higher resolutions, using a customized prescription of platforms, sensors, spatial, radiometric, and temporal resolutions that best conform to specific needs of a particular project (Anderson and Gaston 2013; McGee 2017). sUAS flight operations can be conducted on a chosen day (and time) to take advantage of favorable weather, solar radiance, and other characteristics. Often, sUAS operations can be conducted more safely, as pilots and other personnel are not exposed to undue risk (such as flying low to collect higher resolution imagery) (Lovelace and Wells 2018). Additionally, after completing flight planning processes, data collection operations can be conducted on relatively short notice, and image processing and delivery of both raw and processed imagery can be provided to clients relatively rapidly, especially given advances in cloud computing.

Collecting 'imagery on demand' is in itself a game-changer for researchers, and for the private sector and nonprofit organizations. And sUAS will likely contribute to our society in many other ways. For example, pilot projects that provide proof of concept for 'on-demand' package delivery continue to demonstrate their effectiveness and value (Locascio et al. 2016). In rural Virginia, for example, medical supplies have been delivered by drone to 24 individuals. Through this demonstration project, sUAS cut the delivery time of medical supplies from 'more than a day' to 30 minutes (Virginia Unmanned Systems Commission 2018). Delivery opportunities are not just limited to medical supplies, as sUAS have also been used in the delivery of burritos, pizza, and beer (Reid 2016; Price 2017; Kornatowski et al. 2018). Some countries are using drones to manage livestock (Beach et al. 2017). sUAS has also been used to collect microbes at high altitudes (Jimenez-Sanchez et al. 2018). Such considerations are building the effectiveness of sUAS as a transformative technology (Christie et al. 2016). It is no wonder that, given their impending influence on society, UAS have been described as 'potentially as revolutionary as the introduction of the Internet' (Beach et al. 2017).

4.3 Poised for Takeoff: Equating Industry Trends with Workforce Demands

The adoption of this enabling technology across industry sectors, and geographically across the United States (and abroad) indicates great promise for the future of UAS use (Federal Aviation Administration 2018b). In 2013, the Association for Unmanned Vehicle Systems International (AUVSI) estimated that the economic impact of the integration of

UAS into the national airspace could total more than $13.6 billion in the first three years of integration. This report further projected that more than 100,000 new unmanned aircraft jobs will be created, providing economic impacts of approximately $82.1 billion by 2025 (Jenkins and Vasigh 2013). Corporations are anticipating enormous growth potential in this sector. Industry is endorsing the future of unmanned aircraft operations by extending record levels of venture capital to facilitate research and innovation (Teal Group Corporation 2018). Such developments will likely nurture capabilities of the unmanned aircraft industry to function much differently in the years to come than it does today.

In the United States, UAS operate in public airspace, which is the most complex airspace in the world (Federal Aviation Administration 2018a), and therefore closely regulated by the FAA. Therefore, the technician-level workforce must be trained to obtain UAS-specific knowledge and skills. Under FAA Part 107, an individual flying an sUAS commercially must acquire a remote pilot certification (RPC), or be directly supervised by someone with an RPC. An RPC can be obtained by passing an initial aeronautical knowledge test at an FAA-approved testing center. Given sales trends, it is not surprising that the number of RPC's has been increasing steadily. For example, in September 2017, approximately 60,000 individuals had received an RPC (Bellamy 2017). By February 2018, the number had increased to 70,000. In a 2018 report, the FAA projected a potential need for 300,000 additional RPC's by 2022 to support the burgeoning sUAS industry (Federal Aviation Administration 2018b).

State governments have noticed these trends and are investing in sUAS infrastructure. For example, in 2016, the Virginia Unmanned Systems Commission identified UAS as one of the most important drivers of Virginia's New Economy (Commonwealth of Virginia 2016). It is not surprising, therefore, that one of the recommendations in the Virginia Unmanned Commission Report is to enhance UAS education at several levels, including precollege and higher education programs, and STEM programs in general (Virginia Unmanned Systems Commission 2018).

Current instruction and education efforts have value, but may not address key needs for comprehensive professional preparation. For example, the FAA anticipates a substantial demand for RPC's to legally operate an unmanned aircraft well into the future. The FAA's RPC was designed to ensure that remote pilots acquire a baseline of knowledge to safely and professionally operate unmanned aircraft in the national airspace. The FAA's RPC, while incredibly important, only scratches the surface of knowledge and skills typically associated with operating an sUAS. For example, the FAA's certification does not address any aspect of data collection and processing. There is more to being an sUAS technician than legally operating a drone. As a result, continued availability of updated, properly targeted educational opportunities are paramount to support projected employment and application demands.

4.4 Mapping the Educational and Professional Development Needs of sUAS Industry

Because sUAS represents a technology that is rapidly developing, it is imperative that educational institutions respond to the needs of employers that will actually hire sUAS operators. To be most effective, educational pathways should include outreach educational programs (including 4-H, scouts, and workforce development), high school STEM programs, two-year technical and community colleges, and four-year colleges and universities must provide instruction that reflects the needs of the sUAS industry. In addition, such educational programs should not only provide the education required to support current workforce needs, but must be agile and responsive enough to evolve to meet requirements of the future workforce.

To understand educational requirements of the sUAS workforce, especially given that this industry is so new, it is imperative that employers participate in the educational planning processes. Industry involvement can, and should, take many different forms, including participation on advisory committees, classroom presentations, providing internship opportunities for students (and faculty!), and facilitating real-world educational opportunities, through service-learning projects; an especially challenging facet of expanding the reach of educational and testing facilities in remote/regional locations. Two- and four-year colleges with an emphasis on enabling employees to meet current employer demand and a shorter certificate/degree cycle time are well-suited to quickly enable sUAS operators to meet the current and future demands.

Development of curriculum models and educational pathways can serve as foundations for educators to build upon. These models must mirror the desired skills of industry. Thus, understanding the needs of industry is paramount to developing a successful educational program. One of the most efficient mechanisms to survey the tasks and duties required by employers is to conduct a DACUM (Developing A CurricUluM) analysis.

4.4.1 What Is a DACUM?

While many individuals may consider themselves to be sUAS operators, whether they are operating unmanned aircraft as hobbyists, or as civilian commercial operators, the job title of 'UAS Operator', or 'UAS Pilot' is a bit limiting, as it does not accurately depict actual roles and tasks of the individuals who operate drones for business applications. For example, would a typical sUAS operator only operate (or pilot) an unmanned vehicle, or are there other responsibilities, tasks, and expectations also associated with this job? Developing a consensus of the responsibilities, tasks, duties, and expectations associated with a specific job is paramount to effectively target educational needs of a particular employee.

Information about a specific job can be gleaned from an individual who is currently working in that occupation. Perhaps an even better approach would be to encourage discussions with several individuals, perhaps representing different types of organizations (size, geographic spread, public and private, etc.), and differing priorities. The process of engaging a panel of experienced experts associated with a specific job, brainstorming the tasks, experiences, and expectations associated with the job, and coming to a consensus is paramount to the DACUM method.

A *DACUM* is a process by which a panel, consisting of experienced individuals currently employed in a specific occupation, analyze their own jobs. The DACUM process is based on consensus building by leveraging experience and expertise of individuals who are currently engaged in the specific employment sector. These individuals, through the assistance of a trained facilitator, break down the day-to-day responsibilities associated with their jobs into specific tasks which relate to core competencies for a particular job (Willmett and Herman 1989). The specific tasks are often associated with general categories, or duties. By charting day-to-day expectations and master competencies (or knowledge areas) associated with a specific job, it is possible to begin to identify general skills that are required to complete each task. Educators can utilize information collected through the DACUM process to vet existing course material. This information can also form a framework to support development of a new course, or a new sequence of courses, and to identify educational competencies that students will need in order to respond to the employment needs of industry. The product of this process forms a DACUM chart.

The DACUM chart is a direct reflection of feedback from industry representatives that serve on DACUM panels. Therefore, it is vital that DACUM panels are not just comprised of managers or administrators that oversee employees associated with specific job titles. Often, upper management and administrators, although providing a supervisory role, may not fully understand specific tasks or day-to-day operational responsibilities associated with sUAS operations. Thus, inexperienced supervisors that oversee jobs to be analyzed may have unrealistic expectations of the tasks associated with these roles.

Likewise, academics and educators may not form the best candidates to serve on DACUM panels. The role of the educational community is not to contribute to the job analysis, but to learn from the job analysis, through the findings of the DACUM panel. By gleaning information from the job analysis, educators can develop curriculum and tailor student instruction to better target specific employment demands of industry.

Thus, in order for a DACUM to be successful, it must provide a process driven by needs and demands of an experienced cadre of industry representatives. Such needs, once identified and catalogued, must then be integrated into a comprehensive educational effort either through traditional education at two-year and four-year colleges and universities, or through dedicated workforce development initiatives through competency-based training. A DACUM chart can transcend precollege, two-year college, and higher education. A DACUM panel often consists of 5–12 individuals who are working practitioners and experts in their fields. The panel is facilitated by a professionally trained and certified DACUM facilitator. Typical outcomes of the DACUM process include (DeOnna 2002):

- A DACUM can be used to assess or verify existing curricula and training approaches;
- A DACUM can be used to anticipate and formulate future training curricula for an emerging field;
- A DACUM can be used as a means to better understand the required breadth of knowledge associated with a specific job;
- A DACUM can be used not only to identify tasks and duties associated with a particular job, but it also provides insight to important worker behaviors, worker knowledge sets, tools, equipment and supplies, and potential qualifications or certifications.

4.4.2 Towards an sUAS DACUM

Geospatial DACUMs have been developed for other geospatial related employment tracks, including GIS Technicians and Remote Sensing Analysts (Johnson 2009; 2010). Prior to the sUAS Operations Technician DACUM, there was no evidence of an sUAS-related DACUM. The Geospatial Technician Education for Unmanned Aircraft Systems (GeoTEd-UAS | http://geoted-uas.org/) project, funded through the National Science Foundation, is comprised of a consortium consisting of the Virginia Space Grant Consortium (VSGC), Virginia Tech (VT), Thomas Nelson Community College (TNCC), Mountain Empire Community College (MECC), and the Virginia Community College System (VCCS). This consortium explored options to clarify sUAS industry workforce needs, which resulted in hosting an sUAS DACUM panel session.

Selecting panel members is perhaps the single most important part of the DACUM process. sUAS DACUM panel members consisted of experienced individuals who have worked with small unmanned aircraft (defined by the FAA as weighing less than 55 pounds at takeoff). These professionals, affiliated with an array of companies, organizations, and agencies spanning a range of industrial sectors, provided a breath of sUAS experience. Panel members were affiliated with public agencies, nonprofits, and private companies, and experienced with sUAS operations across the United States and abroad. Selected organizations provided resource management services, general surveying, aerospace, and other sUAS contractual services. A DACUM panel, led by a certified DACUM facilitator, met for two days in 2015 in Roanoke, Virginia. Panel members and their affiliations are listed in Table 4.1.

An initial challenge that sUAS DACUM panel members encountered was that the employment sector was, at the time, so new that many job titles had not been concretely defined and universally accepted by industry. For example, DACUM panel members initially considered the title of *sUAS Technician*. While panel members agreed that the title implied entry level positions, they also maintained that the title was too limiting, as the term conjured images of an entry level technician who repaired and maintained unmanned aircraft. There was a consensus among panel members that an *sUAS Technician* involved more than just drone maintenance and repair. Therefore, DACUM panel members developed a DACUM for *sUAS Operations Technicians* (sUAS OT), a title that extends beyond the limited perceptions associated with an *sUAS Technician*.

The sUAS OT DACUM process took two days to complete, and included discussions led by the facilitator, deliberations, story boarding, and extensive charting. Once the panel adjourned, the DACUM facilitation team, led by Jennifer Stevens, a certified DACUM facilitator, compiled the information into a draft DACUM chart. This draft was disseminated to panel members for further review and comment. The final sUAS-OT DACUM Chart serves as a job analysis that maps tasks, duties, and responsibilities of an sUAS OT. A DACUM Chart is intended to be a living document, as this product will need to evolve as the sUAS industry continues to be embraced by additional sectors of industry, as regulations continue to evolve, and as equipment continues to develop and mature. The sUAS OT DACUM Chart is provided in Figure 4.2, and is available online from the GeoTEd-UAS Website (http://geoted-uas.org/wp-content/uploads/2018/10/SUAS-DACUM-2.pdf).

TABLE 4.1

sUAS DACUM Panel Members and Affiliations

Mark Blanks	Mid-Atlantic Aviation Partnership
Daniel Cross	Conservation Management Institute
Darren Goodbar	Draper Aden Associates
Keith Harris	NASA Langley Research Center
Thomas Jordan	NASA Langley Research Center
Matthew Jungnitsch	Measure UAS, Inc.
Alexander Mirot	Embry-Riddle Aeronautical University
Chase Riley	Sentinel Robotic Solutions
Jeff Sloan	U.S. Geological Survey
Jay Willmott	Nexutech, LLC

4.4.3 Targeting Employer's Needs: sUAS Operation Technician DACUM

The final sUAS-OT DACUM Chart outlined 116 tasks that are associated with a UAS-OT. These tasks were associated with seven major duties, including:

- Planning an sUAS Flight
- Flight Operations
- Process Geospatial Data
- Maintain sUAS
- Coordinate Flight Operation Logistics
- Maintain Proficiency, Currency, and Recency in Professional Knowledge/Skills
- Perform Administrative Tasks

These major duties are summarized as follows.

4.4.3.1 Planning an sUAS Flight

Since one of the major advantages of data collection using an sUAS is the ability to acquire 'data on demand'; proper flight planning is paramount to the success of the operation. DACUM panel members, for example, emphasized that they spend more time (often by several factors) planning the sUAS operation than they actually spend during flight operation. Tasks associated with sUAS flight planning encompass everything from reconnaissance level flight planning (involving a cursory scouting of the region to identify topographic and initial safety and regulatory concerns) and more intensive flight planning (potentially involving acceptable weather parameters, tidal ranges, personnel, equipment requirements, and a specific and detailed report of potential regulatory and safety considerations). An overarching component of flight planning was adherence to safety. Avoiding potential conflicts with people, property, and wildlife is paramount.

Having a comprehensive understanding of federal, state, and local regulations was identified as an essential knowledge area during the flight planning process. Defining objectives of the project is also a critical component of the sUAS planning operation, as this will help to inform flight altitude, aircraft platform (which may include fixed wing vs. rotor) and sensor selection. Other considerations taken into account during sUAS flight planning include temporal considerations (time of day, seasonal variations, and temperature characteristics, especially for thermal imagery), and spatial parameters (shapes and sizes of proposed data collection, terrain characteristics, pixel size, etc.).

4.4.3.2 Flight Operations

The sUAS DACUM chart identifies several important tasks and duties associated with the operation of sUAS. The first, and perhaps the most important, item is that actual operation (piloting) of the sUAS (task B-16) forms a very small component of overall employment expectations and skills associated with an sUAS operations technician. This is not intended to minimize the importance of piloting an unmanned aircraft, but sUAS OT DACUM panel members maintained that there is a much broader range of tasks and requirements that is not often recognized by casual users, and perhaps by the educational community as a whole.

DACUM Job Analysis Research Chart for Small Unmanned Aircraft

DUTIES .... **TASKS**

Duty							
Plan sUAS Flight **A**	A.01 Define operation objective	A.02 Define area of operation	A.03 Assess area of operations (e.g. aviation, chart, map, NOTAMs, weather)	A.04 Define data requirements (e.g. data types, accuracies, frequencies, deliverables, format)	A.05 Plan flight path (e.g. manual or autonomous)	A.06 Obtain regulatory permissions (FAA)	A.07 Obtain landuse authorization (e.g. landowner, state/local permits)
Perform Flight Operation **B**	B.01 Conduct site survey B.15 Launch aircraft	B.02 Check advisory info (NOTAMs, weather, TFR) B.16 Fly sUAS *	B.03 Establish ground control points (GCPs) B.17 Maintain visual contact with aircraft	B.04 Set up flight operations center B.18 Monitor site communications	B.05 Assemble aircraft B.19 Monitor air traffic communications	B.06 Upload flight plan to aircraft B.20 Collect geospatial data	B.07 Calibrate sUAS sensors B.21 Execute emergency procedures
Process Geospatial Data **C**	C.01 Transfer geospatial data	C.02 Prepare data for processing	C.03 Select processing software	C.04 Load data to system	C.05 Create 3-D point cloud	C.06 Create digital model	C.07 Create orthomosiac
Maintain sUAS **D**	D.01 Troubleshoot/repair electrical system D.15 Inspect sUAS for maintenance issues	D.02 Troubleshoot/repair mechanical systems D.16 Inspect aircraft for air worthiness	D.03 Troubleshoot/repair airframe D.17 Integrate pay loads	D.04 Troubleshoot/repair fuel system** D.18 Implement configuration changes (e.g. hardware, software)	D.05 Troubleshoot/repair propulsion/power plant D.19 Document configuration changes (e.g. hardware, software)	D.06 Troubleshoot/repair ground control station D.20 Maintain maintenance logs	D.07 Perform scheduled maintenance on electrical system D.21 Conduct maintenance test flight
Coordinate Flight Operations Logistics **E**	E.01 Develop day of flight schedule E.15 Pack sUAS equipment/supplies	E.02 Communicate flight schedule E.16 Secure physical access to the operations area	E.03 Coordinate mission dependent resources E.17 Secure operations area	E.04 Schedule mission personnel E.18 Transport sUAS equipment/supplies	E.05 Schedule transport vehicles E.19 Prepare operations area	E.06 Establish transport route E.20 Safeguard geospatial data	E.07 Plan travel logistics
Maintain Proficiency, Currency & Recency in Professional Knowledge/Skills **F**	F.01 Practice flying sUAS	F.02 Train on flight simulator	F.03 Attend professional conferences and training	F.04 Maintain membership in professional organizations	F.05 Maintain required certifications	F.06 Review professional literature	F.07 Participate in online forums
Perform Administrative Tasks **G**	G.01 Order equipment/supplies	G.02 Develop operations budget	G.03 Complete expense reports	G.04 Submit timesheets	G.05 Create proposals & presentation	G.06 Present to a variety of audiences	G.07 Submit accident reports

* Denotes that the task may be performed within more than one duty.
** e.g. battery or liquid
* A full-size version of this image is available for download at CRCpress.com/9780367199241

FIGURE 4.2
The sUAS-OT DACUM chart.

The DACUM Panel also recognized that the operator not only requires skills and proficiency with remote piloting of the aircraft, but also needs to be knowledgeable of all components associated with the unmanned aircraft system. Specifics include ground control software, system components, battery technology, knowledge of weather, understanding Global Navigation Satellite Systems (GNSS), and spatial enhancement methods, including Real-Time Kinematics (RTK) systems. Basically, an sUAS OT must be able to diagnose and troubleshoot an entire spectrum of both technical and nontechnical issues.

Systems Operations Technician (sUAS)

A.08	A.09	A.10	A.11	A.12	A.13	
Select sUAS to meet objective	Define aircraft configuration (eg. firmware, payload)	Conduct operational risk assessment	Establish operation communications plan (data links, radio, frequencies)	Assign sUAS and personnel	Rehearse flight operation	
B.08	B.09	B.10	B.11	B.12	B.13	B.14
Conduct flight checks of sUAS	Conduct safety briefing	Conduct mission briefing	Check for FOD	Verify use of PPEs	Communicate with crew and ATC	Secure launch and recovery area
B.22	B.23	B.24	B.25	B.26	B.27	B.28
Adapt flight plan	Recover aircraft	Retrieve geospatial data	Verify data integrity	Conduct post flight inspection	Conduct post flight debrief	Pack sUAS for transport
C.08	C.09	C.10	C.11			
Create derivative products	Generate data report	Deliver aerial images/videos	Disseminate deliverables to customer			
D.08	D.09	D.10	D.11	D.12	D.13	D.14
Perform scheduled maintenance on mechanical systems	Perform scheduled maintenance on airframe	Perform scheduled maintenance on fuel system**	Perform scheduled maintenance on propulsion/power plant	Perform scheduled maintenance on ground control station	Troubleshoot software	Update software
D.22						
Maintain inventory of equipment/ supplies						
E.08	E.09	E.10	E.11	E.12	E.13	E.14
Acquire spare parts	Create pack list	Determine fuel/battery requirements	Facilitate personnel needs in the field (e.g. water, tent, chairs, etc.)	Secure mission supplies	Ship sUAS equipment	Charge UAS batteries
F.08						
Participate in community awareness activities						
G.08	G.09	G.10	G.11	G.12	G.13	
Train peers	Submit budget requests	Submit budget reports	Maintain company vehicle	File flight reports	Maintain flight logs	

FIGURE 4.2
Continued.

4.4.3.3 Process Geospatial Data

Industry representatives have identified image processing as a topic of required knowledge by sUAS technicians. Such skills require careful planning, collecting, and assembling of individual sUAS images to construct accurate maps from the network of aerial images that have been stitched together to prepare a single broad-scale image. Such images can support a variety of applications, including agriculture, forestry, and construction, to name a few. Some sUAS OTs, especially individuals associated with larger organizations, may not be as involved with image processing as sUAS OTs in other (often smaller) organizations.

In this context, however, there was a consensus that sUAS OTs need to understand how image processing works, and be familiar with image acquisitions processes so sUAS OTs can make better decisions about the collection of data while in the field (for example, understanding impacts of solar reflectance from water, resolution issues in hilly terrain, etc.). The panel expected that sUAS OTs be capable of generating orthomosaics, point clouds, and derivative products (including NDVI, etc.). Technicians did not only need proficiency in generating spatial products, but also have a basic understanding of such products, and provide reports associated with these products.

4.4.3.4 Maintain sUAS

Safe operation of an sUAS is not only associated with actual flight operation, but also with proper UAS maintenance. The term *technician* often conjures up a mix of maintenance and repair. This consideration is no different for an sUAS OT, as keeping expensive equipment maintained is essential for safety, operational readiness, retaining data, and minimizing risk and damage to property. DACUM panel members therefore advocated that an sUAS OT should be expected to maintain, troubleshoot, and repair (if necessary) electrical, mechanical, propulsion, fuel, and software systems. The operations technician may require special skills to support these tasks, including, but not limited to, soldering, programming, diagnostic procedures, experience with cable/crimp and harness, experience working with composites, and a basic understanding of physics of flight. sUAS OTs should be adept in use of a voltage meter and experienced in working with varied power sources (battery and liquid fuel). Further, the panel suggested that an sUAS OT should have the knowledge and experience necessary to disassemble and reassemble a basic quadcopter.

In addition to having necessary trouble-shooting and technical skills, an sUAS OT also needs to be adept at maintaining organized records and logs, as appropriate documentation can help to mitigate serious maintenance issues. Battery logs, maintenance logs, software update documentation, and flight logs provide examples of critical elements of record keeping. Such records can then be reviewed in the event that an anomaly occurs during flight testing, pre-flight inspections, or flight operation.

4.4.3.5 Coordinate Flight Operations Logistics

Many of these logistical issues are associated with leadership and the organization of people, equipment, travel (either transportation of people or packing and shipping of equipment). These tasks require systematic organization and planning skills, as well as an attention to detail.

4.4.3.6 Maintain Proficiency, Currency, and Recency in Professional Knowledge/Skills

As with most technical employment opportunities, maintaining currency and pursuing professional development opportunities are essential to supporting long-term career development of sUAS OTs. This is especially critical, given the rapid pace of change associated with flight components, sensors, software, and the evolving regulatory environment. Such tasks include maintaining professional memberships in professional organizations (local, regional, state, and national), maintaining proper certifications, participating in workforce development opportunities including Webinars and forums, and organizing and/or participating in community awareness opportunities (high school drone clubs, university drone clubs, 4-H activities, scouting activities, science festivals, etc.). Another important item is to continue training and gaining experience with all phases of the operation of an sUAS, because every flight and every mission is unique. There is always something to be learned.

4.4.3.7 Perform Administrative Tasks

DACUM panel members also identified an array of administrative tasks often conducted for sUAS OTs. Many skills associated with these administrative tasks are also required throughout other duties categories of the DACUM. These skills include strong oral and written communication skills (proposal writing, report submission, logs, and budgets, and developing and providing education and training to coworkers).

4.4.4 The DACUM Chart: In Summary

The DACUM Panel maintained that sUAS OT's need to be adept and proficient at an array of skillsets in order to meet employer needs. And, as mentioned earlier, operating aircraft is only a very small component in a much wider breadth of required skills. An sUAS OT requires organizational and leadership skills. While many of these skills can be acquired in the classroom, many skills identified by the DACUM Panel are best acquired through valuable first-hand experience and hands-on training in the field. Students should be exposed to networks outside of the classroom, including, but not limited to: robotics clubs, drone clubs, 4-H, and/or scouting drone programs, robotics, and other technical courses, and programming opportunities.

4.5 Launching the DACUM: Developing Curriculum Pathways for sUAS Operation Technicians (UAS OT)

Service-learning courses, designed to take students through all facets of an sUAS project, provide excellent vehicles to augment classroom instruction, with real-world experiences, which introduce many tasks and duties associated with the sUAS DACUM.

Conducting an occupational analysis through a DACUM represents an initial step in the educational process. Perhaps the most challenging endeavor is the transfer of knowledge gained from the DACUM to ensure that educational courses have been properly developed to match the desired skills of industry.

The DACUM provides a comprehensive collection of duties, tasks, and associated skills vetted by industry. These duties, tasks, and skills identified in the sUAS OT DACUM are not intended to be covered in a single course, but rather through a series of courses. The VCCS, through the leadership of GeoTEd-UAS partner institutions, including faculty from Mountain Empire Community College (MECC) and Thomas Nelson Community College (TNCC), and Virginia Space Grant Consortium staff and consultants, developed four new courses that contain course content to teach the skills, knowledge, and competencies aligned with the DACUM chart.* Faculty from Danville Community College developed

---

* Cherie Aukland, Associate Professor, Thomas Nelson Community College; Chris Carter, Deputy Director, Virginia Space Grant Consortium; Fred Coeburn, Instructor, Mountain Empire Community College; and David Webb, Consultant, GeoTEd-UAS Project.

a fifth course, UMS-112, to address the growing need to collect, store, and organize sUAS data. These five new courses are now available to be used by all 23 community colleges in Virginia. The foundation for all of these courses is associated with the needs of industry voiced through the sUAS OT DACUM.

Prior to developing new courses, the GeoTEd-UAS team worked with the VCCS to develop a new unmanned systems (UMS) course prefix. UMS was identified for the new prefix (instead of UAS) as community college institutions and the VCCS recognize that sUAS is a component of a broader umbrella of 'unmanned systems' (which includes air, land, and marine). Air, land, and marine unmanned platforms are anticipated to significantly contribute to Virginia's New Economy (Beach et al. 2017; Virginia Unmanned Systems Commission 2018). These UAS course descriptions are available online (http://geoted-uas.org/resources/) and include:

- UMS-107: Small Unmanned Aircraft System (sUAS) Remote Pilot Ground School
- UMS-111: Aircraft Systems (sUAS) I, UMS 111 Small Unmanned
- UMS-112: Program and Flight Data Management
- UMS-177: Small Unmanned Aircraft System (sUAS) Components and Maintenance
- UMS-211: Small Unmanned Aircraft Systems (sUAS) II

These new UMS course descriptions and content were developed to be broad enough for individual community colleges to tailor them to either meet local needs or for a broader general purpose UAS education (field data collection approaches), to support specific programs (real estate, public safety, forestry, geospatial technologies, surveying, agriculture, biology, environmental studies, etc.). Individual colleges can package the courses with other related courses to develop a variety of flexible pathways to meet workforce demand. The courses can be offered sequentially, alone as part of a certificate program, or as required courses in support of a two-year associate degree program. The sUAS OT DACUM provides the foundation for options by identifying tasks that students will need when entering industry. Examples of courses that have been adopted by community college institutions in Virginia are included in Table 4.2.

After the UMS courses were established, formal academic pathways were developed to prepare the sUAS workforce in Virginia. TNCC developed a Career Studies Certificate (CSC) in sUAS Flight Technician, and plans to develop a CSC in sUAS-OT that are stackable certificates

TABLE 4.2

Summary of New sUAS Courses Adopted by VCCS Member Institutions

Community College	UMS 107	UMS 111	UMS 177	UMS 211	UMS 290/296
Blue Ridge (BRCC)	X	X	X	X	
Dabney S. Lancaster (DSLCC)	X	X	X	X	
Danville CC	X	X	X	X	
Eastern Shore (ESCC)		X			
John Tyler (JTCC)	X				
Mountain Empire (MECC)	X	X	X	X	X
New River (NRCC)	X		X		
Thomas Nelson (TNCC)	X	X	X	X	
Virginia Highlands (VHCC)	X	X	X		

into AAS degrees in Information Technology and Mechanical Engineering Technology at the College (https://tncc.edu/programs/small-unmanned-aircraft-drone-systems-flight-technician-certificate). TNCC also partnered with Peregrine Solutions for an apprentice-ship program in Cyber-UMS, the first registered apprenticeship program of its kind. MECC developed an Associate's of Applied Science degree in Technical Studies with an emphasis on sUAS (www.mecc.edu/programs/unmanned-aerial-systems-uas/). The 68-credit A.A.S. degree combines the UMS courses with other courses designed to prepare a small UAS oper-ations technician for the workforce. Both TNCC and MECC pathways are being shared across the Commonwealth and throughout the nation as models for other colleges.

Both colleges have also developed campus policies for UAS and drone use that have been approved and implemented by the college. Establishing an effective UAS policy to regulate missions and integration is often the first step in establishing a UAS pathway. Through a partnership with Liberty University, three of the UMS courses offered at any Virginia community college can now transfer to Liberty as equivalent UAS courses in their BS in the Aeronautics–Unmanned Aerial Systems Cognate degree program.

Virginia community college students who complete the three-course sequence, UMS 107, UMS 111, and UMS 211 can transfer these courses to Liberty University for equiv-alency with two Liberty courses, AVIA 230 (Unmanned Aerial Systems) and AVIA 235 (Small UAS Ground). This partnership makes it easier for students to transfer community college credit to Liberty University, a four-year institution, which helps them to further their academic pursuits in UAS. It is anticipated that this successful articulation agreement will create a precedent and likely facilitate additional transfer agreements between VCCS colleges and other four-year institutions.

4.5.1 Teaching the Teachers

Creating course options and educational pathways is important. However, it is essential that there are faculty 'in place' who can provide students with access to these courses across the community college system. Faculty development is therefore crucial to provide these academic opportunities and pathways. The GeoTEd-UAS Institute was therefore developed in response to the need for faculty sUAS education.

During spring 2017, spring 2018, and spring 2019, Virginia Tech hosted the GeoTEd-UAS Institute. Supported through NSF ATE funding, the GeoTEd-UAS Institute provides an intensive week of hands-on sUAS education for community college faculty and high school dual enrollment teachers from Virginia and Maryland. The sUAS OT DACUM pro-vides the foundation for instruction at the institute, as tasks and elements from all seven major duties associated with sUAS OT DACUM are covered. Participants receive targeted instruction required to facilitate their efforts to integrate sUAS education in the class-room through hands-on, lab-based exercises supplemented with flight operational expe-rience in the field. Figure 4.3 shows community college faculty learning about different aspects of unmanned aircraft, including maintenance, operation, and image processing. This is a 'train the trainer' model, whereby participants not only acquire the knowledge required for their remote pilot certification, but faculty participants also receive valu-able and experience on an array of subjects that span the sUAS OT DACUM. Participants receive instruction on both fixed wing and rotor-based aircraft, and receive exposure and image processing experience using a variety of sensors. They then become sUAS instruc-tors at their respective two-year colleges and high schools. Faculty are expected to inte-grate sUAS instruction into their respective disciplines (wildlife, agriculture, earth science, forestry, etc.) or implement new sUAS courses and programs at their respective colleges.

FIGURE 4.3
Through the GeoTEd-UAS Institute, community college faculty learn about many of the skills associated with an sUAS Operations Technician, including: aircraft maintenance, aircraft operation, and image processing and interpretation.

These individuals are also encouraged to support outreach activities, which might include drone clubs, drone demonstrations, etc.

Instruction does not cease at the conclusion of the week-long GeoTEd-UAS institute. Institute cohort members receive mentoring by program leaders over the course of the year after each of the spring institutes. Mentoring can include onsite support, in the form of guest presentations, assistance with sUAS-based student service learning (SSL) projects and other outreach programs (including community sUAS youth programs, etc.). In addition to receiving support from their assigned mentors, institute cohort participants also forge a strong community of practice, and support each other. Such networks are vital to support the enrichment and sustainability of courses. This is especially the case with sUAS education, as there is no fail proof formula, or approach to complete a successful mission. Every mission is unique.

GeoTEd-UAS community college partners have been integrating the sUAS OT DACUM not only through traditional classroom-based environments, but through real-world educational experiences as well through SSL opportunities. In SSL courses and missions, community college faculty partner with organizations to plan and operate missions, collect real-world data and use analytical software to solve problems. Faculty-led SSL missions are the primary method through which GeoTEd-UAS partners are teaching sUAS skills and concepts found in the DACUM. By interacting with customers to planning and coordinating a mission, students gain experience with the full breadth of applied skills to use sUAS complex problems using an sUAS.

The GeoTEd-UAS team, in partnership with the Virginia Cooperative Extension (VCE) provided cohort members with valuable sUAS-based image acquisition experience over agricultural operations. Extension agents and agricultural producers have, in turn, gained valuable information associated with the challenges and benefits of sUAS operation, as well as value of imagery associated with sUAS payloads. The team has conducted missions that include inspections of infrastructure, including silos and barns, orthomosaic images of property, NDVI of crops and pastureland, and videography. Figure 4.4 shows GeoTEd-UAS cohort participants and instructors after flying service-learning missions over a commercial agricultural operation in Pulaski County, Virginia.

The GeoTEd-UAS team also partnered with The Nature Conservancy (TNC) to plan and operate SSL missions to support the data collection and analysis needs of TNC.

FIGURE 4.4
Service learning in partnership with Virginia Cooperative Extension. (Photo credit: Stacey Kuhar Photography.)

The partnership has been piloted with missions conducted on Virginia's Eastern Shore, but the goal is to expand the partnership to include all college service regions in Virginia. GeoTEd-UAS has established a student service-learning model for all community colleges to work with TNC across the state. The team has operated more than 20 missions and mapped more than 3,000 acres producing high resolution imagery, maps, 3-D models, and other products that have been shared with TNC to support their decision making. Figure 4.5 shows faculty participating in a service-learning course at the Eastern Shore, Virginia.

FIGURE 4.5
Service-learning courses with The Nature Conservancy integrates classroom knowledge with real-world experience. (Photo credit: Chris Carter.)

FIGURE 4.6
Real-world service-learning opportunities can provide engaging and rewarding experiences for both students and educators. (Image credit: Stacey Kuhar Photography.)

The GeoTEd-UAS team, led by high school teacher and GeoTEd-UAS Institute cohort member Veronica Spradlin, partnered with Virginia Tech to conduct UAS operations to map and model Lane Stadium, Virginia Tech's football stadium which seats more 66,000 attendees. Spradlin, her students, and other operators led several UAS operations, and collected over 2,000 images of the stadium resulting in a variety of geospatial products, images, and digital surface models. The student service-learning missions were some of the first multiple UAS operations in Virginia Tech's campus airspace outside the FAA test site area. These operations were a proof of concept and demonstrated that UAS missions can be safely and efficiently conducted on campus and will result in future service-learning partnerships. Figure 4.6 shows a high school teacher and student conducting service-learning flights in Lane Stadium; Figure 4.7 provides an oblique perspective of the stadium, and an orthomosaic that was generated by the high school students is provided in Figure 4.8. These resources are now used to help recruit students into STEM related programs at the high school. As a result, the high school has started offering a Career and Technical Education (CTE) program drone course.

Individual community colleges across the state are starting to implement SSL models to further reinforce the skills in the sUAS OT DACUM. MECC's summer 2018 UMS-177 class conducted five missions to collect multispectral data from University of Virginia sponsored hemp fields in three different locations in rural southwest Virginia. Using a DJI Matrice equipped with a Parrot Sequoia sensor, the class was able to generate a normalized difference vegetation index (NDVI) to help identify the best variety of hemp to grow in southwest Virginia. MECC is currently meeting with the leaders of the Appalachia Sustainable Development (ASD) organization to add more service-learning opportunities in precision agriculture at the farms sponsored by ASD to the MECC UMS program. Faculty from

FIGURE 4.7
An oblique image of Lane Stadium. (Image credit: Veronica Spradlin and GeoTEd-UAS.)

Germanna CC conducted an SSL project over a large farm in Fredericksburg, Virginia. Faculty from TNCC worked with a sustainable strawberry farm outside of Williamsburg to facilitate the farmer's efforts to establish a new drip irrigation system.

The GeoTEd-UAS team has led the development of new student outreach materials and programs for students in middle school through college. The programs included hands-on flying time with competitions, UAS kits, drone-building opportunities, access to a variety of vehicles and sensors, and exposure to professionals working the field. These hands-on experiences are essential, as they not only provide students with invaluable knowledge and skills, but also can serve as a 'marketing hook' to initially engage students. After an initial introduction, students often gain an appreciation and further interest to explore other STEM areas, including some remote sensing image processing. Figure 4.9 shows a middle school student, who has just successfully built his first drone.

Examples of outreach initiatives can include the following:

- Hosting STEM day events with demonstrations, hands-on flying, and competitions;
- Organizing UAS Tracks during the conferences and faculty peer group meetings;
- Partnering with scouts, 4-H, and other information educational programs to provide an introduction to data collection using drones;
- Working directly with CTE teachers, high school drone clubs, and governor's schools to disseminate information about drone-supported applications;
- Identifying opportunities for student scholarships to encourage participation and support recruitment in UAS pathway courses.

FIGURE 4.8
An orthomosaic of Lane Stadium captured with multiple unmanned aircraft. (Image credit: GeoTEd-UAS, Veronica Spradlin, and Daniel Cross.)

FIGURE 4.9
Building an unmanned aircraft is a great way to learn how they work. This is also a great marketing hook to encourage younger students to become more involved in STEM related fields. This student has since joined his school's robotics team. (Photo credit: Stacey Kuhar Photography.)

4.6 Summary

sUAS represents an emergent, transformative technology anticipated to permeate across many sectors of the economy. While the unmanned aircraft industry is currently in its infancy, long-term economic prospects associated with this industry are extremely promising. New instruction, courses, curriculum models, and educational pathways must directly target industry needs, and those of our broader society. Thus, understanding the needs of industry is paramount to developing successful courses, educational programs, and academic pathways. A DACUM was established to chart the tasks and duties associated with sUAS operation technicians.

The sUAS OT DACUM serves as the foundation to develop curriculum models for courses, identify proper course sequences, design new degree and certificate programs, and establish overall educational pathways for high school dual enrollment teachers and two-year community college faculty in Virginia. Workforce development programs for educators and other professionals seeking to enter the sUAS domain are using the DACUM chart to validate curriculum and educational initiatives. Faculty are further utilizing the sUAS OT DACUM to help better facilitate informal learning, through outreach programs, and by providing service-learning options for students. While the sUAS DACUM chart has been used extensively in Virginia, the sUAS DACUM chart can be easily be modified to take into account local contexts, and can be applied to other geographic areas to support sUAS education.

Acknowledgments

The authors wish to acknowledge the following individuals for their contributions:

Cherie Aukland (Thomas Nelson Community College), Fred Coeburn (Mountain Empire Community College), Chris Carter (Virginia Space Grant Consortium), and David Webb (John Tyler Community College) for content contributions to this chapter. Jennifer Stevens (Virginia Advanced Studies Strategies) for her expertise as a certified DACUM facilitator. The sUAS OT DACUM Panel members (identified in the text) for the time and expertise during the development of the sUAS OT DACUM. Stacey Kuhar, for her photographic expertise. Chris Bruce (The Nature Conservancy) and Morgan Paulette (Virginia Cooperative Extension) for their assistance with Student Service Learning (SSL) projects. Daniel Cross (Virginia Tech Conservation Management Institute) for his contributions and leadership with sUAS. The National Science Foundation Division for Undergraduate Education for their continued support to support for GeoTEd-UAS (NSF DUE #1601614), which provided funding for the DACUM, and the GeoTed-UAS Faculty Development Institute.

References

Anderson, K., and K.J. Gaston. 2013. "Lightweight Unmanned Aerial Vehicles Will Revolutionize Spatial Ecology." *Fronteirs in Ecology and the Environment* 11(3):136–146.

Barden, J.P., B.S. Valentine, and B.S. Culliver. 2017. "Seeing is Believing – Improving Spaceport Planning & Public Outreach through Visualization." AIAA SPACE and Astronautics Forum and Exposition, Orlando, FL.

Beach, R.P., V.L. Stouffer, G.J. Belcher, and P.M. Schultz. 2017. "Commonwealth of Virginia Unmanned Aerial Systems Strategic Economic Development Plan." Plan: Commonwealth of Virginia Unmanned Aerial Systems Strategic Economic Development.

Bellamy, W. 2017. "US Now Has 60,000 Part 107 Drone Pilots." *Avionics.*

Christie, K.S., S.L. Gilbert, C.L. Brown, M. Harfield, and L. Hanson. 2016. "Unmanned Aircraft Systems in Wildlife Research: Current and Future Applications of a Transformative Technology." *Fronteirs in Ecology and the Environment* 14(5):241–251.

Colomina, I., and P. Molina. 2014. "Unmanned Aerial Systems for Photogrammetry and Remote Sensing: A Review." *ISPRS Journal of Photogrammetry and Remote Sensing* 92:79–97.

Commonwealth of Virginia. 2016. "Unmanned Systems Commission Final Report."

DeOnna, J. 2002. "DACUM-A Versitile Competency-Based Framework for Staff Development." *Journal for Nurses in Staff Development* 18(1):5–13.

Eninger, R. 2015. "Unmanned Aerial Systems in Occupational Hygiene—Learning from Allied Disciplines." *The Annals of Occupational Hygiene* 59(8):949–958.

Federal Aviation Administration. 2015. "Operation and Certification of Small Unmanned Aircraft Systems: Notice of Proposed Rulemaking (NPRM)."

Federal Aviation Administration. 2018a. "FAA 2018 Summary of Performance and Financial Information."

Federal Aviation Administration. 2018b. "FAA Aerospace Forecast 2018–2038." Federal Aviation Administration.

Floreano, D., and R.J. Wood. 2015. "Science, Technology and the Future of Small Autonomous Drones." *Nature* 521:460–466.

Freeman, P. and Freeland, R. 2015. "Agricultural UAVs in the U.S.: Potential, Policy, and Hype." *Remote Sensing Applications: Society and Environment* 2 (December):35–43.

Gallik, J., and L. Bolesova. 2016. "sUAS and the Application in Observing Geomorphological Processes." *Solid Earth* 7:1033–1042. doi:10.5194/se-7-1033-2016

Gibbens, D. 2014. "Integrating UAS into the Oil and Gas Industry." *Journal of Unmanned Vehicle Systems* 2(2):vi–vii.

Jenkins, D., and B. Vasigh. 2013. "The Economic Impact of Unmanned Aircraft Systems Integration in the United States."

Jimenez-Sanchez, C., R. Hanton, K.A. Aho, C. Powers, C.E. Morris, and D.G. Schmale III. 2018. "Diversity and Ice Nucleation Activity of Microorganisms Collected with a Small Unmanned Aircraft System (sUAS) in France and the United States." *Fronteirs in Microbiology* 9. doi:10.3389/fmicb.2018.01667

Johnson, J. 2009. "DACUM Research Chart for Remote Sensing Specialist – California In *Adobe pdf.*" edited by ATE CENTRAL: National GeoTEch Center. www.geotechcenter.org/uploads/2/4/8/8/24886299/remotesensingdacum_mc_2011.pdf. Accessed 8 August 2019.

Johnson, J. 2010. "What GIS Technicians Do: A Synthesis of DACUM Job Analyses." *Journal of the Urban and Regional Information Systems Association* 22(2):31–40.

Kornatowski, P.M., A. Bhaskaran, G.M. Heitz, S. Mintchev, and D. Floreano. 2018. "Last-Centimeter Personal Drone Delivery: Field Deployment and User Interaction." *IEEE Robotics and Automation Letters* 3(4):3813–3820.

Kuhn, K. 2016. "Small Unmanned Aerial System Certification and Traffic Management Systems." Rand Corporation.

Locascio, D., M. Levy, K. Ravikumar, B. German, S.I. Briceno, and D.N. Mavris. 2016. "Evaluation of Concepts of Operations for sUAS Package Delivery." AIAA Aviation Technology, Integration, and Operations Conference, Washington, DC, June 13–17.

Lovelace, B., and J. Wells. 2018. "Improving the Quality of Bridge Inspections Using Unmanned Aircraft Systems (UAS)." Minnesota Department of Transportation, Collins Engineers, Inc.

McGee, J. 2017. "Unmanned Aircraft Systems for Natural Resource Managers." *Virginia Forests* LXXIII(3):6–11.

Pasztor, A., and R. Wall. 2016. "Drone Regulators Struggle to Keep Up with the Rapidly Growing Technology." *Wall Street Journal*, July 10. www.wsj.com/articles/drone-regulators-struggle-to-keep-up-with-the-rapidly-growing-technology-1468202371. Accessed 8 August 2019.

Preble, B. 2015. "A Case for Drones." *Technology and Engineering Teacher* 74(7):24–29.

Price, E. 2017. "Beer Delivery Drones have Arrived in Iceland." *Fortune*.

Reid, D. 2016. "Domino's Delivers World's First Ever Pizza by Drone." *CNBC*.

Tang, L., and G. Shao. 2015. "Drone Remote Sensing for Forestry Research and Practices." *Journal of Forestry Research* 26(4):791–797.

Teal Group Corporation. 2018. "Teal Group Predicts Worldwide Civil Drone Production Will Soar Over the Next Decade." www.tealgroup.com/index.php/pages/press-releases/54-teal-group-predicts-worldwide-civil-drone-production-will-soar-over-the-next-decade. Accessed 17 February 2019.

Virginia Unmanned Systems Commission. 2018. "Unmanned Systems Industry in Virginia." Virginia Unmanned Systems Commission.

Willmett, J., and G. Herman. 1989. "Which Occupational Analysis Technique: Critical Incident, DACUM, and/or Information Search?" *The Vocational Aspect of Education* 41(110):79–88.

5

Federal Government Applications of UAS Technology

Bruce Quirk and Barry Haack

CONTENTS

5.1 Introduction

Many Federal agencies are investigating how Unmanned Aircraft Systems (UAS) technology, especially small UAS, can be applied to their missions. Remotely-sensed data and the information derived from these data have played a key role in the scientific and operational activities of many agencies. The rapid development of UAS technology, particularly smaller, light-weight, less-expensive sensors that require less power, have provided new opportunities to acquire aerial data repeatable over an area at a low cost and more safely than with other methods. UAS also provide access to hazardous or inaccessible areas with fewer disturbances to flora and fauna, and are ideal for imaging transient events, such as floods, or providing rapid response. Both rotary and fixed wing UAS bring different capabilities to meet Federal requirements, especially when combined with other remote sensing and *in situ* data sets. These digital data sets can be archived to create a baseline for future multi-temporal studies.

The United States' economic security and environmental vitality rely on continual observations of the Earth's land surface to understand changes on the landscape at local, regional, and global scales. Improving the ability to monitor, analyze, and permanently record these changes promotes sustained, informed economic growth, well-managed natural resources, environmental awareness, national security, and the advancement of scientific knowledge. This information is critical to the formulation of successful public policy and is used by officials and decision-makers in fulfilling their public-service responsibilities and UAS technology contributes to meeting these responsibilities.

This chapter will focus on the different applications that have been and are being developed across the Federal government with this rapidly changing technology. The types of applications include monitoring wildland fire, agriculture, coastal and river studies, cryosphere, geologic and mining, mapping and surveying, public safety, marine, water and hydrology, wildlife monitoring, climate change, damage assessment, and forestry. Not since the introduction of the airplane or GPS has a technology had such a disruptive or transformational impact on the remote-sensing community.

5.2 UAS for Agriculture

One of the most promising areas for applying UAS technology is in agriculture. The potential uses for UAS are in the areas of soil and field analysis, planting, crop spraying and monitoring, irrigation (identifying problems or when and where to irrigate), vegetation health assessment, identifying and managing disease, insect and weed problems, rangeland management, measuring the temperature of animals, and identifying diseased/sick animals in pastures. UAS sensors can provide even more valuable information in the form of very high spatial resolution data. As an example, color infrared imagery from UAS sensors can be used to identify very specific locations with problems, such as invasive plants, pests, or weeds, resulting in a more precise application of fertilizers, irrigation, herbicides, and pesticides. This type of precision application benefits cash-crop practices and overall range management with chemical-cost savings, reduced impacts to the environment, improved yields, and overall vegetation health.

There has been a significant effort applying UAS technology for agricultural applications within the Federal government and most of those activities have been within the U.S. Department of Agriculture (USDA). Among the various USDA organizations using UAS are the National Institute of Food and Agriculture (NIFA), the Agricultural Research Service (ARS), and the Animal and Plant Health Inspection Service (APHIS). USDA scientists conduct both basic and applied research using UAS, as well as funding relevant studies with other organizations, especially universities. In addition, USDA scientists publish reports, book chapters, and journal articles on the topic for the broader scientific community. The following is a review of some of the UAS activities in agriculture being conducted by Federal agencies.

5.2.1 Pest Control

The USDA APHIS Plant Program and Quarantine (PPQ) Science and Technology program has multiple activities on the use of UAS for both their ongoing and planned activities. These UAS activities include releasing sterile insects; conducting remote pest surveys; assessing plant health over wide areas; and collecting samples of seeds, grain, water, air, spores, and insects (Rosenthal 2017). Since 2014, PPQ has been working under various cooperative agreements with other governmental organizations, the private sector, and universities to develop UAS technology for plant-health protection. One project involves using both fixed wing and rotary UAS, which are safer, less expensive, and more flexible, to release sterile insects to suppress invasive pests. One disadvantage is the limited payload capacity and short flight times and consequently inability to cover large areas. One solution to these issues is to fly a group of UAS or "swarm" to deliver the sterile insects or employ larger and longer duration UAS that have increased payload capacity to cover larger areas.

In addition to the sterile insect release activities, PPQ has at least four other application areas for UAS technology. One potential activity is tree surveys, specifically for the Asian longhorned beetle (ALB). UAS are used to detect ALB-infested trees, instead of ground observers using binoculars to identify possible infected trees. The second activity involves using UAS to inspect U.S. bound vessels for invasive pests, such as the Asian gypsy moth (AGM). UAS are employed to look for AGM egg masses, instead of manually inspecting these vessels. PPQ also conducts rangeland inspections for outbreaks of grasshoppers and Mormon crickets (GHMC). The approach uses UAS data to check on the health of the rangeland grasses the GHMC impact.

PPQ and the USDA Grain Inspection Packers and Stockyards Administration (GIPSA) work together to certify the quality of U.S. grain exports. That effort requires the collection and inspection of grain exports to confirm the absence of pests and diseases. This is typically a difficult, expensive, and time-consuming activity. To assist in this effort PPQ evaluated the capability of a UAS to gather grain samples before export. The UAS was equipped with a sampling device suspended from a nylon line and a nadir-facing camera, and successfully collected grain samples (Rosenthal 2017).

5.2.2 Weed Management

The USDA ARS is employing UAS technology for weed management. One study uses a UAS equipped with multispectral or hyperspectral sensors to locate glyphosate-resistant (GR) weeds. Glyphosate is the primary ingredient in the most widely used herbicide in the United States and there is an increase in resistant weeds. Two GR weeds of concern are pigweed in soybeans and Italian ryegrass in cotton (AgResearch Magazine 2017). Huang et al. (2018) explored UAS technology to provide low-altitude, high spatial resolution data for precision weed management. Their studies included collection of color and color infra-red imagery over fields to identify weed species, determine crop damage from different doses of herbicide, and GR weed detection. Vegetative indices from these sensors were very effective in monitoring damaged crops from different herbicide application rates.

5.2.3 Rangeland Monitoring

The USDA ARS Jornada Experimental Range (JER) facility in New Mexico has incorpo-rated UAS into their studies of rangeland mapping and monitoring. JER scientists have developed processes for the acquisition, processing, and analysis of high spatial resolution UAS data and integrating the data with *in situ* measurements. These efforts have been used for improving field-sampling efforts, to derive parameters for hydrologic models, support phenologic rangeland studies, and to assist in the evaluation of rangeland distur-bance experiments (Jornada 2018).

In another study, multispectral data acquired by UAS have been used in rangeland veg-etation classifications and have the potential for integration with satellite data (Laliberte et al. 2011). Laliberte et al. (2010) also showed that small UAS can be used to inventory, assess, and monitor rangeland environments, and these data may complement or replace other measurements, such as ground-based measurements. In other studies, researchers found that including texture information and using object-based analysis on high spatial resolu-tion UAS data increased its value in rangeland monitoring (Laliberte and Rango 2009) and for classifying rangeland vegetation (Laliberte and Rango 2011).

5.2.4 Improved Crop Yields

The USDA NIFA funds basic research that advances agriculture-related sciences and technologies. One funded Kansas State University study employed UAS to assist in the breeding of better wheat varieties. The role of the UAS was to provide "precision mea-surements of plant characteristics in the field" that when combined with other informa-tion could model in-season yield predictions (K-State Research and Extension News 2016). NIFA has also funded Texas A&M to develop analytical tools using UAS data to determine if they can be used to select superior corn varieties. This study incorporates both fixed wing and rotary UAS in coordination with ground vehicle data. The analytical tools will measure the physical characteristics of hybrid corn varieties during the growing season (Figure 5.1). The goal is to identify the corn hybrid varieties that will increase yields for farmers (Fannin 2017).

The USDA also funded a study at the University of Nebraska to assess UAS for moni-toring and improving farmers' center pivot irrigation systems. The study collected both

FIGURE 5.1
Dr. Lonesome Malambo, a doctoral student in Dr Sorin Popesu's research program, flies a quadcopter over the 2017 Genomes to Fields experiment to collect images for phenotypic measurement (Texas A&M AgriLife photo by Colby Ratcliff).

(Source: Fannin 2017)

multispectral and thermal sensor data from UAS flying over crops during the growing season. In parallel to the UAS flights, there were in-field sensors, which collected soil water content and temperature. The data from these combined systems are able to create water management recommendations for center pivot irrigation systems, which maximize water use efficiency and increase crop production (Daugherty Global Institute 2018).

Hunt and Daughtry (2018) identified higher ground resolutions, light sensors for data calibration, and canopy height models from structure from motion as three advantages that UAS have when compared to manned aircraft. They also categorized UAS applications into three niches: scouting for problems, monitoring to prevent yield losses, and planning crop management operations. In addition, there is information available on crop monitoring; biomass and yield potential; and weeds, pests, and diseases (Hunt et al. 2010).

Hunt and Rondo (2017) demonstrated how multispectral imagery acquired from low-flying UAS could be used for detection of Colorado potato beetle damage in agricultural fields. They also found that that plant height calculated from structure from motion was related to the damage, but required extensive operator involvement when processing of UAS data.

Precision aerial applications of pesticides and other agro-chemicals from UAS provide farmers with the capability to target specific areas of fields, lowering the cost, and benefiting the environment (Lan et al. 2010). Huang et al. (2013) contend that because of their limited payload and endurance, UAS should focus on inspecting and treating small agricultural fields or areas not easily reached by traditional equipment and for monitoring crop health at higher resolutions.

UAS imagery has also been used to delineate tile-drain networks in agricultural fields. Williamson et al. (2019) demonstrated how UAS multispectral and thermal data assisted in delineating tile-drain networks in agricultural fields and facilitated identifying cracked tiles, which helped clarify the differences in water quantity and quality between two adjacent fields.

5.3 UAS for Natural Vegetation

As there are many types of natural vegetation and communities, there are many Federal agencies that have employed UAS to collect information on these communities. Those efforts have included different sensors and multiple informational needs. Small UAS have been used as an important source of reference or calibration data for satellite product validation. NASA has developed the Multi AngLe Imaging Bidirectional Reflectance Distribution Function small-UAS (MALIBU) that allows multi-angular remote-sensing measurements of surface anisotropy and albedo. MALIBU acquires high-resolution measurements from two multispectral sensors with different viewing geometries, which can be used to validate land products from satellites like Terra and Landsat (Wang et al. 2016).

5.3.1 Wetlands

Wetlands are one of the most complex natural vegetation communities, as well as one of the most important. There has been significant loss in the extent of wetlands both nationally and globally. One of the long-standing efforts by the U.S. Government has been the mapping of wetlands and specifically the National Wetlands Inventory conducted by the U.S. Fish and Wildlife Service (USFWS).

The U.S. Army Corps of Engineers funded a study by the Florida Cooperative Fish and Wildlife Research Unit at the University of Florida for UAS to delineate high-scale wetland vegetation communities (Zweig et al. 2015). The UAS collected true color images from a height of 150 m for a study site in the Everglades. The original imagery had a spatial resolution of 5 cm, which was too detailed for the analysis method so the images were resampled to 5 m. The resampled imagery was classified into nine wetland classes with an overall accuracy of 69%. More spectral bands and a multi-temporal approach or fewer classes were options to improve the classification results. In another wetland classification study, Pande-Chhetri et al. (2017) used UAS natural color imagery acquired at 8 cm and down-sampled to 30 cm to compare three different classifiers. The results concluded that UAS imagery has potential, but still has issues, such as image quality, that need to be addressed. Jones and Hartley (2018) have used UAS to collect high spatial resolution imagery to document current land and water fragmentation, degradation, and distribution in marsh habitats that are critical for mottled ducks in Texas and Louisiana coastal areas.

The National Oceanic and Atmospheric Administration (NOAA) has been working with the National Park Service (NPS), Pacific Northwest National Laboratory, and the commercial UAS community on a wetland restoration project within the Lewis and Clark National Park in Oregon. The UAS collected hyperspectral imagery that can be used to map vegetation, including invasive species, in the wetland environment (Lillian 2017).

5.3.2 Forestry

The U.S. Forest Service (USFS) in cooperation with Michigan Technological University are using UAS to provide data and advise on land management efforts in the Hiawatha National Forest in the Upper Peninsula of Michigan. The UAS provided data will be utilized to maintain habitat for near-shore fisheries and migrating birds; monitor water quality, quantity, and flow conditions; manage non-native invasive species that threaten the health of the lakes; and monitor the impacts of topographic features such as roads, bridges, and levees (Crooks 2017).

Dandois et al. (2015) created three-dimensional multispectral point clouds of temperate deciduous forests using structure from motion processing of UAS data. The estimates of tree canopy heights compared well to actual field measurements and can be used as part of an integrated (ground, aerial, and satellite) approach to measuring vegetative structure.

5.4 UAS for Public Safety

5.4.1 Public Safety

UAS platforms, sensors, and data-analysis systems are improving quickly and generally all aspects of these systems are becoming more economical. The costs are much more viable as applications migrate from large to small UAS. These improvements have increased interests within both the Federal and other governmental levels in the employment of UAS for public-safety applications.

According to a 2009 Bureau of Justice (BOJ) Statistics report only about 350 U.S. law enforcement agencies had aviation programs in active use (Langton 2009). The number was low because of the substantial costs of operating fixed wing and rotary aircraft. The Department of Justice (DOJ) estimated in 2013 that the cost of operating a UAS was about US$25 per hour versus US$650 per hour for manned aircraft. Thus, there is an increased use of and interest in UAS for various public safety purposes and to augment the manned aircraft activities (Office of the Inspector General 2013).

In addition to cost advantages, UAS are capable of performing a number of missions that are either too dangerous or beyond the technical capability of manned aircraft. Current public safety uses include response to and assessment of hazardous materials spills and incidents, explosive ordnance disposal incidents, search and rescue missions, barricaded subject surveillance, active-shooter incidents, execution of search warrants, and disaster response and recovery. Local law enforcement agencies often respond to assistance requests from local and State fire authorities to deliver real-time awareness on fires (Figure 5.2). Thermal sensors can be particularly advantageous for determining hot spots and flare-ups at a fire scene. Increasingly, fire agencies are deploying their own UAS and sensors. Other potential law enforcement UAS applications include crime scene mapping,

FIGURE 5.2
UAS oblique view for fire agencies.

(Source: U.S. Department of Justice 2016).

damage assessment, patrolling of critical infrastructure, three-dimensional mapping of transportation accidents, missing persons, and crowd monitoring (US DOJ 2016).

With some exceptions, these current UAS activities are conducted by local law enforcement agencies and not Federal agencies. However, the DOJ has been very involved in supporting these local agencies by publishing reports and organizing workshops and seminars for local agencies. Among these reports are guidelines for operating UAS, regulations for UAS compliance, and interactions with the public especially relative to concerns on privacy. Other Federal agencies have also created guidelines for UAS relative to concerns on privacy. These include a U.S. Department of Homeland Security (DHS) Privacy Impact Assessment for Aircraft Systems in 2013 and more specifically the U.S. Customs and Border Protection Office of Air and Marine UAS Operations and Privacy Policy (US DHS 2013; Scharnweber 2018).

5.4.2 Customs and Border Patrol

One Federal agency directly involved in the use of UAS for public safety is the DHS Custom and Border Protection (CBP). CBP employs multiple types of aircraft including manned and unmanned. The manned aircraft include both helicopters and fixed wing. Both large and small UAS are being operated. All aircraft have some type of imaging capabilities including video, natural color cameras, thermal, and radar. Increasingly they are expanding the range of more sophisticated sensors, such as multispectral, hyperspectral, and laser technology.

CBP is responsible for guarding nearly 7,000 miles of land border with neighboring countries and 2,000 miles of coastal waters surrounding the Florida peninsula and off the coast of Southern California. In addition, CBP, in partnership with the U.S. Coast Guard (USCG), protects 95,000 miles of coastal border. CBP aircraft are primarily used to: (1) patrol the border; (2) conduct surveillance for investigative operations; (3) conduct damage assessment in disaster situations; and (4) respond to officer safety scenarios (US DHS 2013). These border patrol operations are to detect, identify, apprehend, and remove individuals and contraband from illegally entering the United States between ports of entry.

5.4.3 Disasters

The various CBP aircraft also support the Federal Emergency Management Agency (FEMA), NOAA, and first responders to conduct reconnaissance missions during natural disasters. State and local law enforcement officials may also request CBP support in emergency situations. This assistance is typically when officer safety is implicated and other data sources may be inadequate because of factors, such as difficult terrain (US DHS 2013). In 2018, DHS created a program to expand their UAS sensors. The DHS Science and Technology Directorate began a study to examine commercially available sensors, other than cameras, that may be useful in search and rescue, surveillance, active shooter response, hostage situations, and other scenarios. The intent of the study is to identify new technology that will improve component operations, as well as educate suppliers on the needs of CPB, Immigration and Customs Enforcement (ICE), and the USCG so they can incorporate those needs into their products and activities (Rockwell 2018).

There has been an interesting evolution by the Federal Aviation Authority (FAA) in their support of UAS relative to disasters. At one time, their concern was unauthorized UAS potentially interfering with sponsored efforts for data collection and relief. However, with the assistance provided by UAS for hurricane response in Florida and Texas in 2017, the FAA stated that this was a landmark in UAS usage in supporting response and recovery efforts (McNabb 2017).

After Hurricane Irma in Florida, many agencies requested FAA authorization to fly UAS in affected areas. Most of these agencies were from the Federal government including FEMA, CBP, and the Air National Guard, but also private companies, such as Jacksonville Electric requested airspace access. The FAA issued over 130 authorizations for mapping of infrastructure and damage assessment, as well as search and rescue. Similarly, after the impact of Hurricane Harvey in Houston, there were many authorizations for similar purposes (Figure 5.3).

One of the important applications in these situations was to restore electrical power and the FAA authorized electrical companies to use UAS to survey areas not accessible

FIGURE 5.3
UAS view of impact from Hurricane Harvey.

(Source: McNabb 2017).

to vehicles to plan strategies. The rapid authorization by the FAA for these purposes was critical because most local airports were either closed or dedicated to emergency relief flights and the fuel supply was low. In effect, every UAS flight meant that a traditional aircraft was not adding strain to the situation (McNabb 2017).

5.4.4 Construction Safety

The Centers for Disease Control (CDC) has published information on how UAS can make construction safer. CDC stated that UAS have the potential to prevent injury and death in the construction industry where nearly 1,000 workers died in the United States in 2015. The most common UAS applications in construction include monitoring, inspection, and maintenance (Howard et al. 2017).

Monitoring a large site is a challenge for construction management. A UAS can be deployed to send video footage of site conditions to management faster and more efficiently than on-the-ground personnel. The acquired data can be converted into a three-dimensional image of the site and compared to architectural plans for progress evaluation. UAS can be equipped with other sensors, such as LiDAR, thermal, or other specialized cameras to increase monitoring capabilities. UAS can inspect a large worksite more efficiently through aerial photography and other sensors than ground inspection. The UAS can potentially detect hazardous conditions, materials, and dangerous structures. UAS can also inspect difficult locations such as on and under bridges or along highways, which can reduce cost and risk. UAS inspection of tall or difficult structures such as skyscrapers, bridges, or towers can also reduce cost and risk. UAS have also been proposed as material handling vehicles to transport tools, equipment, and materials at a construction site (Howard et al. 2017).

5.5 UAS for Volcanoes

Researchers have been using remote-sensing systems, primarily satellites and manned aircraft, for many years to monitor volcanoes. These studies have included using thermal imagery from helicopters and satellites to monitor lava flows (Patrick et al. 2017; Patrick et al. 2016; Wright et al. 2008) to integrating satellite data and models for quantitative hazard assessment (Del Negro et al. 2016). Satellite data have also been used to measure elevation changes over time caused by lava flows by comparing digital elevation models (Poland 2014). Using UAS to monitor and collect data over volcanoes is a safe, low-cost option being explored by many organizations. During the October 2004 eruption of Mount St Helens, volcanologists at the USGS accelerated the level of monitoring using a variety of ground-based sensors deployed in and on the flanks of the volcano using manned helicopters. They also developed additional unmanned sensing methods that could be used in potentially hazardous and low visibility conditions, which resulted in a UAS experiment being conducted during the ongoing eruption early in November 2004. The UAS was flown over and inside the crater to perform routine observation and data gathering, as well as to demonstrate whether UAS technology could collect real-time data at a reduced risk to scientists (Patterson et al. 2005).

Many UAS experiments have been conducted over volcanoes in Costa Rica. In 2013 scientists at the Universidad Politécnica de Madrid collaborated with the University of Costa

Rica and used UAS to explore and monitor active volcanoes in the central mountains of Costa Rica to locate hot spots and areas of hydrothermal alteration (UAS Vision 2018a). In another 2013 study of the Turrialba Volcano in Costa Rica, researchers collected sulfur dioxide (SO_2) concentrations with a UAS and used the data to constrain a SO_2 degassing flux model. The study successfully demonstrated the utility of UAS for analyzing volcanic gas compositions (Xi et al. 2016).

In another experiment, small mass spectrometers were flown with other sensors on a UAS to collect temperature, pressure, relative humidity, SO_2, hydrogen sulfide (H_2S), and carbon dioxide (CO_2) data to generate real-time, three-dimensional concentration maps of the Turrialba Volcano plume in Costa Rica. These maps were compared to data collected simultaneously from satellites (Diaz et al. 2015). In January 2018, researchers flew a UAS equipped with sensors designed to measure CO_2 and water vapor being emitted by an active volcano in Costa Rica. By monitoring and measuring the occurrence of volcanic gases emitted from the vents and fractures of active volcanoes, NASA/Jet Propulsion Laboratory (JPL) scientists expect to understand how volcanoes work and improve volcano eruption planning and warning capabilities (sUAS News 2018).

In Hawaii, researchers from the University of Hawaii at Hilo successfully mapped the active flow front of a Kīlauea lava flow with a UAS in October 2014. In a collaborative partnership with Hawaii County Civil Defense and the USGS Hawaiian Volcano Observatory (HVO), the flight team used a UAS with a visible camera to collect high spatial resolution images to create a photomosaic for use by civil defense emergency planners (UAS Vision 2018b).

Starting on May 17, 2018, the Department of the Interior (DOI) Office of Aviation Services, the USGS, and the Bureau of Land Management (BLM) deployed equipment and trained personnel to provide remote-sensing data acquisition over the impacted area of the Kīlauea eruption in Hawaii. The operations of the Multi-Bureau UAS Response Team (Team) were conducted in accordance with established authorizations from the FAA, the DOI, the National Guard, and local emergency operations organizations coordinated by the Hawaii County Emergency Operations Center (EOC). The Team flew over 1,200 flights totaling more than 300 hours of aerial geospatial and gas emission data collection. A USGS volcanologist guided mission planning in coordination with the UAS operations team lead to ensure tight integration of the Team's efforts and the needs of the USGS HVO, FEMA, and local emergency managers. The initial mission was to use visible and thermal imagery to monitor lava flows moving through residential areas and threatening critical infrastructure and to construct up-to-date digital elevation models of areas where lava flows had dramatically changed the topography. Measuring concentrations of hazardous gases were also a critical concern to emergency managers so the UAS payload was modified to collect gas data utilizing three different sensors (Figure 5.4).

On May 27, 2018, the Team was conducting mapping missions of Kīlauea's active lava flows in the Lower East Rift Zone to monitor flow directions and velocities in support of the emergency response being coordinated by the EOC. That evening the Team identified a new outbreak of lava that was moving swiftly through a residential subdivision. They immediately notified the EOC of the situation and need for immediate evacuation of residents remaining in the area. The Team also provided live video coverage of the flow's progress, which the EOC used to dispatch police and fire units and guide their emergency evacuation actions. While preparing another UAS for flight, the Team overheard radio communications about a civilian trapped by the lava flow at his residence. They confirmed the location of the residence and flew their drone to the area to lend assistance. In the darkness, the individual was spotted and instructed to follow the drone to safety.

FIGURE 5.4
USGS UAS monitoring of Kīlauea Volcano, Hawaii 2018.

(Source: USGS NUPO 2018a).

The individual began making his way through the forest toward the hovering drone. While he was making his way through the forest, the Team and EOC personnel tracked his progress and relayed location information to the first responders on the ground. After ten minutes of providing directional guidance to both the stranded person and the search team, the searchers found the man and guided him to safety.

The Team was routinely asked to provide 24×7 surveillance including separate UAS operations (requiring three shifts) to cover FEMA and HVO data requests to provide situational awareness missions for emergency managers to assess spillovers and new lava channel breakouts. All collected data, approved by government and emergency managers for release, have been organized and made available to the emergency managers and Volcano Science Centers for immediate access and investigations are underway for long-term archiving, distribution, and analysis (USGS NUPO 2018a; Zoeller et al. 2018).

The EPA National Risk Management Research Laboratory has used multiple UAS sensors for emissions analysis at a number of volcanoes in addition to Kīlauea. The primary emission is SO_2, but also CO_2 and H_2S. Multiple missions were flown for the Turrialba Volcano in Costa Rica. There were over 40 deployments over a two-year time frame for this site. Similar UAS sensors were deployed to study the La Solfatara Crater, in Italy. In addition to emissions analysis, UAS have been used to obtain high spatial resolution topography data of volcanoes for hazardous flow modeling, change detection, deformation studies, instrumentation siting, and geologic and hazard mapping (Gullett 2017).

5.6 UAS for Hydrologic Features

There are numerous examples of applying UAS technology to hydrologic applications, such as monitoring flood events, river flow and structure, ground water discharge, shoreline erosion, and river restoration.

5.6.1 Flooding

The USGS State Liaison for New Hampshire and the New Hampshire Department of Environmental Services requested a rapid response UAS mission to acquire low altitude high spatial resolution UAS imagery of flooding caused by an ice dam (jam) occurring in and near the towns of Plymouth and Holderness, New Hampshire. Rapid delivery of these data was crucial for the situational awareness and emergency response activities taking place to support any before and after flooding of the Pemigewasset River associated with the release of the ice dam.

The USGS immediately employed existing processes to establish an end-product contract with a local commercial UAS operator, who was able to start collecting the requested data within four hours and delivered the data, as well as a mosaicked image of the area, by late afternoon of the same day. Data processing of the imagery, which covered a four-mile stretch of the Pemigewasset River, resulted in the creation of a georeferenced image mosaic of the area with 3-cm ground resolution. Both the collected and processed data products were delivered to the New Hampshire Department of Environmental Services and the Cold Regions Research and Engineering Laboratory (CRREL) for briefings to the Governor, first responders, and news outlets (Figure 5.5) (USGS NUPO 2018e).

The U.S. Army Corps of Engineers (USACE) 2017 Duck Unmanned Aircraft Systems (UAS) Pilot Experiment was designed to evaluate existing and new UAS survey and monitoring techniques for the USACE Flood Risk Management Program. The diverse array of UAS sensors (LiDAR, multispectral, and high spatial resolution cameras) can collect data to estimate topography, bathymetry, terrain, land cover, vegetation, and structures at high temporal and spatial resolutions (Bruder et al. 2018).

NOAA has been conducting a River Watch program in Alaska for several years. The purpose of the program is to conduct visual monitoring and evaluation of river ice and rough volumetric snowpack changes in specific target areas prior to the breakup events using UAS technology (NOAA 2017).

FIGURE 5.5
UAS derived image of an ice dam on the Pemigewasset River in New Hampshire.

(Source: USGS NUPO 2018e).

5.6.2 Bathymetry and Elevation

The National Centers for Coastal Ocean Science tested the utility of UAS technology to provide land elevation and water depth data along the coastline and near shore waters of St. Croix in the U.S. Virgin Islands. These areas are difficult to access or are remotely located making them challenging and expensive to map with existing technologies. UAS imagery offers a potential inexpensive and accurate method to meet these requirements (Battista 2018). The NOAA Office of Coastal Management evaluated the quantitative spatial accuracy of both UAS imagery and LiDAR products from commercial UAS operators to map marsh habitat. A critical component of understanding marsh vulnerability to sea level rise requires good elevation data. The potential of LiDAR to penetrate to the ground with a small ground sampling footprint and higher point density than is available from other remote-sensing systems was investigated (Waters 2018).

Schumann et al. (2016) concluded that digital elevation models generated using structure from motion processing on UAS imagery could complement LiDAR data for floodplain mapping and modeling, especially in cases where the areas of interest are small and collecting of LiDAR data may be too costly or impractical.

5.6.3 River Change Monitoring

The central Platte River valley near Kearney, Nebraska is an internationally recognized site, which supports several million waterfowl that seasonally migrate including half a million sandhill cranes. There is concern about the sustainability of the migratory and resident birds, as well as other biota, because of changes in water and land use that have influenced the river channel and altered adjacent wetland and riparian habitats. One of the significant concerns is how river flow influences sandbar height used by nesting birds. Because of the dynamic changes in river flows, determining the sedimentation process at high temporal scales and high spatial resolution using traditional methods are challenging and costly. Kinzel et al. (2015) conducted UAS missions to ascertain if that platform could more effectively map the spatial extent and elevation of emergent sandbars along segments of the Platte River. The imagery was collected with a high spatial resolution digital camera. This sensor provides data with very accurate horizontal and vertical information, which support multiple types of hydrologic modeling and measurements applicable for the waterfowl issues in this landscape.

5.6.4 Ground Water Discharge

Red Rock Lakes National Refuge in Montana is a unique marshland, which hosts waterfowl including the rare trumpeter swan. The water budget for this refuge is significantly impacted by ground water discharge thus understanding the volume, distribution, and temporal variability of the groundwater discharge is the basis for management at the refuge. Refuge managers were aware that during the summer, the ground water discharge temperature is about 15 degrees cooler that the existing surface water. Based upon this information, a UAS flight with a thermal sensor was flown to determine if that system could map the extent of the ground discharge areas within the lakes and near the shoreline. From an initial mapping, fieldwork could more accurately be directed to map the rates of discharge and quantification of the amount of discharge. The UAS-based thermal sensor was successful in detection of temperature differences in the water areas. In addition, the thermal imagery was able to locate springs that flowed into the lake, which contributed to the water budget (Cress et al. 2015).

5.6.5 Shoreline Erosion

The upper Missouri River in South Dakota has a history of rapid shoreline erosion, as much as 8 feet per year. The erosion may affect sites of cultural value on the Lower Brule Sioux Tribe Reservation, as well as agricultural and recreational activities. The USGS flew multiple UAS missions to determine if this technology could accurately map areas of shoreline erosion. The UAS were launched from both the shoreline and from an off-shore powerboat to map the areas of interest. The primary sensor was video from which still-frame photographs were extracted and georeferenced to existing base imagery. These rectified images were then mosaicked to create a seamless shoreline image over several years. This time series of images can detect change, evaluate rates of erosion, and assist in land-management decisions (Cress et al. 2015).

The USGS has also been collecting low-altitude digital imagery over multiple locations along the Lake Ontario shoreline in New York with a camera mounted on a small UAS. These data were collected to document and monitor effects of high lake levels, including shoreline erosion, inundation, and property damage and were collected with the support of FEMA and the State of New York Departments of State and Environmental Conservation (Sherwood et al. 2018). Sturdivant et al. (2017) found that structure from motion point clouds used to produce orthomosaics and digital elevation models could extract geomorphic features with a high degree of accuracy and also improve land cover classification at a beach and wetland site near Buzzards Bay, Massachusetts.

5.6.6 Dam Removal

The Elwha River Restoration Project is the largest dam removal project in U.S. history and is coordinated by the DOI in collaboration with other Federal, State, local, and Tribal organizations. A high priority activity of the removal was the erosion and transport of sediment and the geomorphic changes that occurred to the reservoir deltas, downstream river, and shorelines. A sediment monitoring program was established that included UAS flights to track the sediment. The Bureau of Reclamation, NPS, and the USGS flew multiple UAS missions in June and September 2012 to obtain data that would provide information on the rates and patterns of change during the river restoration. One of the more critical concerns was how quickly the sediment from the reservoir was removed and how and where it was deposited downstream. Digital elevation models were created from the UAS data to monitor changes in the reservoirs and river channel, and in particular the redeposited sediment, which may affect salmon habitat and the height of flood stage. This effort provided an orthorectified imagery base of the reservoir, showing the restored river basin with the rapid change in sediment movement and deposition (Clark et al. 2013).

5.6.7 Water Management

Yueh et al. (2018) have developed a P-band Signals of Opportunity (SoOp) sensor for a small UAS that gathers data on Snow Water Equivalent (SWE) and Root Zone Soil Moisture (RZSM). The UAS will collect high-resolution observations of both SWE and RZSM to improve the estimation of terrestrial water storage for water management, crop production, and forecasts of natural hazards. The sensor integration has been completed and field testing will begin in early 2018.

Remote-sensing measurements from satellite-, aerial-, and ground-based platforms are being used to compute river discharge or streamflow, a key component of the water cycle. These systems provide measurements on river width, hydraulic grade, and water velocity. A Doppler radar system, called QCam, has been developed to fly on a small UAS and measures the along-track river surface velocity at set heights above the river channel. Field tests along the South Platte River in Colorado demonstrated surface water velocities derived from conventional and UAS data provided similar results (Fulton 2019).

5.7 UAS for Wildland Fires

Wildland fires occur in many types of terrain and may be difficult to reach due to their remote locations or rugged terrain. The consequences of wildland fires differ greatly, but in extreme cases they encompass vast expanses of land, destroy extensive wildlife and wildlife habitat, cause irreparable damage to wilderness areas, and may cause the loss of human lives and damage to property. Wildland fires may occur any time during the year, but summer in the Western states is when they are most frequent and intense, resulting in 10 to 20 or more fires being reported daily.

Variable terrain, high winds, and complex support logistics are just a few factors that contribute to the high degree of danger associated with wildfire management. These same factors also complicate the data collection activities needed to supply the real-time, critical information about fire location, perimeters, direction of spread, burn intensity, and smoke plumes.

The DOI and the USFS, working within the National Interagency Fire Community, are conducting wildland fire UAS data collection exercises and simulations. The results of these pilots indicate that small UAS are easily transportable and can be equipped for rapid deployment, can provide situational awareness, serve as a communications relay, can provide continuous weather coverage, and assist the Burned Area Rehabilitation Teams and other post-fire activities, such as aerial seeding and assessing soil erosion and debris flow management (Cress et al. 2015).

There are multiple informational needs for wildfires in grasslands, shrublands, and forests. Those needs include fire prediction, real-time monitoring, and post fire assessment. Different remote-sensing platforms and sensors have roles in these needs. More recently, UAS have served as a platform for several of these applications. The National Wildfire Coordinating Group has developed an interagency fire guide for UAS operations that standardizes the processes and procedures for using UAS in support of fire management objectives (National Wildfire Coordinating Group 2017).

The USFS and the NPS used a UAS to gather thermal infrared imagery of an active fire in Olympic National Park in 2015. That information allowed firefighters to pinpoint the fire perimeter and identify areas with intense heat. The UAS use little fuel and present a very low risk to employees in contrast to manned aircraft. Also in 2015, the DOI used a UAS to collect imagery of the Tepee Spring Fire in Idaho. The resulting information provided firefighters with near real-time information about the movement and intensity of the fires (Markiewicz and Nash 2016).

Another application of UAS for fire is to employ the UAS for prescribed fire actions. The DOI used a UAS to ignite and burn 26 acres of restored tall grass prairie at the Homestead National Monument of America in Nebraska. This successful effort determined that UAS have the potential to reduce direct risk to firefighters by doing aerial ignition work while reducing costs and making an aerial resource more widely accessible to wildland firefighting efforts (US DOI 2018a; 2018b; Johnson 2016).

NOAA is investigating using a small UAS as a platform for measuring biomass burning emissions, plume distribution, fire extent and perimeter, and supporting meteorological data, especially at night when manned aircraft typically do not operate. The Nighttime Fire Observations eXperiment (NightFOX) project objective is to develop and deploy a UAS with two modular and easily exchangeable payloads. One payload will provide measurements of CO_2, carbon monoxide (CO), and high- and coarse-mode aerosol size distributions in biomass burning plumes and the second payload will be flown over the fire to make fire perimeter and fire radiative power measurements using visible and short-, mid-, and long-wavelength infrared observations. The UAS data will be used to provide information for comparison with satellite observations, and along with measured meteorological parameters, will be used in fire-atmosphere modeling (NOAA 2018).

5.8 UAS for Wildlife

There are many examples of the use of UAS for wildlife studies by different Federal agencies either directly by Federal scientists or by others with Federal support. Some of the Federal agencies that have activities on this topic include NOAA, the USDA, and various bureaus of the DOI, such as the USGS, USFWS, and NPS. The applications for UAS and wildlife include population counts, habitat assessment, and damage caused by wildlife. There are many animal species and/or their habitats that have been studied including sandhill cranes, pygmy rabbits, white pelicans, salmon, and turtles.

5.8.1 Animal Population Counts

The first USGS operational UAS animal population count project took place in 2011 at the Monte Vista National Wildlife Refuge (NWR) in cooperation with the USFWS. The purpose of the project was to test the feasibility of using the sensors on a UAS to survey the sandhill cranes during their migratory stopover at the Colorado site. These surveys traditionally used either fixed wing aircraft, which can place both birds and staff at risk of mid-air collisions, or time-consuming ground methods, where biologists visually estimate the number of birds. The main objective was to determine whether a UAS thermal sensor could identify the sandhill crane heat signature in enough detail to allow biologists to obtain accurate counts. Equally important was determining how the cranes would react to the UAS, especially because of concerns that the cranes would see the UAS as a predator, and flush or fly away when it approached. These flights, at different altitudes, demonstrated that the UAS could be flown without disturbing the cranes and that the thermal sensor could detect the cranes. The sensor images were mosaicked and the crane population estimates were within five percent of the USFWS ground survey results. These results established that UAS can make safe, cost-effective, and accurate population counts of sandhill cranes (USGS NUPO 2018b; Cress et al. 2015).

In another waterfowl population study, the USGS conducted proof-of-concept UAS missions to collect imagery for use in performing ground-nesting colonial waterbird surveys at the Chase Lake National Wildlife Refuge in south central North Dakota. This NWR is home to one of the largest colonies of ground-nesting waterbirds in the northern Great Plains supporting 20,000–30,000 nests of 40,000–60,000 breeding birds and over 12 species.

Access to reliable population data is crucial for effective conservation and management of the Chase Lake waterbird colony and other colonies in the region.

Ground surveys by USFWS observers were used in the 1970s, but ground surveys are not recommended because they are known to cause disturbance and nest abandonment by American White Pelicans and other waterbirds. Annual aerial surveys of nesting pelicans with manned aircraft have proven critical in documenting the increases and decreases in American White Pelican populations at Chase Lake; however, these manned aerial surveys have potential complications, including safety, costs, and logistics. The UAS missions, flown in 2014, collected imagery for use in performing ground-nesting colonial waterbird surveys (Figure 5.6). The high spatial resolution color and infrared images provide detection of individual nesting pelicans and other ground-nesting waterbirds and proved to be a useful alternative to manned aircrafts and have the potential to be more economical, less obtrusive, safer, and a more efficient and versatile means to survey nesting pelicans and other ground-nesting waterbirds (USGS NUPO 2018d; Cress et al. 2015).

Christie et al. (2016) found UAS to be a cost-effective, safe, relatively quiet, and an effective alternative to traditional survey methods, especially over small areas to support wildlife research. Regulatory issues, data-processing times, and short flight times were recognized as issues, but the technology will become more popular as these issues are addressed and resolved. Future activities, such as detecting and monitoring nests, locating GPS or radio-tagged wildlife, and collecting biological samples were identified.

The USFWS Marine Turtle Conservation Fund supported a study of monitoring nesting Kemp's Ridley sea turtles at a site near Rancho Nuevo, Mexico during two nesting seasons. The UAS were equipped with a video camera for identification of both adult and hatchling turtles. The results indicated that the UAS provides a practical and effective method of conducting daytime surveys in nearshore waters for monitoring sea turtle abundance and movements. Another advantage is the low risk of the UAS disturbing the turtles. UAS technology can complement and enhance ongoing conservation programs and can collect data that would normally be difficult to acquire (Bevan et al. 2015).

FIGURE 5.6
Image of pelicans on Chase Island taken from a Sony S100 on a Raven UAS.

(Source: USGS NUPO 2018d).

In another study, Rees et al. (2018) determined UAS are becoming increasing popular for gathering of high spatial and temporal data due to their low-cost, safe operations, and ease of use. UAS technology is improving existing methods of studying turtle nesting, at-sea distribution, and behaviors, as well as for public outreach and engagement. Data storage and analysis of the large amounts of UAS data can be a constraint, but future advances will enhance UAS technology in several areas, such as collecting data directly from sensors on turtles.

Durban et al. (2015) conducted 60 flights of a vertical takeoff and landing (VTOL) UAS launched from a boat to collect over 18,900 images of killer whales. The VTOL was quiet enough to collect detailed images of the whales that could be used to differentiate individual whales using their natural markings and for estimating whale lengths. The VTOL was a safe and low-cost approach for collecting data and will have utility in other wildlife studies.

In another NOAA study, a VTOL UAS was used to investigate the distribution and abundance of predator populations in remote areas, such as Antarctica. The VTOL was flown in remote, windy locations with frequent cloud cover making satellite observations impractical. Images from over 60 flights were used to estimate abundance, colony area, and density of krill dependent predators in Antarctica. The VTOL characteristics of portability, flight stability, launch, and recovery from small areas, safety, and low noise or minimal wildlife disturbance make it a good fit for wildlife studies (Goebel et al. 2015). Also since 2014, NOAA has successfully implemented UAS to supplement Stellar sea lion surveys and has expanded this approach to surveying the abundance of northern fur seals. These surveys are based on capturing both thermal and high-resolution imagery of the rookery and investigate the benefits of spectral measurements (Sweeney 2018).

Johnston et al. (2017) compared both fixed wing and rotary UAS estimates of gray seal populations with data from manned flights. The population estimate results from the fixed wing UAS and manned aircraft were similar; however, the rotary UAS data were also found to be very useful for assessing the seal molt-stage. The UAS provided a flexible, low-cost, and safe data collection platform and are a reliable tool for conducting seal population studies.

Angliss et al. (2018) conducted manned and unmanned aerial surveys for monitoring cetacean distribution and density in the Arctic. The paired aerial surveys near Barrow, Alaska helped identify components that contributed to successful UAS data collection, such as Internet service and a portable weather station, and changes in UAS flight operations, such as climate-controlled storage of UAS equipment and automated aircraft position broadcasting that will improve future UAS data acquisitions.

Hanson et al. (2014) demonstrated that small UAS could efficiently detect and provide estimates of breeding greater sage-grouse on lek sites. The location of leks (breeding sites) and accurate population counts for each lek are needed to successfully manage greater sage-grouse. This proof-of-concept mission objectives included how the greater sage-grouse responded to the UAS flights, what type of sensor (visible or thermal) would best detect greater sage-grouse, can the sensors detect greater sage-grouse that are obscured by the sagebrush, and whether the sensor can distinguish males from females. The project demonstrated that the greater sage-grouse were not disturbed by the electric UAS and could provide the needed information. It also provided the beginning of a historical archive that could be used to build sequential time-series records for future research. Overall, the mission proved that UAS could provide a low-cost, high spatial resolution, and mobile platform for determining greater sage-grouse population counts. As visible and thermal sensor technology continues to advance, more of the mission objectives will be met.

5.8.2 Habitat Studies

In addition to animal census, UAS have been used by Federal agencies to evaluate the habitat of various species. The USGS partnered with Boise State University and the University of Idaho to determine if data collected by UAS could help in the analysis of the landscape habitat of pygmy rabbits. The UAS data were successfully used to map vegetation cover and document how the rabbit's habitat was becoming increasingly fragmented by development, agriculture, fire, and rangeland improvements by replacing sagebrush with grasses. The UAS data were used with other parameters to develop and test spatial models of the rabbit's habitat (Olsoy et al. 2017; USGS NUPO 2018c).

Scientists at the USGS Northern Rocky Mountain Science Center have used UAS to model talus, microclimate, and vegetative characteristics of montane landscapes. These landscapes are subject to rapid changes in community assemblages from temperature increases. Surface and subsurface temperatures may vary greatly thus affecting heat sensitive species, such as the American pika. UAS data from multiple sensors were successfully collected to better understand and protect the pika and other species (Johnson and Preston 2017).

NOAA has funded the use of UAS to quantify the restoration of juvenile salmon habitat in the National Park Service's Colewort Creek Wetland Restoration Site along the Columbia River in Oregon. The UAS collected data with a 120-band hyperspectral sensor and three-band orthophotography at 2-cm spatial resolution. The effort noted the increased affordability, flexibility, data resolutions, and weather adaptability of the UAS for this application. The project collected and processed an extensive amount of imagery and was supported by field collection of wetland vegetation species. These combined activities were considered very successful and resulted in a completely operational UAS for this application (Roegner et al. 2016).

NOAA researchers also investigated the feasibility of UAS to provide regional systematic ship-based surveys for seals in seasonal Arctic ice. A fixed wing UAS collected over 27,000 images from 10 flights that were analyzed for the presence of seals. These images indicated a reduction in seal disturbance by the UAS compared to manned aircraft data collection. These results suggest that as UAS technology improves, surveys of arctic habitats can be accomplished at reduced costs and safer than traditional methods (Moreland et al. 2015).

5.8.3 Landscape Damage

The USDA APHIS National Feral Swine Program's goal is to minimize the damage caused by feral swine and to protect agriculture and livestock, natural resources, property, and human health and safety. Suppressing feral swine populations in states where they are well established and eliminating them in states where they are new or just getting established accomplishes this goal. Training on UAS technology began in 2016 and both fixed wing and rotary UAS are being used for detecting and mapping feral swine damage and surveillance to locate feral swine (Lutman 2017; Ravindranath 2017). Another USDA study used UAS to assess damage to rice crops by feral swine in Louisiana. Data from these systems can be used to accurately identify and assess damage to rice fields and once verified, be extrapolated to estimate the economic damage over larger areas to verify where feral swine control projects need to focus, whether crop damage is increasing or decreasing, and the success or failure of the feral swine program (RiceTec 2018).

5.9 UAS for Public Health

There is considerable interest and application in the employment of UAS for different aspects of public health. Those applications typically are either for information collection or for transport of materials. The information collection is primarily for disaster assessment and relief, and is also considered in the section of this chapter on public safety. The transport of materials is for movement of medical supplies and also samples for testing. Several Federal agencies are either actively employing UAS for these purposes or are promoting their use by increased awareness and funding. Leaders among these agencies are the United States Agency for International Development (USAID), but there is also interest by NASA and the CDC.

USAID has produced comprehensive reports on UAS for public health to broaden the awareness of these applications, as well as document specific examples (USAID 2017a; 2017b). Those reports describe many of the past, current, and potential UAS roles in public health. There are many pilot projects ongoing to further these activities. Additionally, a Humanitarian Unmanned Aerial Vehicles (UAV) Network has created a Code of Conduct to ensure safe and effective use of these platforms in humanitarian and development settings. The potential of UAS is being further explored in a testing corridor established in 2017 in Malawi through coordination of UNICEF and the Government of Malawi (USAID 2017b).

Much of the role of USAID and other similar organizations is to distribute information and develop guidelines for successful inclusion of UAS in public health. Their approach is to develop example case studies, test those opportunities, evaluate the results and identify challenges, and finally develop an investment plan for broader implementation. Multiple UAS-based sensors have been activated for humanitarian missions primarily for surveillance following natural disasters. One UAS application is vector research for malaria. UAS sensors also have been employed in Haiti for flood mitigation and refugee camp management. In Tanzania, the World Bank has funded a UAS mapping effort of informal settlements. Those maps have led to faster response during cholera outbreaks in Tanzania (USAID 2017b). Another interesting application being examined is the use of UAS sensors to assist in demining efforts; unfortunately, there are still many areas of land mines in many countries.

5.9.1 Material Transport

The USAID Global Health Supply Chain Program-Procurement and Supply Management functions to ensure uninterrupted supplies of health commodities in support of U.S. Government funded public health initiatives globally. One aspect of that program is to explore the potential of incorporating UAS for moving health commodities through public health supply chains. The expectation is that UAS can circumvent problematic infrastructure on the landscape, speed up delivery processes, and make supply chains more responsive.

The demand for UAS for transport is increasing in many sectors and especially for public health. With the ability to fly considerable distances and over difficult terrain, they can accelerate delivery of needed, often live-saving, medical supplies. These systems can also improve access to non-emergency supplies for locations where health worker visits and commodity deliveries are infrequent. Perhaps the first application of UAS for delivery of medicine and other supplies for public health occurred in the Dominican Republic and Haiti after the major earthquake in 2013 (USAID 2017b).

Among the transport of materials for public health opportunities is the delivery of blood for life-saving transfusions. Another example is to deliver medical diagnostic testing kits and then return samples to a central laboratory for analysis. A third, but very different, transport is vector release programs often of sterile male insects, such as mosquitos for disease control. Examples of these efforts are Zika-carrying mosquitoes in Brazil and Tsetse fly in Ethiopia. There are examples of this by USDA in this chapter's discussion on agriculture, but for crops rather than people or animals.

Another example of UAS for sample delivery was by Médecins Sans Frontières in 2014 for tuberculosis testing in Papua New Guinea (USAID 2017b). The results of this effort were very positive and there have been increased numbers of similar applications. These platforms can connect remote health centers to central hospitals, labs, and distribution centers. A significant advantage is to reduce the time necessary for testing. USAID supported a similar program in Madagascar for transporting of blood and stool samples in 2016. The samples were transported from rural villages to a research station where they could be properly stored and analyzed. This initial demonstration creates opportunities for bringing health care to those living in very remote areas. The study indicated that UAS could also be used to deliver vaccines (USAID 2017b).

Another international example of UAS for medical transport involved delivery of blood samples and other medical items in Rwanda. The Government of Rwanda has commissioned a commercial firm for blood deliveries to 21 transfusion facilities and to place all citizens within a 30-minute delivery zone for essential medical products. A similar example from Sweden has tested UAS to deliver automated external defibrillators to out of hospital cardiac arrests. USAID in their 2017b report and in its efforts to improve awareness of UAS for public health has an Annex with a list of 42 humanitarian UAV projects organized by country.

Domestically, in the United States, a clinic in southwest Virginia, Health Wagon, partnered with NASA, Virginia Tech's Institute for Critical Technology and Applied Science, and others to fly the first UAS mission approved by the FAA to deliver critical medications to pharmacists that were given to patients in isolated pockets of Appalachia. The program was very successful, but experimental because of FAA restrictions preventing ongoing regular deliveries. Those restrictions are likely to be changed in the future (Gardner 2016).

5.10 UAS for Environmental Monitoring

There are multiple examples of the use of remote sensing for monitoring and inspection of varied environmental issues. For these applications, various Federal agencies have conducted basic and applied remote-sensing science with UAS as described in the following sections.

5.10.1 Oil Spills

The NOAA Bureau of Safety and Environmental Enforcement have funded and assisted research on the use of UAS for tactical oil spill response operations. These studies have not only examined UAS for mapping the location and extent of an oil spill, but also methods to determine the oil thickness (Garcia-Pineda 2018).

5.10.2 Contour Surface Mines

Inspecting mine sites to monitor water quality, hazardous conditions, terrain, wildlife habitats, land use after mining, and the safety of cultural features is typically done by inspectors walking the site, but the DOI has increasingly used UAS for these inspections because of safety, lower costs, and flexibility. For example, coal mines are required to have drainage-control devices to control mine runoff and to prevent flooding, and verifying of these structures is a very expensive, hazardous, and time-consuming process. In addition, information on the number, extent, and location of underground mine fires is important to establish for monitoring purposes. The USGS, in coordination with the DOI Office of Surface Mining Reclamation and Enforcement (OSMRE), have tested UAS for supporting inspections of mine drainage and underground fires. Multiple missions with different sensors were flown in these tests. The data from these sensors were used to produce three-dimensional models of the mining areas to make perimeter and volume measurements and included thermal data for locating coal seam fires. These missions determined that UAS are an efficient tool for monitoring extensive mining locations (Cress et al. 2015).

5.10.3 Abandoned Mines

Abandoned mines have multiple potential concerns including landslides, erosion, and burning coal refuse. The Federal government is required to inspect abandoned mine sites for any necessary remediation. A UAS experimental mission was conducted at the Coal Basin mining operation in Pitkin County, Colorado to identify potentially dangerous aspects of an abandoned mine, such as portals or other openings.

The Coal Basin complex includes five adjacent underground mines, a rock tunnel entry, several waste piles, and numerous ancillary facilities. A UAS mounted camera provided stereo images to produce three-dimensional photogrammetric models of the site. The mission was successful for monitoring and inspecting abandoned mine areas (Cress et al. 2015).

5.10.4 Abandoned Solid Waste

The USGS flew UAS missions in the Mojave National Preserve in California to determine if the high spatial resolution imagery could locate and map abandoned solid waste, aid the difficult task of identifying abandoned materials, determine historical significance, and support removal and clean up. The mission successfully demonstrated that UAS could be a very useful reconnaissance and monitoring tool for identification of abandoned materials and provide coordinates of the materials for follow-on visits from ground personnel (Figure 5.7). The imagery also could distinguish various vegetation types, such as Joshua trees, for inventory and monitoring activities (Cress et al. 2015).

5.11 UAS for Ice Applications

The Department of Energy's (DOE) Atmospheric Radiation Measurement (ARM) UAS Program was started in 1991 and has conducted a series of flight campaigns that have demonstrated that UAS technology can contribute to the understanding of cloud and

FIGURE 5.7
Image of abandoned solid waste in Mojave National Preserve, California. Image taken by a GoPro Hero camera on a Raven UAS.

(Source: Cress et al. 2015 USGS).

radiative processes. Specific examples include the evaluation of cloud parameterizations and the development and evaluation of cloud remote-sensing techniques (Stephens et al. 2000). The DOE has been applying UAS technology to collect observations that expand the understanding of the Earth system processes in the Arctic. UAS can provide researchers data, such as soil moisture, surface temperature, and elevation, from locations that are not easily or safely assessable by manned aircraft. The Marginal Ice Zone Observations and Processes Experiment (MIZOPEX) conducted in 2013 exploited UAS technology to examine conditions in the marginal ice zone off Oliktok Point, Alaska to assist in understanding the loss of ice in the Arctic Ocean (Ivey et al. 2013).

Zaugg, et al. (2010) tested the MicroASAR on a NASA UAS to study sea ice roughness and break-up as part of the Characterization of Arctic Sea Ice Experiment 2009 (CASIE-09). The study demonstrated the value of a compact, low-power synthetic aperture radar on a UAS for this and many other applications.

5.12 UAS for Archeology

There is a very long history and a very interesting series of applications of remote sensing from multiple platforms and multiple sensors in archeology also known as cultural resources management. Those applications include discovery of new features, as well as documentation during excavation or exploration. It is quite common to obtain imagery from an overhead perspective during the various stages of excavation. This would frequently involve using a tethered platform for repeat imaging, such as a balloon, but more recently with a UAS.

Many of the early efforts of discovery would employ soil, vegetation, or shadow marks from aerial photography. Other interesting examples are bathymetric remote sensing for sunken ships and other artifacts and in very dry environments, radar sensors that can penetrate soils to reveal buried features. LiDAR has also been employed to map tropical rainforest to locate features of archeological interest below the tree canopy, such as in Central America. The availability of UAS has created more opportunities for archeological studies. Among the Federal agencies, these applications have primarily been accomplished by the BLM and the NPS.

5.12.1 Cultural Resource Inventories

The BLM employed UAS to determine the extent of archeological site features in the Lower Salmon River canyon in Idaho and potential impacts to these sites from livestock, high river flows, or recreational use related to rafting and off-road vehicle activities. Among the archeological sites located with UAS were prehistoric pit house depressions, open surface prehistoric sties, abandoned mine features, and foundations from historic residents from the 19th century (US BLM 2017). Some of the historic features such as building rock foundations and mine ditches were easier to locate with low altitude, oblique UAS views than traditional vertical views from aircraft. The UAS were able to identify prehistoric pit houses because they have higher organic content, which retain moisture and thus provide more vigorous vegetation, which UAS sensors at lower altitudes can detect.

The Henry Smith site is in the BLM Milk Cultural Area of Critical Environmental Concern and covers about 2,000 acres in North Central Montana. The location has buffalo kill locations, prehistoric driveways, both anthropomorphic and zoomorphic ground figures, habitation sites, and medicine wheels. BLM conducted about 20 hours of UAS flights to accurately map the features of interest on this site, which will allow for better management and protection of these unique national Register of Historic Places eligible features. This effort clearly established the effectiveness of UAS for cultural resource inventories. The UAS accomplished this task more accurately, faster, and at a lower cost than traditional survey methods (US BLM 2017).

The BLM also employed UAS for reconnaissance and mapping in the Miller Creek drainage of the Skull Creek Wilderness Study Area in Colorado. The rough, steep, and rocky terrain in this location largely prohibits traditional cultural resource inventories. The primary UAS purpose was to locate rock overhangs, which might contain Formative Era granaries, rock art, or other cultural material. The UAS imagery was successful in locating several features of interest and was cost effective and safer than traditional methods (Figure 5.8) (US BLM 2017).

5.12.2 Fossil Animal Tracks

Several DOI bureaus employed UAS to locate and document the presence of exposed dinosaur tracks at the paleontological track site at White Sands National Monument, New Mexico. The pilot study prototyped the methodologies for an aerial survey for photogrammetric documentation of extremely fragile and ephemeral fossilized footprints from the late Pleistocene time period. The area of study includes a portion of a Late Pleistocene megatrack site within and around White Sands National Monument. Thousands of Ice Age fossil vertebrate tracks and trackways, which date to approximately 20,000 years ago, have been documented within this protected area (USGS NUPO 2018f).

FIGURE 5.8
Photo from UAS of granary by BLM in the Miller Creek drainage of the Skull Creek Wilderness Study Area in Colorado.

(Source: U.S. Bureau of Land Management 2017).

5.13 UAS for Other Applications

The U.S. Department of Transportation sponsored research into monitoring of unpaved road conditions with a UAS. The UAS demonstrated the capability to determine the unpaved road's condition and the UAS data were suitable for extracting many of the road condition parameters needed for monitoring of these surfaces (Zhang 2008). High-resolution imagery acquired by small UAS can be used to generate three-dimensional models. The USGS, in cooperation with the NPS, flew multiple UAS flights over Devils Tower at the Devils Tower National Monument in Wyoming. Both vertical (nadir) and oblique imagery, with a ground resolution of less than 5 cm, were acquired and used to create the three-dimensional model. The model supports assessing rock quality conditions, potential rock fall areas, vegetative health, and peregrine falcon nesting sites. This model also serves as a baseline for the NPS for rescue operations and delineating recreational climbing routes (Bauer 2017).

5.14 Summary

Many Federal agencies have actively embraced the employment of UAS for multiple basic and applied science and operational activities. The variation in the types of sensors and the many different applications is very impressive. The disciplinary topics have included agriculture, public health and safety, hydrologic features, wildlife, environmental monitoring, wildfires, and volcanoes among others. The Department of the Interior has one of the most active UAS programs in the Federal government. In 2018 the Department logged

10,342 UAS flights compared to 4,976 flights in 2017 and estimated it saved US$14.8 million over the cost of "traditional ground-based methods." It reported 359 certified pilots operating over 530 UAS to perform a wide variety of missions (US DOI 2018c).

UAS are helping Federal agencies meet the challenges across a wide range of applications, especially when a quick, safe approach is required to collect data for traditional applications, such as mapping or to support emergency response activities. The flexibility of operations and relative low cost to purchase and operate sUAS enhances the ability to track long-term landscape and environmental change, and promotes the integration of these data with *in situ* and satellite data sets. As sUAS technology advances, longer duration flights, etc., and more complex sensors, such as ground penetrating radars, become available for these platforms, coupled with new regulations that allow nighttime flights, swarms, and beyond visual line of sight (BVLOS) flights, the future applications of sUAS in the Federal community is limitless.

In addition to the direct involvement by Federal agencies in UAS, they have also promoted the awareness and employment of these platforms by funding other organizations, especially universities, convening workshops, making presentations, and publishing reports and scientific articles. Federal agencies have also been foremost in developing standards for UAS use and regulations for safety and privacy. It is evident that UAS will continue to expand in utilization and that Federal agencies will maintain leading roles, but also increasingly promote this technology in State, local and Tribal governments, and the private and commercial sectors.

References

AgResearch Magazine, 2017. Weed spotting by drone, USDA Agriculture Research Service, https://agresearchmag.ars.usda.gov/2017/dec/drone (accessed December 13, 2018).

Angliss, R., M. Ferguson, P. Hall, V. Helker, A. Kennedy, and T. Sfromo, 2018. Comparing manned and unmanned aerial surveys for cetacean monitoring in the Arctic: Methods and operational result, *Journal of Unmanned Vehicle Systems*, 6(3):109–127. doi:10.1139/juvs-2018-001.

Battista, T., 2018. NOAA evaluates using drones to map coastline and nearshore waters, https://coastalscience.noaa.gov/news/noaa-evaluates-drones-to-map-coastline-and-nearshore-waters/ (accessed January 18, 2019).

Bauer, M., 2017. Modeling Devils Tower, Wyoming using small Unmanned Aerial Systems (UAS) for baseline comparisons, highlighting UAS Data acquisition methods, American Association of Geographers (AAG) Annual Meeting, Boston, MA.

Bevan, E., T. Wibbels, B. Najera, M. Martinez, L. Martinez, F. Martine, J. Cuevas, T. Anderson, A. Bonka, M. Hernandex, L. Pena, and P. Burchfield, 2015. Unmanned aerial vehicles (UAVs) for monitoring sea turtles in near-shore waters, *Marine Turtle Newsletter*, 145:19–22.

Bruder, B., A. Renaud, N. Spore, and K. Brodie, 2018. Evaluation of unmanned aircraft systems for flood risk management – Field experiment conspectus, Coastal and Ocean Data Systems Program Flood and Coastal Systems Research and Development Program, U.S. Army Corps of Engineers, Engineer Research and Development Center, ERDC/CHL SR-18-2, 66 p.

Christie, K., S. Gilbert, C. Brown, M. Hatfield, and L. Hanson, 2016. Unmanned aircraft systems in wildlife research: Current and future applications of transformative technologies, *Frontiers of Ecology and the Environment*, 14(5):241–251, doi:10.1002/fee.1281.

Clark, D., A. Bell, J. Sloan, M. Bauer, and S. Goplen, 2013. Aerial missions with small unmanned aircraft systems to monitor sediment flow and changing topography resulting from the removal of dams on the Elwha River, Bureau of Reclamation Technical Memorandum 86-68260-13-03, 33 p.

Cress, J., M. Hutt, J. Sloan, M. Bauer, M. Feller, and S. Goplen, 2015. U.S. geological survey unmanned aircraft systems (UAS) roadmap 2014, U.S. Geological Survey Open-File Report 2015-1032, 60 p., doi:10.3133/ofr20151032.

Crooks, J., 2017. Drones collect information to benefit great lakes, United States Forest Service, www.usda.gov/media/blog/2017/12/28/drone-collects-information-benefit-great-lakes (accessed December 8, 2018).

Dandois, J., M. Olano, and E. Ellis, 2015. Optimal altitude, overlap and weather conditions for computer vision UAV estimates of forest structure, *Remote Sensing*, 7:13895–13920, doi:10.3390/rs71013895.

Daugherty Global Institute, 2018. Drones are buzzing toward increased crop production, University of Nebraska, https://waterforfood.nebraska.edu/news-and-events/news/2018/03/drones-are-buzzing-toward-increased-crop-production (accessed December 13, 2018).

Del Negro, C., A. Cappello, and G. Ganci, 2016. Quantifying lava flow hazards in response to effusive eruption, *Bulletin of the Geological Society of America*, 128:1–13, doi:10.1130/B31364.1.

Diaz, J., D. Pieri, K. Wright, P. Sorensen, R. Kline-Shoder, C. Arkin, M. Fladeland, G. Bland, M. Buongiorno, C. Ramirez, E. Corrales, A. Alan, O. Alegria, D. Diaz, and J. Linick, 2015. Unmanned aerial mass spectrometer systems for in-situ volcanic plume analysis, *Journal of the American Society for Mass Spectrometry*, 26(2):292–304, doi:10.1007/s13361-014-1058-x.

Durban, J., H. Fearnbach, L. Barrett-Lennard, W. Perryman, and D. Leroi, 2015. Photogrammetry of killer whales using a small hexacopter launched at sea, *Journal of Unmanned Vehicle Systems*, 3:131–135, doi:101139/juvs-2015-0020.

Fannin, B., 2017. AgriLife research receives USDA-NIFA grant for rhenotyping tool development, Texas A&M, https://today.agrilife.org/2017/06/02/agrilife-research-receives-usda-nifa-grant-phenotyping-tool-development/ (accessed December 13, 2018).

Fulton, J., 2019. Radar on Drones, https://www.usgs.gov/centers/co-water/science/radar-drones?qt-science_center_objects=0#qt-science_center_objects (accessed August 23, 2019).

Garcia-Pineda, O., 2018. Using UAS for tactical oil spill response operations, https://gulfseagrant.org/wp-content/uploads/2018/09/5.-Garcia-Pineda-FOR-WEB.pdf (accessed December 12, 2018).

Gardner, T., 2016. Drone-delivered health care in rural Appalachia, http://thehealthwagon.org/hwwp/2016/12/07/drone-delivered-health-care-in-rural-appalachia/ (accessed January 8, 2019).

Goebel, M., W. Perryman, J. Hinke, D. Krause, N. Hann, S. Gardner, and D. Leroi, 2015. A small unmanned aerial system for estimating abundance and size of Antarctic predators, *Polar Biology*, 38:619–630, doi:10.1007/s00300-014-1625-4.

Gullett, B., 2017. Emission sampling using UAS, UAS Workshop, NASA, Ames, March 28–30, 2017.

Hanson, L., C. Holmquist-Johnson, and M. Cowardin, 2014. Evaluation of the Raven sUAS to detect and monitor greater sage-grouse leks within the Middle Park population, U.S. Geological Survey, Open-File Report 2014–1205, 20 p., doi:10.3133/ofr20141205.

Howard, J., V. Murahov, and C. Branche, 2017. *Centers for Disease Control. Can Drones Make Construction Safer?*, https://blogs.cdc.gov/niosh-science-blog/2017/10/23/drones-constr (accessed December 8, 2018).

Huang, Y., K. Reddy, R. Fletcher, and D. Pennington, 2018. UAS low-altitude remote sensing for precision weed management, *Weed Technology*, 32(1): 2–6.

Huang, Y., S.J. Thomson, W.C. Hoffman, Y. Lan, and B.K. Fritz, 2013. Development and prospect of unmanned aerial vehicle technologies for agricultural production management, *International Journal of Agricultural and Biological Engineering*, 6(3):10 p. 26.

Hunt, R., Jr., and C. Daughtry, 2018. What good are unnamed aircraft systems for agricultural remote sensing and precision agriculture?, *International Journal of Remote Sensing*, 39(15–16): 5345–5376, doi:10.1080/01431161.2017.1410300.

Hunt, R., Jr. W. Hively, S. Fujikawa, D. Linden, C. Daughtry, and G. McCarty, 2010. Acquisition of NIR-green-blue digital photographs from unmanned aircraft for crop monitoring, *Remote Sensing*, 2:290–305, doi:10.3390/rs2010290.

Hunt, R., Jr., and S. Rondo, 2017. Detection of potato beetle damage using remote sensing from small unmanned aircraft systems, *Journal of Applied Remote Sensing*, 11(2):10 p., doi:10.1117/1.jrs.11.026013.

Ivey, M., R. Petty, D. Desilets, J. Verlinde, and R. Ellingson, 2013. Polar research with unmanned aircraft and tethered balloons, A Report from the Planning and Operational Meeting on Polar Atmospheric Measurements Related to the U.S. Department of Energy ARM Program Using Small Unmanned Aircraft Systems and Tethered Balloons, DOE/SC-ARM-TR-135, 36 p.

Johnson, A., and T. Preston, 2017. Talus and microclimate mapping with UAS to Identify Mechanisms of Mammalian Distribution, http://uas.usgs.gov/mission/MT_TalusSlopes.shtml (accessed December 13, 2018).

Johnson, M., 2016. *Small Unmanned Aircraft Used for Prescribed Fire at Homestead National Monument of America*, National Park Service, www.nps.gov/articles/homestead-prescribed-fire-uses-unmanned-aircraft.htm (accessed December 7, 2018).

Johnston, D., J. Dale, K. Murray, E. Josephson, E. Newton, and S. Wood, 2017. Comparing occupied and unoccupied aircraft surveys of wildlife populations: Assessing gray seal (Halichoerus grypus) breeding colony on Muskeget Island, USA, *Journal of Unmanned Vehicle Systems*, 5(4):178–191, doi:10.1139/juvs-2017-0012.

Jones, W., and S. Hartley, 2018. *Using Unmanned Aerial Systems (UAS) Capabilities to Help Identify Hummock-Hollow Formation and Fragmentation in Critical Marsh Habitat for Mottled Ducks*, www.usgs.gov/centers/wetland-and-aquatic-research-center-warc/science/using-unmanned-aerial-systems-uas?qt-science_center_objects=0#qt-science_center_objects (accessed December 7, 2018).

Jornada Rangeland Research Programs, 2018. *Unmanned Aircraft Systems (UAS) for Remote Sensing*, ARS New Mexico State University, https://Jornada.NMSU.edu/remote-sensing/UAS (accessed December 15, 2018).

Kinzel, P., M. Bauer, M. Feller, C. Holmquist-Johnson, and T. Preston, 2015. Experimental flights using a small unmanned aircraft system for mapping emergent sandbars, *Great Plains Research*, 25(1):39–52, doi:10.1353/gpr.2015.0018.

K-State Research and Extension News, 2016. *$975,000 Grant Will Help Scientists Employ UASs to Improve Wheat Breeding*, www.ksre.k-state.edu/news/stories/2016/12/drone-nifa-grant.html (accessed December 16, 2018).

Laliberte, A., M. Goforth, C. Steele, and A. Rango, 2011. Multispectral remote sensing from unmanned aircraft: image processing workflows and applications for rangeland environments, *Remote Sensing*, 3(12):2529–2551, doi:10.3390/rs3112529.

Laliberte, A., J. Herrick, A. Rango, and C. Winters, 2010. Acquisition, orthorectification, and object-based classification of unmanned aerial vehicle (UAV) imagery for rangeland monitoring, *Photogrammetric Engineering and Remote Sensing*, 6(6):661–672.

Laliberte, A., and A. Rango, 2009. Texture and scale in object-based analysis of subdecimeter resolution unmanned aerial vehicle (UAV) imagery, *IEEE Transactions on Geoscience and Remote Sensing*, 47(3):761–770, doi:10.1109/TGRS.2008.2009355.

Laliberte, A., and A. Rango, 2011. Image processing and classification procedures for analysis of sub-decimeter imagery Acquired with an unmanned aircraft over arid rangelands, *GIScience & Remote Sensing*, 48(1):4–23, doi:10.2747/1548-1603.48.1.4.

Lan, Y., S. Thomson, Y. Huang, C. Hoffmann, and H. Zhang, 2010. Current status and future directions of precision aerial application for site-specific crop management in the USA, *Computers and Electronics in Agriculture*, 74(1):34–38, doi:10.1016/j.compag.2010.07.001.

Langton, L., 2009. Aviation Units in Large Law Enforcement Agencies, 2007, Bureau of Justice Statistics, Washington, DC, www.bjs.gov/content/pub/pdf/aullea07.pdf (accessed December 8, 2018).

Lillian, B., 2017. *NOAA Teams with Drone Services Company RYKA UAS for Wetland Restoration Work*, https://unmanned-aerial.com/noaa-teams-drone-services-company-wetland-restoration-work (accessed December 16, 2018).

Lutman, M., 2017. The use of unmanned aircraft systems in the national feral swine damage management program, presentation to DOI UAS User Group.

Markiewicz, A. and L. Nash, 2016. *Small Unmanned Aircraft and the U. S. Forest Service: Benefits, Costs, and Recommendations for Using Small Unmanned Aircraft in Forest Service Operations*, U.S. Forest Service White Paper. U.S. Department of Transportation, John A. Volpe National Transportation Systems Center, 28 p.

McNabb, M., 2017. FAA: Hurricane response a "landmark in the evolution of drone usage," Dronelife, https://dronelife.com/2017/09/19/faa-hurricane-response-landmark-evolution-drone-usage (accessed December 8, 2018).

Moreland, E., M. Cameron, P. Angliss, and P. Boveng, 2015. Evaluation of a ship-based unoccupied aircraft system (UAS) for surveys of spotted and ribbon seals in the Bering Sea pack ice, *Journal of Unmanned Vehicle Systems*, 3:114–122, doi:101139/juvs-2015-0012.

National Oceanic and Atmospheric Administration (NOAA), 2017. Center for Weather and Climate Prediction, Arctic Domain Awareness Unmanned Aircraft Systems (UAS) Workshop Report, College Park, MD from January 31 to February 1, 2017, 64 p.

National Oceanic and Atmospheric Administration (NOAA), 2018. Nightime fire observations eXperiment (NightFOX), unmanned aircraft systems program, https://uas.noaa.gov/News/ArtMID/6699/ArticleID/797/Nighttime-Fire-Observations-eXperiment-NightFOX (accessed November 25, 2018).

National Wildfire Coordinating Group, 2017. *Interagency Fire Unmanned Aircraft Systems Operations Guide*, PMS 515, 24 p., www.nwcg.gov/sites/default/files/publications/pms515.pdf (accessed November 18, 2018).

Office of the Inspector General, 2013. *Interim Report on the Department of Justice's Use and Support of Unmanned Aircraft Systems*, Report 13-37, U.S. Department of Justice, Audit Division, Washington, DC, September 2013, https://oig.justice.gov/reports/2013/a1337.pdf (accessed December 8, 2018).

Olsoy, P., L. Shipley, J. Rachlow, J. Forbey, N. Glenn, M. Burgess, and D. Thorton, 2017. Unmanned aerial systems measure structural habitat features for wildlife across multiple scales, *Methods in Ecology and Evolution*, 9(3):594–604, doi:10.1111/2041-210X.12919.

Pande-Chhetri, R., A. Abd-Elrahman, T. Liu, J. Morton, and V. Wilhelm, 2017. Object-based classification of wetland vegetation using very high-resolution unmanned air system imagery, *European Journal of Remote Sensing*, 50(1):564–576, doi:10.1080/22797254.2017.1373602.

Patrick M., J. Kauahikaua, T. Orr, A. Davies, and M. Ramsey, 2016. Operational thermal remote sensing and lava flow monitoring at the Hawaiian Volcano Observatory, *Geological Society of London*, Special Publication SP426.17, doi:10.1144/SP426.17.

Patrick M., T. Orr, G. Fisher, F. Trusdell, and J. Kauahikaua, 2017. Thermal mapping of a pāhoehoe lava flow, Kīlauea Volcano, *Journal of Volcanology and Geothermal Research*, 332:71–87, doi:10.1016/j.jvolgeores.2016.12.007.

Patterson, M., A. Mulligan, J. Douglas, J. Robinson, and J. Pallister, 2005. *Volcano Surveillance by ACR Silver Fox*, Infotech@Aerospace, 26–29 September 2005, Arlington, Virginia, American Institute of Aeronautics and Astronautics, AIAA 2005-6954.

Poland, M., 2014. Time-averaged discharge rate of subaerial lava at Kīlauea Volcano, Hawaii, measured from TanDEM-X interferometry: Implications for magma supply and storage during 2011–2013, *Journal of Geophysical Research*, 119:5464–81, doi:10.1002/2014JB011132.

Ravindranath, M., 2017. *Government Using Drones to Track Down Dangerous Pigs*, Nextgov, www.nextgov.com/cio-briefing/2017/08/government-using-drones-track-down-wild-pigs/140389/ (accessed December 16, 2018).

Rees, A., L. Avens, K. Ballorain, E. Bevan, A. Broderick, R. Carthy, M. Christianen, G. Duclos, M. Heithaus, D. Johnston, J. Mangel, R. Paladino, K. Pendoley, R. Reina, N. Robinson, R. Ryan, S. Sykora-Bodie, D. Tilley, M. Varela, E.Whitman, P. Whittock, T. Wibbels, and B. Godley, 2018. The potential of unmanned aerial systems for sea turtle research and conservation: a review and future directions, *Endangered Species Research*, 5:81–100, doi:10.3354/esr00877.

RiceTec, 2018. *USDA Seeks Louisiana Growers with Hog Damage for Drone Study*, www.ricefarming. com/departments/feature/usda-seeks-louisiana-growers-with-hog-damage-for-drone-study (accessed December 8, 2018).

Rockwell, M., 2018. DHS seeks drones to patrol the border, FCW, https://fcw.com/ articles/2018/05/02/dhs-border-drones.aspx (accessed December 1, 2018).

Roegner, C., A. Borde, A. Colemand, J. Aga, R. Erdt, G. Pierce, and C. Cole, 2016. *Quantifying Restoration of Juvenile Salmon Habitat with an Unmanned Aerial Vehicle System*, https://uas.noaa. gov/Portals/5/Docs/Projects/2016/Quantify_Juvenile_Salmon_Restoration_Mission%20 Concept%20Review%20UAS_2016.pdf (accessed August 10, 2019).

Rosenthal, G., 2017. *Plant Protection Today – Meet PPQ's Plant Health Guardians in the Sky*, www.aphis. usda.gov/aphis/ourfocus/planthealth/ppq-program-overview/plant-protection-today/arti-cles/unmanned-aircraft-systems (accessed December 12, 2018).

Scharnweber, A., 2018. *Privacy Impact Assessment Update for the Aircraft Systems*, DHS/CBP/PIA-018(a), U.S. Customs and Border Protection, 18 p., www.dhs.gov/sites/default/files/publica-tions/privacy-pia-cbp018a-aircraftsystems-april2018.pdf (accessed December 16, 2018).

Schumann, G., J. Muhlhausen, and K. Andreadis, 2016. The value of a UAV-acquired DEM for flood inundation mapping and modeling, *Geophysical Research Abstracts*, 18, EGU2016-10158-1.

Sherwood, C., S. Brosnahan, S. Ackerman, J. Borden, E. Montgomery, E. Pendleton, and E. Sturdivant, 2018. *Aerial Imagery and Photogrammetric Products from Unmanned Aerial Systems (UAS) Flights over the Lake Ontario Shoreline at Sodus Bay, New York*, July 12 to 14, 2017: U.S. Geological Survey data release, doi:10.5066/P9XQYCD0 (accessed December 12, 2018).

Stephens, G.L., R.G. Ellingson, J. Vitko, W. Bolten, T.P. Tooman, P.J. Valero, P. Minnis, P. Pilewskie, G.S. Phipps, S. Sekelsky, J.R. Carswell, S.D. Miller, A. Benedetti, R.B. McCoy, R.F. McCoy, A. Lederbuhr, and R. Bambha, 2000. The department of energy's atmospheric radiation measure-ment (ARM) unmanned aerospace vehicle (UAV) program, *Bulletin of the American Meteorological Society*, 18:2915–2937.

Sturdivant, E., E. Lentz, E. Thieler, A. Farris, K. Weber, D. Remsen, S. Miner, and R. Henderson, 2017. UAS-sfm for coastal research: Geomorphic feature extraction and land cover classification from high-resolution elevation and optical imagery, *Remote Sensing*, 9(1020):21 p., doi:10.3390/ rs9101020.

sUAS News, 2018. Black Swift Technologies and NSAS's Jet Propulsion Laboratory (JPL) demonstrate effective use of sUAS for volcano research, www.suasnews.com/2018/02/black-swift-tech-nologies-nasas-jet-propulsion-laboratory-jpl-demonstrate-effective-use-suas-volcano-research/?mc_cid=21ad50b88c&mc_eid=888e960df6 (accessed November 25, 2018).

Sweeney, K., 2018. Advanced UAS sensor development for marine mammal monitoring, https:// uas.noaa.gov/News/ArtMID/6699/ArticleID/795/Advanced-UAS-Sensor-Development-for-Marine-Mammal-Monitoring (accessed January 18, 2019).

UAS Vision, 2018a. RPAS used in exploration of active volcanoes, www.uasvision.com/2013/04/10/ rpas-used-in-exploration-of-active-volcanoes/?utm_source=Newsletter&utm_ campaign=dbef83b684-RSS_EMAIL_CAMPAIGN&utm_medium=email (accessed November 25, 2018).

UAS Vision, 2018b. Hawaii University uses UAS to map lava flow, www.uasvision.com/2014/10/31/ hawaii-university-uses-uas-to-map-lava-flow/?utm_source=Newsletter&utm_ medium=email&utm_campaign=fdd0e9d213-RSS_EMAIL_CAMPAIGN&utm_ term=0_799756aeb7-fdd0e9d213-297542849#sthash.sjRE1BeF.dpuf (accessed November 25, 2018).

USAID Center for Accelerating Innovation and Impact, 2017a. *UAS in Global Health: Defining a Collective Path Forward*, 29 p., www.usaid.gov/cii/uavs-global-health (accessed December 2, 2018).

USAID Global Health Supply Chain Program-Procurements and Supply Management, 2017b, *Unmanned Aerial Vehicles Landscape Analysis: Applications in the Development Context*, Washington, DC: Chemonics International Inc., 65 p., www.ghsupplychain.org/resource/unmanned-aerial-vehicles-landscape-analysis (accessed December 2, 2018).

U.S. Bureau of Land Management, 2017. Unmanned aerial systems (UAS) cultural resources cases, 2014–2017, www.doi.gov/sites/doi.gov/files/uploads/blm_uas_cultural_summary.pdf (accessed December 28, 2018).

U.S. Department of Homeland Security, 2013. *Privacy Impact Assessment for the Aircraft Systems*, www.dhs.gov/sites/default/files/publications/privacy-pia-cbp-aircraft-systems-20130926.pdf (accessed December 2, 2018).

U.S. Department of Justice, Office of Justice Programs, 2016. *Considerations and Recommendations for Implementing an Unmanned Aircraft Systems (UAS) Program*, www.ncjrs.gov/pdffiles1/nij/250283.pdf (accessed December 16, 2018).

U.S. Department of the Interior, 2018a. Aerial ignition UAS and payload development, Office of Aviation Services Briefing Paper, www.doi.gov/sites/doi.gov/files/uploads/doi_uas_aerial_ignition_and_payload_development_08_2018.pdf (accessed December 7, 2018).

U.S. Department of the Interior, 2018b. UAS aerial ignition UAS aerial ignition operational test and evaluation, Office of Aviation Services Briefing Paper, www.doi.gov/sites/doi.gov/files/uploads/doi_uas_aerial_ignition_ote_field_report_08_2018.pdf (accessed December 7, 2018).

U.S. Department of the Interior, 2018c. U.S. Department of the Interior Unmanned Aircraft Systems (UAS) program 2018 use report, www.doi.gov/sites/doi.gov/files/uploads/doi_fy_2018_uas_use_report.pdf (accessed April 15, 2019).

USGS NUPO (National UAS Project Office), 2018a. Rapid response: Volcano monitoring Kīlauea Volcano, Hawaii, https://uas.usgs.gov/mission/HI_KilaueaVolcano.shtml (accessed November 25, 2018).

USGS NUPO, 2018b. Sandhill crane population estimates, https://uas.usgs.gov/mission/CO_SandhillCranesMonteVistaNWR.shtml (accessed December 16, 2018).

USGS NUPO, 2018c. Pygmy rabbit habitat study, https://uas.usgs.gov/mission/ID_PygmyRabbitLandscape.shtml (accessed December 16, 2018).

USGS NUPO, 2018d. Census of ground-nesting pelicans, https://uas.usgs.gov/mission/ND_ChaseLakeNWRPelicans.shtml (accessed December 16, 2018).

USGS NUPO, 2018e. Rapid response: UAS data collection for flood monitoring – Pemigewasset River near Plymouth, New Hampshire, https://uas.usgs.gov/mission/NH_PlymouthIceJamER.shtml (accessed November 25, 2018).

USGS NUPO, 2018f. Surveying and document paleontological pleistocene tracks: White Sands National Monument in New Mexico, https://uas.usgs.gov/mission/NM_FossilTraceWhiteSands.shtml. (accessed December 28, 2018).

Wang, Z., M. Roman, N. Pahlevan, M. Stachura, J. McCorkel, G. Bland, and C. Schaaf, 2016. MALIBU: A high spatial resolution multi-angle imaging unmanned airborne system to validate satellite-derived BRDF/Albedo products, American Geophysical Union, Fall Meeting 2016, abstract #B31B-0471.

Waters, K., 2018. NOAA evaluates using drones for lidar and imagery in the national estuarine research, https://uas.noaa.gov/News/ArtMID/6699/ArticleID/796/NOAA-Evaluates-Using-Drones-for-Lidar-and-Imagery-in-the-National-Estuarine-Research (accessed January 18, 2019).

Williamson, T., E. Dobrowolski, S. Meyer, J. Frey, and B. Allred, 2019. Delineation of tile-drain networks using thermal and multispectral imagery – Implications for water quantity and quality differences from paired edge-of-field sites, *Journal of Soil and Water Conservation*, 74:1–11, doi:10.2489/jswc.74.1.1.

Wright, R., H. Garbeil, and A. Harris, 2008. Using infrared satellite data to drive a thermo-rheological/stochastic lava flow emplacement model: A method for near-real-time volcanic hazard assessment, *Geophysical Research Letters*, 35:L19307, doi:10.1029/2008GL035228.

Xi, X., M. Johnson, S. Jeong, M. Fladeland, D. Pieri, J. Diaz, and G. Bland, 2016. Constraining the sulfur dioxide degassing flux from Turrialba Volcano, Costa Rica using unmanned aerial system measurements, *Journal of Volcanology and Geothermal Research*, 35:110–118, doi:10.1016/j.jvolgeores.2016.06.023.

Yueh, S., R. Shah, X. Xu, K. Elder, S. Margulis, G. Liston, M. Durand, C. Derksen, and J. Elston, 2018. UAS-based P-band signals of opportunity for remote sensing of snow and root zone soil moisture, *Proceedings Volume 10785, Sensors, Systems, and Next-Generation Satellites* XXII: 107850B (2018), 8 p., doi:10.1117/12.2325819.

Zaugg, E., D. Long, M. Edwards, M. Fladeland, R. Kolyer, I. Crocker, J. Maslanik, U. Herzfeld, and B. Wallin, 2010. Using the microASAR on the NASA SIERRA UAS in the characterization of Arctic Sea Ice Experiment, Radar 2010: IEEE International Radar Conference, May 10–14, 2010, Arlington, VA, pp. 271–276, doi:10.1109/RADAR.2010.5494611.

Zhang, C., 2008. An UAV-based photogrammetric mapping system for road condition assessment, *The International Archives of the Photogrammetry, Remote Sensing and Spatial Information Sciences*, Vol. XXXVII, Part B5, Beijing, China, pp. 627–632.

Zoeller, M., M. Patrick, and C. Neal, 2018. Crisis remote sensing during the 2018 lower rift zone eruption of Kīlauea Volcano, *Photogrammetric Engineering and Remote Sensing*, 84(12):749–751.

Zweig, C., M. Burgess, H. Percival, and W. Kitchens, 2015. Use of unmanned aircraft systems to delineate fine-scale wetland vegetation communities, *Wetlands*, 35(2):303–309, doi:10.1007/s13157-014-0612-4.

6

sUAS for Wildlife Conservation – Assessing Habitat Quality of the Endangered Black-Footed Ferret

Donna M. Delparte, Kristy Bly, Travis Stone, Sarah Olimb, Michael Kinsey, Matthew Belt, and Thomas Calton

CONTENTS

6.1 Introduction

The functional application of sUAS is rapidly becoming a valuable tool for wildlife conservation. sUAS offers researchers and wildlife managers the ability to determine species distribution, population density, status and trend monitoring, species presence or absence, and habitat classification (Christie et al. 2016; Linchant et al. 2015; Mulero-Pázmány et al. 2015). These efforts are especially valuable in remote or inaccessible areas when studying imperiled habitats or species and over large expanses of land that is time consuming and expensive for pedestrian surveys (van Gemert et al. 2015). This chapter provides a case study for using sUAS for a black-footed ferret (*Mustela nigripes*) habitat assessment based on the spatial distribution and population density of its primary prey species – the prairie dog (*Cynomys spp.*). In this study, automated feature extraction of individual prairie dog burrows from small Unmanned Aircraft Systems (sUAS) multispectral imagery provided assessments of prairie dog colony size and population density. Based on counts collected by ground survey and through manually digitized burrow counts at 27 plot sites, automated feature extraction using sUAS imagery provided a novel approach to efficiently collect accurate extent and density metrics to assess and monitor black-footed ferret habitat. The benefits of sUAS versus traditional on-the-ground biological surveys are to minimize environmental/behavioral impact on prairie dogs, offer a less time-consuming survey option, reduce count error and bias, provide the ability to cover larger areas, and conduct more frequent repeat surveys. sUAS can explore other potential advantages such as the detailed assessment of vegetation species composition and structure with multispectral and advanced sensors such as hyperspectral and LiDAR.

6.1.1 Monitoring the Endangered Black-Footed Ferret

Once found throughout the Great Plains and Intermountain West (Hillman and Clark 1980), the black-footed ferret (*Mustela nigripes*) is one of North America's most endangered terrestrial mammals. They are obligate predators of prairie dogs (*Cynomys spp.*) and rely on their burrows for shelter and denning (Hillman 1968; Biggins et al. 2006); consequently, their fate is directly linked to that of prairie dogs. Conversion of native grasslands to cropland, anthropogenic control of prairie dogs, and disease decimated black-footed ferret populations to the point of extinction in the 20th century (Cully 1993; Biggins et al. 2006; Fagerstone and Biggins 2011; Jachowski and Lockhart 2009; U.S. Fish and Wildlife Service 2013). Concerted efforts from many federal and state agencies, tribal governments, zoos, conservation organizations, and private landowners are facilitating recovery of the black-footed ferret. Although great strides have been made to recover the species through captive breeding and reintroductions, habitat loss, and sylvatic plague (*Yersinia pestis*) – a non-native disease lethal to black-footed ferrets and prairie dogs – is a key threat (Jachowski and Lockhart 2009). To remove the black-footed ferret from the Federal List of Endangered and Threatened Wildlife, national recovery objectives call for the establishment of 30 or more geographically distinct wild populations throughout the historical range of the species with no fewer than 30 breeding adults in any population (U.S. Fish and Wildlife Service 2013). Currently, less than 300 black-footed ferrets live in the wild, which is well below the 3,000 breeding adults needed for delisting of the species. To reach this goal, additional reintroduction sites are needed and existing populations need to be enhanced (U.S. Fish and Wildlife Service 2013).

Once black-footed ferrets are reintroduced in an area, annual assessments of their population and prairie dog colony size and density are needed. Monitoring prairie dog populations is important for assessing habitat suitability for black-footed ferrets, population dynamics, reintroduction potential, and the success or failure of reintroduction efforts (Bevers et al. 1997; Eads et al. 2014; Caldwell 2015). This data informs the U.S. Fish and Wildlife Service of progress toward delisting criteria and helps managers to decide if black-footed ferret population augmentation or sylvatic plague mitigation is warranted. Currently, two monitoring metrics of black-footed ferret habitat are obtained annually: (1) prairie dog colony size, which is estimated by recording the active perimeter of each prairie dog colony via walking or driving with a Global Positioning System (GPS) unit (Hoogland 1995); and (2) estimating approximate densities of prairie dogs from strip transect samples of active prairie dog burrows (Biggins et al. 1993; Biggins et al. 2006). Although effective, collecting these data for large areas of prairie dog colonies within an individual black-footed ferret reintroduction site are costly in terms of labor and time and are also subject to observer bias and colony disturbance (Severson and Plumb 1998; Grenzdörffer 2013).

In addition, personnel at many black-footed ferret reintroduction sites lack the financial and technical capacity to conduct such labor-intensive ground monitoring. Advancements in small Unmanned Aircraft Systems (sUAS) remote-sensing technology, such as the advantages of rapid deployment and efficiency, when compared to ground surveys, offer the potential to improve population monitoring of both black-footed ferrets and their prairie dog prey.

6.1.2 sUAS in Wildlife Conservation

sUAS technologies specific to habitat monitoring involve acquiring high-resolution data for areas of interest with either small fixed wing or multirotor craft at varying resolutions and extents. Fixed wing sUAS offers advantages of longer flight times and larger coverage areas per flight, while multirotor platforms are generally less expensive and can collect data from lower flight altitudes and at higher resolutions. Typical sensors for wildlife habitat assessment and/or surveys include commercial off-the-shelf cameras (Chabot et al. 2015; Beck et al. 2014), video cameras (Jones et al. 2006), thermal sensors (McCafferty 2013), and multi/hyperspectral sensors in the near to mid-infrared range (Grenzdörffer 2013; Gillette et al. 2015). sUAS also has the potential to advance global wildlife conservation efforts (Watts et al. 2010; Chabot and Francis 2016) by monitoring spatial and temporal changes in habitat (Anderson and Gaston 2013), conducting counts of individual animals (Sardà-Palomera et al. 2012; Wich et al. 2015), and identifying wildlife poachers (Koh and Wich 2012). Another unique benefit of sUAS is that it can be flown safely at night to thermal-track (Witczuk et al. 2018) nocturnal animals, such as the black-footed ferret.

sUAS missions generate hundreds of overlapping individual images (typically subdecimeter resolution) and require subsequent processing into mosaics for inspection and analysis. Once processed into orthomosaics with advanced photogrammetric software, this imagery is analyzed with Geographic Information Systems (GIS) and remote-sensing software to yield spatially accurate information that is extracted in an efficient manner. 3D terrain (Digital Surface Models) can also be derived from sufficiently overlapping images, allowing incorporation of features (e.g. raised prairie dog burrows) into spatial analysis algorithms that isolate and automate the count or identification of features of interest (Grenzdörffer 2013).

Automated processing of sUAS derived data products is essential for rapidly assessing large volumes of images collected during an sUAS mission. Visual perusal of imagery by an analyst to manually detect and digitize thousands of wildlife observations or habitat features is time consuming and negates the time efficiency benefit of utilizing sUAS. Although operator counts of sUAS imagery with ground nesting or roosting birds for smaller areas has been successful (Watts et al. 2010; Sardà-Palomera et al. 2012; Hutt 2011), automated counts using feature extraction algorithms that employ pattern recognition based on spectral characteristics of shape and size parameters, for example, have shown encouraging results for estimates of features in the thousands and higher (Abd-Elrahman et al. 2005; Grenzdörffer 2013; Descamps et al. 2011; van Gemert et al. 2015).

As traditional methods of monitoring black-footed ferret habitat are costly in terms of both labor and time, this study examined the ability of fixed wing sUAS equipped with a multispectral sensor and automated analysis to derive two primary metrics of black-footed ferret habitat: (1) prairie dog colony size and (2) prairie dog density within colonies. To derive these metrics, an automated approach to burrow counts was applied to 376 ha of prairie dog colonies and validated against ground survey counts and manually digitized counts interpreted from a visual inspection of imagery.

6.2 Methods

6.2.1 Study Area

Figure 6.1 highlights the study area located in north-central Montana on the Fort Belknap Reservation. Established in 1888 as the homeland for the Gros Ventre (Aaniiih) and the Assiniboine (Nakota) Tribes, Fort Belknap encompasses 273,300 ha in Blaine and Phillips Counties. In addition, there are another 12,032 ha of tribal land outside of the Reservation boundaries. The northern portion of the Reservation is bounded by the Milk River; the southern portion drains into the Missouri River through the Little Rocky Mountains. As part of the North American Great Plains region, this semi-arid area is characterized by flat to undulating prairie overlaying glacial till and alluvial bottomlands, which provides ideal

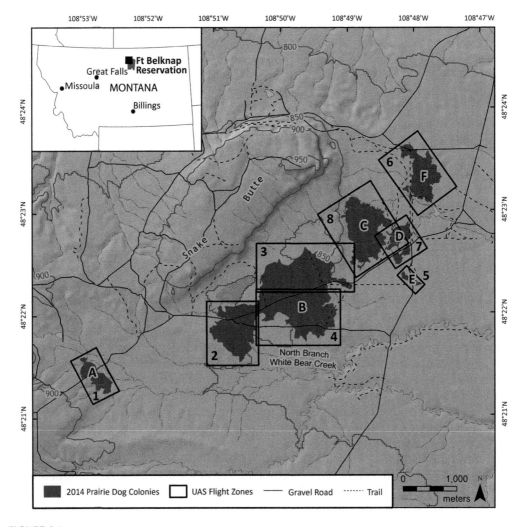

FIGURE 6.1
sUAS habitat survey on the Fort Belknap Reservation, Montana, USA. Six prairie dog colonies are labeled A–F. Flight areas are numbered 1–8.

habitat for prairie dog colonies. The mixed grass prairie consists predominantly of western wheatgrass (*Pascopyrum smithii*), blue grama (*Bouteloua gracilis*), and needle and thread (*Hesperostipa comata*) grasses interspersed with silver sagebrush (*Artemisia cana*) (Natural Resources Conservation Service 2006). The predominant land use is buffalo grazing and dryland agriculture. Yearly precipitation averages 30 cm and temperature averages 9°C (National Oceanic Atmospheric Administration 2016). Elevations within the Reservation range from 700 m to 1,525 m (Goodwin and Longknife 2013).

Six black-tailed prairie dog colonies, labeled A-F on Figure 6.1, were surveyed for this study in 2015, comprising approximately 376 ha. Individual colonies ranged in size from 3.4 ha to 227 ha. The colonies were split into eight survey zones, marked with numbers 1–8 on Figure 6.1 for flight planning and provided a total coverage of 806 ha of prairie dog colonies and grassland habitat (Table 6.1).

The Fort Belknap Reservation hosts one of only 29 populations of the endangered black-footed ferret in North America. By the early 1980's, black-footed ferrets were presumed extirpated in Montana (Knowles et al. 2002) due to extensive prairie dog control programs. In 1997, the Reservation became the sixth federally designated black-footed ferret reintroduction site, and notably, the first reintroduction to occur on tribal land. During 1997–1999, 167 black-footed ferrets were released in prairie dog colonies at the base of Snake Butte and elsewhere on the Reservation. The reintroduced black-footed ferret population was thriving and reproducing until an epizootic outbreak of sylvatic plague (*Yersinia pestis*), a nonnative disease lethal to both black-footed ferrets and prairie dogs, swept through prairie dog colonies in 1999 and decimated populations of both species. Since then, prairie dog populations have rebounded, plague mitigation tools are in place, and partners such as the World Wildlife Fund have been working with the Tribes to re-establish a black-footed ferret population to the Reservation. During 2013–2015, 67 black-footed ferrets were reintroduced and with annual sylvatic plague mitigation in place, this population is growing. At the time of this study, approximately ten black-footed ferrets occupied the 806 ha study area.

6.2.2 sUAS Image Acquisition

sUAS image acquisition of prairie dog colonies within the study area occurred on 9–10 June 2015 between the hours of 9 am to 4 pm under Federal Aviation Administration certificate

TABLE 6.1

Prairie Dog Colony Areas, Flight Zones, and Associated Flight Data

Zone		Colony	
ID	ha	ID	ha
1	49.4	A	18.5
2	111.4		
3	159.2	B	227.0
4	157.6		
5	11.8	E	3.4
6	111.1	F	41.3
7	50.0	D	17.1
8	155.7	C	67.7
Total	806.2		375.0

of authorization 2015-WSA-37-COA. A Topcon Sirius Pro sUAS (Figure 6.2) equipped with a 16MP Panasonic DMC-GX1 digital photo camera with a 14 mm lens modified to collect color infrared imagery (Near-infrared, Red, Green) was deployed at each of the eight survey zones (Figure 6.1). The Sirius Pro is a small fixed wing, hand launched, electrically powered platform with GNSS RTK (global navigation satellite systems, real-time kinematic) onboard hardware for accurate real time positioning that provides georeferenced aerial mapping results within 2–5-cm accuracy. It has a wing span of 163 cm, weighs 2.7 kg, has an optimal cruising speed of 65 km h^{-1}, and can operate in winds with gusts up to 65 km h^{-1}. Depending on wind conditions, the Sirius Pro's flight time is up to 45 minutes. The ground control station consisted of a laptop with MAVinci desktop flight planning and management software to create and upload automated flight plans to the Sirius Pro and to monitor flights during operation. Flights were conducted at an altitude of 120 m with a resulting ground sample distance of 3 cm and forward image overlap of 70% and side overlap of 65%. During the data collection period, the maximum wind gusts were up to 55 km h^{-1}, necessitating flight plan adjustments to fly perpendicular to the wind; downwind flight speed would have exceeded the camera's ability to capture data at the desired overlap. A total of 11 flights were completed during 350 minutes of flight time, resulting in the capture of 5,455 images to cover the entire 806 ha of interest. Acquisitions ranged from 14 to 42 minutes in duration and 128 to 676 individual photos were collected per flight. To conduct quality assurance of sUAS flight lines as well as image quality reports, image overlap and photo exposure were generated on site with a field laptop using Pix4D Mapper software.

During flight operations, a separate Topcon Hiper V GPS base station collected data points each day during flight operations for a minimum of four hours to allow for correction of sUAS image reference coordinates. These base station reference points were post-processed using the Online Positioning User Service from the National Oceanic and Atmospheric Administration. Rover coordinates for the three positions used for the sUAS external RTK base station locations were recorded and post-processed with corrected base station data. MAVinci desktop software facilitated the post GPS correction processing to update photo center coordinates for each image.

FIGURE 6.2
MAVinci SIRIUS Pro Unmanned Aircraft System. (Photo credit: Simeon Kateliev).

6.2.3 Ground Surveys

Ground surveys of prairie dog colony size and density were conducted in June and July 2015. Using traditional methods, we estimated prairie dog colony size by recording the active perimeter of each prairie dog colony via walking or driving using a GPS unit. An index of colony-wide prairie dog density was derived per the detailed methods as described in Biggins et al. (1993; 2006); density was estimated by walking with a meter wheel and transect pipe and counting active and inactive prairie dog burrows along 3-m wide strip transects, spaced at 60 m apart across Colony F. The calculation for estimating prairie dog density from burrow counts is as follows (Biggins et al. 1993)

$$\text{Prairie Dog Density} = (0.179 \times \text{active burrow density})/0.566 \tag{6.1}$$

Prairie dog density was also estimated by counting prairie dog burrows within 27 randomly selected, georeferenced ground plots that ranged in size from 0.05 ha to 0.13 ha. We used a compass and meter wheel to establish plot boundaries, which yielded imperfectly shaped plots and subsequently varying plot sizes. A survey grade GPS unit recorded plot corners. All burrows within each of the 27 plots were counted as well as the time it took two observers on foot to collect the data. The same two observers (and sometimes two additional observers) independently counted burrows in each plot to address observer bias and confirm burrow numbers. This counting process was later used to determine accuracy for counts manually digitized from sUAS imagery and from computer automated counts.

6.2.4 Image Processing and Analysis

Once the acquired images and photo center GPS coordinates were differentially corrected, we followed a standard workflow commonly used across sUAS image processing softwares. RAW images captured with the modified near-infrared 16MP Panasonic DMC-GX1 were pre-processed using Adobe Camera Raw 9.1 with a manual lens vignetting set to +25%. RAW images were exported to TIFF format with 16 bits/channel, no compression, and colorspace of sRGB IEC6 1966–2.1. Individual frames were stitched together to create orthomosaics and digital surface models (DSM) using Agisoft PhotoScan software. In Agisoft PhotoScan, the image alignment, dense cloud, and mesh settings were set to high accuracy. Height field was calculated during mesh generation to produce a DSM. Both orthoimagery and DSM products were exported with a pixel resolution of 3 cm.

The analysis workflow consisted of a five-step process: (1) the orthomosaics were pre-processed, (2) segmented, and (3) rules surrounding computer identification of burrows were built. Burrows within the images were then (4) classified and (5) verified (Figure 6.3). Orthomosaic image pre-processing involved applying a high pass convolution filter to enhance edge detection and enhance differences between brightness values. Segmentation and rule-building used eCognition software (Trimble). Imagery segmentation parameters were interactively adjusted based on visual interpretation of the accuracy of delineation of key features such as prairie dog burrows and land cover (Figure 6.3B). These values varied between the eight separate orthomosaic scenes processed due to variations in scene brightness from lighting conditions specific to the time and day of data collection. For verification, computer-generated counts of burrows within the plots were compared to visual counts of burrows from the processed images and ground surveyed counts of burrows. Density calculations made by ground observers from 17 transects in Colony F (Table 6.2) were also compared to the computer-generated density calculations.

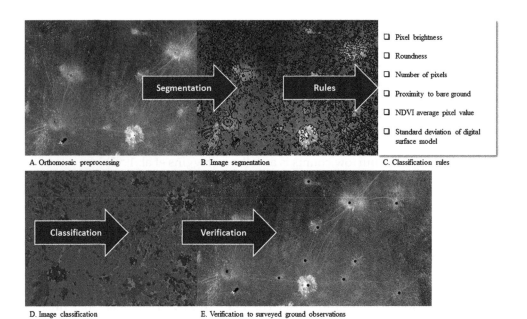

FIGURE 6.3

Analysis workflow consisting of five steps. A. Orthomosaic pre-processing, B. Image segmentation, C. Classification rules, D. Image classification, and E. Verification.

Prairie dog burrows have specific characteristics that make their identification possible using a rule-based approach (Figure 6.4). These characteristics were translated to rules in an algorithm that included specific variation in pixel brightness values, proximity to bare ground classed pixels, number of pixels, normalized difference vegetation index (NDVI) values, averaged pixel values of the standard deviation of the digital surface model within a segmented section, and roundness of shape (Figure 6.3 C). The algorithm was processed on each set of orthomosaic images to create a classified image with main classes of burrows, bare earth, shrubs, and grass (Figure 6.3 D). The classified imagery was brought into ArcGIS and burrow pixels converted to polygons and subsequently, polygons to point values. Thus, each image classified burrow was assigned a coordinate point. Each set of plot boundaries was converted to polygons and displayed with burrow point locations on top of the orthomosaics (Figure 6.3 E). Counts of rule-generated burrows were automatically determined based on location within plot polygons. As a final check, a computer operator visually counted burrows within each plot and recorded the difference between number of burrows visually recognized on the imagery and computer-generated counts. Esri ArcGIS kernel density tool, with a search radius of 1 ha, calculated density of automated burrow locations for comparison to the density of burrows obtained through ground surveys.

6.2.5 Sensor Limitations

Although the sUAS platform and sensor systems worked well for data collection, the camera was not calibrated to account for changing light conditions. This is a common problem when using cameras that are modified to capture near-infrared imagery. Other camera systems are equipped with light sensors to compensate for changing light conditions (e.g. Parrot Sequoia+, Tetracam ADC Micro). Reflectance targets placed on the ground during flight may have helped compensate for this issue. The problem this created was that imagery brightness

TABLE 6.2

Counts of Prairie Dog Burrows Within Plots by Observers in the Field, Visually from the Imagery and by Computer Feature Detection

Plot_ID	Ground Survey Count	Manually Digitized Count	Automated Computer Count	Plot Area (ha)
6-6	9	9	8	0.061
6-3	9	9	8	0.064
6-20	14	14	12	0.064
6-12	10	10	9	0.055
6-15	7	7	8	0.068
4-17	7	4	5	0.066
4-20	5	5	5	0.067
5-4	4	4	5	0.065
4-15	8	8	7	0.064
3-10	6	5	5	0.066
3-9	4	4	3	0.061
3-19	6	5	3	0.059
3-4	6	6	6	0.068
3-12	4	4	5	0.067
5-1	4	4	4	0.062
5-20	8	7	7	0.064
5-10	4	4	4	0.064
5-13	6	6	6	0.062
4-10	4	4	4	0.062
1-11	11	11	10	0.126
1-14	12	12	11	0.126
1-20	15	16	12	0.125
2-1	11	10	10	0.056
2-11	12	11	12	0.057
2-4	10	10	10	0.064
2-17	10	12	10	0.062
2-18	10	10	10	0.060
Total	214	211	199	2.15

and NDVI values were not consistent across the flight area and thus the algorithm values had to be refined for each of the eight flights. Other confusion issues arose during automation of burrow counts, where prairie dog shadows were interpreted as burrows (Figure 6.5). Shadows cast from rocks initially posed a problem in confusion with burrow detection but were eventually eliminated by using proximity features to distinguish them from burrows. Similar confusion between detection of species and background were also found in other studies due to similarities of shape and size (Grenzdörffer 2013; Chabot et al. 2015).

6.3 Results

The six prairie dog colonies in the study area totaled 375.0 ha and ranged in size from 8.3 ha to 561.0 ha (Table 6.1), as estimated by traditional GPS mapping methods. Ground

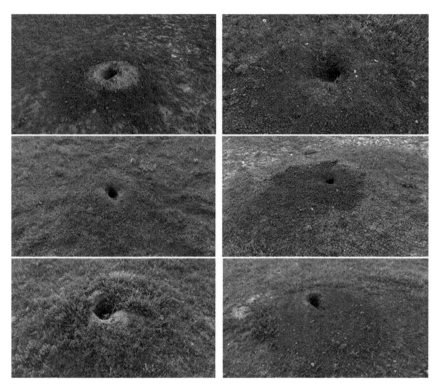

FIGURE 6.4
Prairie dog burrows photographed from the ground within the study area. Unique characteristics of the bur-rows facilitated the development of rules that included variation in pixel brightness values, proximity to bare ground classed pixels, number of pixels, and roundness of shape. (Photo credit: Jessica Alexander).

based prairie dog density for Colony F (also using the traditional method) was calculated using 17 surveyed transects, aligned east to west at 3-m wide and spaced at 60 m apart. The 17 transects ranged from 152.44 m to 685.96 m in length, containing counts ranging from 8 to 34 active burrows (Figure 6.6). The transect survey average number of burrows was estimated at 135.4 ha^{-1} which yielded 44.6 prairie dogs ha^{-1}. Five of the comparison plots (Table 6.2) were also located within Colony F, with an average burrow density of 158.3 ha^{-1} and prairie dog density of 50.1 ha^{-1}. To create a comparison metric to the automated bur-row count across the colony, a kernel density operation generated burrow locations using a 1 ha search radius. The average burrow density was 110 ha^{-1} with a density of 35 prairie dogs ha^{-1} across the entire colony (Figure 6.6).

Within the 27 plot sites, observer counts of burrows (214), those manually digi-tized from sUAS images (211), and computer automated burrow locations (199) were recorded (Table 6.2). Computer-generated burrow counts indicated a 93% accuracy when compared with ground-truthed burrow counts. Manually digitized counts of burrows from the processed images indicated a 98% accuracy when compared to the ground survey burrow counts. The mean average error (MAE) and root mean square error (RMSE) for the computer-generated burrow counts, in comparison to the on-the-ground counts were MAE = 0.74 and RMSE = 1.08 burrows ha^{-1}. For all plots, the ground survey average burrow density was 117.5 ha^{-1} with an average prairie dog density of 37.2 ha^{-1}.

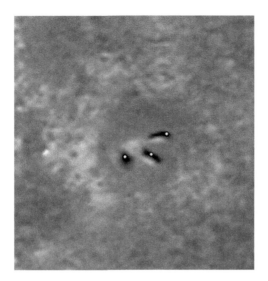

FIGURE 6.5
Prairie dog population densities for Colony F. Yellow dots represent burrow locations auto-detected from the imagery. Black squares are field plots for the colony.

6.4 Discussion

As traditional methods of monitoring black-footed ferret habitat are costly in terms of both labor and time, this study examined the ability of sUAS equipped with a multispectral sensor and automated analysis to derive two primary metrics of black-footed ferret habitat: (1) prairie dog colony size and (2) prairie dog density within colonies. We demonstrated that sUAS-acquired imagery combined with remote-sensing analysis can accurately collect estimates of prairie dog colony size and density. sUAS data collection also has the added benefit of minimal disturbance of prairie dog behavior when compared to ground survey methods. Furthermore, sUAS surveys are less prone to double and missed counts as well as observer bias. sUAS technologies thus show great promise for assessing black-footed ferret habitat for potential re-introductions and long-term monitoring of existing ferret reintroduction sites.

6.4.1 Cost Comparison Between Ground Surveys and sUAS

For this study, mapping prairie dog colonies and conducting density estimates on the 806.2 ha study area by all-terrain vehicles and on foot took two people two weeks with two GPS units, meter wheels, and tally counters. The approximate cost of obtaining estimates of prairie dog colony size and density using traditional ground survey methods was $9.54 ha^{-1}. This cost per ha includes downloading GPS track files onto ArcGIS and subsequent analysis of colony size, as well as time required to pre-populate strip transect start and end points and translation of burrow counts to prairie dog density. These costs are largely representative of costs incurred annually to conduct this work on the Fort Belknap Reservation.

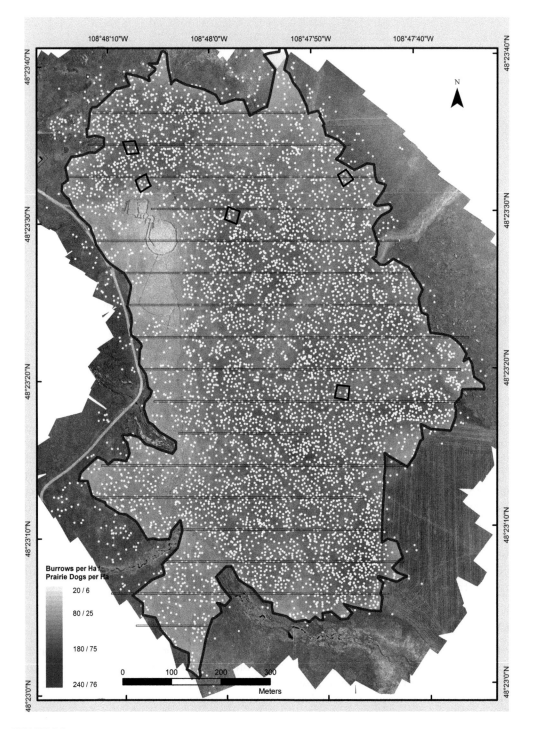

FIGURE 6.6
Example of a classification error based on the sUAS imagery. Two prairie dogs are exiting their burrow which is correctly identified by the placement of a point, yet the shadows cast from their bodies are incorrectly identified as burrows as indicated by additional points.

The total financial cost of obtaining metrics of colony size and density using a fixed wing sUAS platform in the study area was more difficult to ascertain. This project leveraged partnerships that eliminated costs associated with sUAS data collection, image processing, analysis work flow, creating the burrow algorithm, and travel expenses. Time wise, sUAS data collection flights were much faster than traditional methods – three days versus two weeks respectively – yet sUAS image processing and analysis took longer (approximately two months) for this study. Labor and time costs are often higher when initiating a new project versus later efforts to replicate the work because equipment has already been purchased, workflow and details are established, and increased operational efficiencies are realized. To replicate this study using an sUAS approach, we estimate a cost of $9.94 ha^{-1}, in comparison to the $9.54 ha^{-1} cost for a ground survey. We are estimating the sUAS costs based on the lower end of the per ha flying costs ($7.41) and adding in the burrow detection image analysis ($2.53 ha^{-1}). The sUAS estimate, although higher, includes the cost associated with developing an algorithm; subsequent iterations would see a reduction in automated burrow detection expenses.

For large area coverage, as is the case in this study (806.2 ha), a fixed wing sUAS platform is preferred for faster data collection, thereby reducing personnel time in the field. Fixed wing sUAS can fly up 202 ha per flight, while rotor sUAS platforms can cover around 40 ha per flight. In general, commercial sUAS operators charge-out fixed wing data collection on a $7.41–12.35 ha^{-1} basis for 3–6-cm resolution imagery. This includes labor costs in the field and post-processing of data products to include a deliverable set of orthomosaics for the flight areas (but not necessarily travel expenses to a remote location). Further processing, at an additional charge (~$2.53 ha^{-1}), would be required to extract burrow point locations from the orthomosaics. With a burrow algorithm detection application, depending on the agency's GIS capabilities, this task could be completed in-house. For smaller areas under approximately 121 ha, a multirotor solution may be preferable and many low-cost sUAS such as the DJI Phantom 4 could be purchased and equipped with a low-cost sensor for <$1,500 and used for data collection. Individual images from the flight could then be uploaded to cloud-based sUAS image processing services such as Pix4d Cloud, Datamapper, and DroneDeploy for orthoimage creation with starting rates of approximately $99/month for these services.

6.4.2 Long-term Monitoring and Transferability to Other Burrow Counting Studies

Our sUAS captured orthoimagery revealed burrow locations beyond the 2015 colony boundaries surveyed by biologists in the same year and is represented by the grey outline boundary in Figure 6.6. Based on these additional points detected by the sUAS outside the on-the-ground surveyed colony boundary, an adjustment to update the boundary to include the proximal burrow locations and recalculate the burrow density could be readily made using GIS tools. The long-term plan for this site is to repeat sUAS flights to monitor prairie dog populations within Fort Belknap's black-footed ferret reintroduction area and provide colony growth and density estimates. In addition, the authors are exploring agent-based modeling approaches to forecast and predict future colony growth patterns and population sizes.

The approach in this study could be used for other burrowing animals. A key criterion in the transferability of this approach is the size of the burrow and vegetative cover. For small burrow openings (less than 5 cm in size) lower and slower sUAS flights would be required to capture higher resolution data, such that the opening would encompass at least more than 3–4 pixels in the captured imagery. Further detection challenges would include the amount of obscuring vegetation and burrow position on a slope angled away from typically nadir flight paths. These smaller burrows would likely best be captured with rotary sUAS. They can fly at lower altitudes to collect higher resolution imagery with the added capability to alter camera angles, thereby allowing oblique photography to capture burrows on side hills. These capabilities are not currently feasible with faster flying fixed wing platforms.

This study advanced and improved our ability to assess and understand black-footed ferret habitat, and therefore the ability to support populations of the species. The results of this study provide a comprehensive map of both the extent and density of burrows within prairie dog colonies on the Fort Belknap Reservation. In addition to the immediate value of this information, these maps will be used over time to quantify changes in prairie dog colony size and density as the black-footed ferret population grows, which aids managers in determining if augmentation or plague mitigation is warranted.

An additional benefit of interest to the Tribes of Fort Belknap was the discovery of the extensive off-road vehicle tire tracks, which has the potential to enable non-native plant encroachment that, in turn, may reduce forage quality and quantity for prairie dogs. sUAS imagery thus provides baseline information that traditional methods alone cannot.

sUAS acquisition and processing technologies continue to improve at a rapid pace with respect to ease of use and the overall regulatory environment has become considerably more favorable to research-related sUAS projects. Furthermore, with an established work-flow model, future iterations will be far simpler, and thus less time consuming and expensive, to employ. By utilizing sUAS and automated processing, additional black-footed ferret reintroduction sites can be identified, monitored, and established, which ultimately, will contribute to advancing the recovery of the species.

6.4.3 Challenges and Opportunities for Habitat Monitoring Using sUAS Hyperspectral Imaging

Spectral imaging is widely used for the detection and classification of vegetative communities. Vegetation emits a spectral response to incoming electromagnetic energy based on plant health, species, and phenology. Newer sUAS multispectral imagers are typically three to five bands that include the visible range (~400–700 nm), the red-edge (~670–780 nm), and near-infrared (~700–860 nm) portions of the electromagnetic spectrum (e.g. Parrot Sequoia+, Micasense RedEdge-MX). Compared to multispectral (wideband) remote sensing, hyperspectral (narrowband) remote sensing provides numerous spectral bands (>10 to hundreds of bands) within the visual to near-infrared and even into the shortwave radiation range (350–2500 nm). This ability to capture a larger spectral range with narrowband precision allows hyperspectral imagers to better discriminate between species of vegetation for long-term monitoring.

As hyperspectral sensors have become lighter and smaller for integration onto sUAS platforms, they still remain costly and generate large amounts of data suitable for processing by remote-sensing professionals or researchers with appropriate hardware resources, software tools, and programming libraries (Adão et al. 2017). Specific examples of hyperspectral remote sensing for habitat monitoring in arid environments include work done by Sankey et al. (2018) that fused sUAS hyperspectral imagery with LiDAR to characterize 3D canopy structure and used spectral signatures to differentiate between plant species. Mitchell et al. (2012) employed unsupervised and supervised classifications to spectrally distinguish between grass, shrubs, and bare ground for dryland vegetation monitoring. Although adoption of sUAS hyperspectral sensing is low, there is an increasing number of sensors available on the market that will likely become more cost-effective with future technological advances (Adão et al. 2017).

In Figure 6.7, we highlight a shrub-steppe vegetation classification generated from an imaging system composed of two sensors onboard an sUAS platform. We equipped

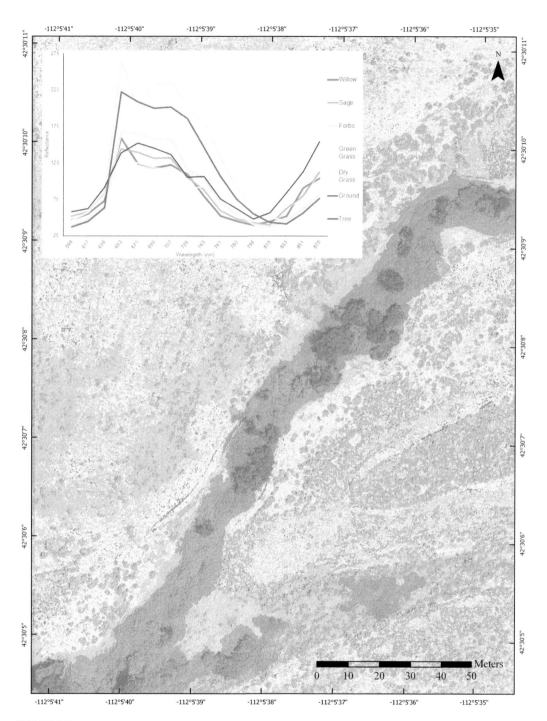

FIGURE 6.7
Columbia Sharp-tailed Grouse habitat classification with spectral characterization between willow (orange), sagebrush (green), grasses (light green/yellow), and forbs (grey), with some areas of larger deciduous vegetation (dark green) and bare soil (brown).

our Matrice 600 Pro hexacopter produced by DJI Technology Co., Ltd with the Rikola Hyperspectral Imager (HSI) and a digital camera. The HSI is a snapshot type imager that provides a real spectral response in each pixel over a range from 500 to 900 nm. Spectral bands (16–25) were programmed into the sensor at 10–20 nm bandwidths. A GR II 16-megapixel digital camera, manufactured by Ricoh Imaging Company, Ltd., captured photos during the flight to produce a natural color image orthomosaic and digital surface model using Agisoft Metashape software. The orthomosaic, DSM, and spectral bands from the HSI provided inputs for classification in eCognition software (Trimble, Inc.). The purpose of this study is to classify the foraging foodscape of brooding Columbian Sharp-tailed Grouse (*Tympanuchus phasianellus columbianus*) on lands removed from agricultural production and returned to natural vegetative cover. Our preliminary results reveal a micro-habitat classification at an individual plant scale that quantifies forage availability between spectrally similar vegetative species.

6.4.4 The Future of sUAS and Wildlife Conservation

The use of small Unmanned Aerial Systems (sUAS) for wildlife conservation has increased dramatically over the last decade (Linchant et al. 2015). Wildlife managers, researchers, and professionals associated with data accumulation in natural systems have begun widely adopting the use of sUAS for research and monitoring purposes. Changes in the regulatory environment has made it easier for researchers to be able to integrate sUAS into their wildlife and habitat studies. sUAS now are easily deployable and are decreasing in cost with the added benefit of onboard geo-referencing and an increasing number of sensors that can be deployed on a variety of airframes. sUAS offer survey options that can be less disruptive to wildlife and safer for researchers compared to traditional surveys from low-flying manned aircraft (Christie et al. 2016). Some barriers to adoption include processing time for large datasets, expertise required for advanced image analysis, and technological limitations for covering large areas due to sUAS limited battery life. With future technological advances and opportunities for sUAS remote-sensing specific training, these drawbacks are likely to be overcome and the continued use of sUAS will be transformative for wildlife conservation and habitat monitoring.

Literature Cited

Abd-Elrahman, Amr, Leonard Pearlstine, and Franklin Percival. 2005. "Development of Pattern Recognition Algorithm for Automatic Bird Detection from Unmanned Aerial Vehicle Imagery." *Surveying and Land Information Science* 65 (1): 37.

Adão, Telmo, Jonáš Hruška, Luís Pádua, José Bessa, Emanuel Peres, Raul Morais, and Joaquim João Sousa. 2017. "Hyperspectral Imaging: A Review on UAV-Based Sensors, Data Processing and Applications for Agriculture and Forestry." *Remote Sensing* 9 (11). doi: 10.3390/rs9111110.

Anderson, Karen, and Kevin J. Gaston. 2013. "Lightweight Unmanned Aerial Vehicles Will Revolutionize Spatial Ecology." *Frontiers in Ecology and the Environment* 11 (3): 138–46. doi:10.1890/120150.

Beck, Jeffrey L., D. Terrance Booth, and Carmen L. Kennedy. 2014. "Assessing Greater Sage-Grouse Breeding Habitat With Aerial and Ground Imagery." *Rangeland Ecology & Management* 67 (3): 328–32. doi:10.2111/REM-D-12-00141.1.

Bevers, M., J. Hof, D. W. Uresk, and G. L. Schenbeck. 1997. "Spatial Optimization of Prairie Dog Colonies for Black-Footed Ferret Recovery." *Operations Research* 45 (February 2015): 495–507. doi:10.1287/opre.45.4.495.

Biggins, D., B. Miller, L. Hanebury, B. Oakleaf, Adrian H. Farmer, Ron Crete, and Arnold Dood. 1993. "A Technique for Evaluating Black-Footed Ferret Habitat. Management of Prairie Dog Complexes for the Reintroduction of the Black-Footed Ferret." *US Fish and Wildlife Service Biological Report* 13: 73–88.

Biggins, D. E., Jerry L. Godbey, Marc R. Matchett, L. R. Hanebury, T. M. Livieri, and Paul E. Marinari. 2006. "Monitoring Black-Footed Ferrets during Reestablishment of Free-Ranging Populations: Discussion of Alternative Methods and Recommended Minimum Standards." *Recovery of the Black-Footed Ferret: Progress and Continuing Challenges* 2006: 155–74.

Caldwell, Rachel A. 2015. "Ecological Status of Black-Tailed Prairie Dogs on Boulder, Colorado Open Space and Mountain Parks Land: An Analysis of Select Indicators." University of Montana. 83 pp.

Chabot, Dominique, S. R. Craik, and D. M. Bird. 2015. "Population Census of a Large Common Tern Colony with a Small Unmanned Aircraft." *PLoS ONE* 10: 1–14. doi:10.1371/journal.pone.0122588.

Chabot, Dominique, and Charles M. Francis. 2016. "Computer-Automated Bird Detection and Counts in High-Resolution Aerial Images: A Review." *Journal of Field Ornithology*. doi:10.1111/jofo.12171.

Christie, Katherine S., Sophie L. Gilbert, Casey L. Brown, Michael Hatfield, and Leanne Hanson. 2016. "Unmanned Aircraft Systems in Wildlife Research: Current and Future Applications of a Transformative Technology." *Frontiers in Ecology and the Environment* 14 (5): 241–51. doi:10.1002/fee.1281.

Cully, J. F. 1993. "Plague, Prairie Dogs, and Black-Footed Ferrets. Management of Prairie Dog Complexes for the Reintroduction of the Black-Footed Ferret." *US Fish and Wildlife Service Biological Report* 1: 38–48.

Descamps, Stig, Arnaud Béchet, Xavier Descombes, Antoine Arnaud, and Josiane Zerubia. 2011. "An Automatic Counter for Aerial Images of Aggregations of Large Birds." *Bird Study* 58 (3): 302–08. doi:10.1080/00063657.2011.588195.

Eads, David A., Dean E. Biggins, Travis M. Livieri, and Joshua J. Millspaugh. 2014. "Space Use, Resource Selection and Territoriality of Black-Footed Ferrets: Implications for Reserve Design." *Wildlife Biology* 20 (1): 27–36. doi:10.2981/wlb.13070.

Fagerstone, Kathleen A, and Dean E. Biggins. 2011. "Black-Footed Ferret Areas of Activity during Late Summer and Fall at Meeteetse, Wyoming." *Journal of Mammalogy* 92 (4): 705–09.

Gillette, Gifford L., Kerry P. Reese, John W. Connelly, Chris J. Colt, and Jeffrey M. Knetter. 2015. "Evaluating the Potential of Aerial Infrared as a Lek Count Method for Prairie Grouse." *Journal of Fish and Wildlife Management* 6 (2): 486–97. doi:10.3996/022015-JFWM-008.

Goodwin, K, and D. Longknife. 2013. "Fort Belknap Indian Community, Noxious Weed Management Strategic Plan 2013–2018." *Center for Invasive Species Management*, Montana State University. http://weedcenter.org/wpa/docs/Ft. Belknap_FINAL.pdf

Grenzdörffer, G. J. 2013. "UAS-Based Automatic Bird Count of a Common Gull Colony." *International Archives of Photogrammetry and Remote Sensing* XL–1 (W 2): 169–74. doi:10.5194/isprsarchives-XL-1-W2-169-2013.

Hillman, Conrad N. 1968. "Field Observations of Black-Footed Ferrets in South Dakota." *Transactions of the North American Wildlife Conference and National Resources Conference* 33: 433–43.

Hillman, Conrad N, and Tim W. Clark. 1980. "Mustela Nigripes." *Mammalian Species* 126: 1–3. doi:10.2307/3503892.

Hoogland, John L. 1995. *The Black-Tailed Prairie Dog: Social Life of a Burrowing Mammal.* University of Chicago Press, 571 pp.

Hutt, Mike. 2011. "USGS Takes to the Sky." *Earth Imaging Journal* 8 (5): 54–55.

Jachowski, David S, and J. Michael Lockhart. 2009. "Reintroducing the Black - Footed Ferret Mustela Nigripes to the Great Plains of North America." *Small Carnivore Conservation* 41: 58–64.

Jones, George Pierce, Leonard G. Pearlstine, and H. Franklin Percival. 2006. "An Assessment of Small Unmanned Aerial Vehicles for Wildlife Research." *Wildlife Society Bulletin* 34 (3): 750–58. doi:10.2193/0091-7648(2006)34[750:Aaosua]2.0.Co;2.

Knowles, Craig J, Jonathan D. Proctor, and Steven C. Forrest. 2002. "Black-Tailed Prairie Dog Abundance and Distribution in the Great Plains Based on Historic and Contemporary Information." *Great Plains Research* 54: 219–54.

Koh, Lian Pin, and Serge A. Wich. 2012. "Dawn of Drone Ecology: Low-Cost Autonomous Aerial Vehicles for Conservation." *Tropical Conservation Science* 5 (2): 121–32. doi:WOS:000310846600002.

Linchant, Julie, Jonathan Lisein, Jean Semeki, Philippe Lejune, and Cédric Vermeulen. 2015. "Are Unmanned Aircraft Systems (UASs) the Future of Wildlife Monitoring? A Review of Accomplishments and Challenges." *Mammal Review* 45 (4): 239–52. doi:10.1111/mam.12046.

McCafferty, Dominic J. 2013. "Applications of Thermal Imaging in Avian Science." *Ibis* 155 (1): 4–15.

Mitchell, Jessica J, Nancy F. Glenn, Matthew O. Anderson, Ryan C. Hruska, Anne Halford, Charlie Baun, and Nick Nydegger. 2012. "Unmanned Aerial Vehicle (UAV) Hyperspectral Remote Sensing for Dryland Vegetation Monitoring." In *Hyperspectral Image and Signal Processing: Evolution in Remote Sensing (WHISPERS), 2012 4th Workshop*, 1–10. IEEE.

Mulero-Pázmány, Margarita, Jose Ángel Barasona, Pelayo Acevedo, Joaquín Vicente, and Juan José Negro. 2015. "Unmanned Aircraft Systems Complement Biologging in Spatial Ecology Studies." *Ecology and Evolution* 5 (21): 4808–18. doi:10.1002/ece3.1744.

National Oceanic Atmospheric Administration. 2016. "1981–2010 U.S. Climate Normals." Climate Normals. www.ncdc.noaa.gov/data-access/land-based-station-data/land-based-datasets/climate-normals

Sankey, Temuulen T., Jason McVay, Tyson L. Swetnam, Mitchel P. McClaran, Philip Heilman, and Mary Nichols. 2018. "UAV Hyperspectral and Lidar Data and Their Fusion for Arid and Semi-Arid Land Vegetation Monitoring." *Remote Sensing in Ecology and Conservation* 4 (1): 20–33. doi:10.1002/rse2.44.

Sardà-Palomera, Francesc, Gerard Bota, Carlos Viñolo, Oriol Pallarés, Víctor Sazatornil, Lluís Brotons, Spartacus Gomáriz, and Francesc Sardà. 2012. "Fine-Scale Bird Monitoring from Light Unmanned Aircraft Systems." *Ibis* 154 (1): 177–83. doi:10.1111/j.1474-919X.2011.01177.x.

Severson, Kieth E, and Glenn E. Plumb. 1998. "Comparison of Methods to Estimate Population Densities of Black-Tailed Prairie Dogs." *Wildlife Society Bulletin* 26: 859–66.

U.S. Fish and Wildlife Service. 2013. "Black-Footed Ferret Recovery Plan." U. S. Fish and Wildlife Service, Denver, Colorado. 157 pp.

van Gemert, Jan C., Camiel R. Verschoor, and Pascal Mettes. 2015. "Nature Conservation Drones for Automatic Localization and Counting of Animals." *Computer Vision - ECCV 2014 Workshops* 8925: 255–70.

Watts, Adam C, John H. Perry, Scot E. Smith, Matthew A. Burgess, Benjamin E. Wilkinson, Zoltan Szantoi, Peter G. Ifju, and H. Franklin Percival. 2010. "Small Unmanned Aircraft Systems for Low-Altitude Aerial Surveys." *The Journal of Wildlife Management* 74 (7): 1614–19. doi:10.2307/40801523.

Wich, Serge, David Dellatore, Max Houghton, Rio Ardi, and Lian Pin Koh. 2015. "A Preliminary Assessment of Using Conservation Drones for Sumatran Orang-Utan (Pongo Abelii) Distribution and Density." *Journal of Unmanned Vehicle Systems* 4 (1): 45–52.

Witczuk, Julia, Stanisław Pagacz, Anna Zmarz, and Maciej Cypel. 2018. "Exploring the Feasibility of Unmanned Aerial Vehicles and Thermal Imaging for Ungulate Surveys in Forests-Preliminary Results." *International Journal of Remote Sensing* 39 (15–16): 5504–21.

7

Multi-View, Deep Learning, and Contextual Analysis: Promising Approaches for sUAS Land Cover Classification

Tao Liu and Amr Abd-Elrahman

CONTENTS

7.1 Introduction

The methods used to analyze sUAS images have not been significantly changed to accommodate the wide adoption of these images in the natural resource management field. Currently, traditional pixel-based and object-based image classification of an orthoimage produced through photogrammetric processing of hundreds or thousands of sUAS images is still the most common way for sUAS image classification. Images captured by sUAS differ from those captured by other remote sensing platforms since they tend to have smaller extent, higher spatial resolution, and large image-to-image overlap with varying object-senor geometry compared to satellite or piloted aircraft images. Unlike satellite images, in a typical image acquisition mission, many overlapped sUAS images are captured within a very short time from different viewing angles, potentially, facilitating a way to study

133

the bi-directional reflectance distribution function (BRDF) of the land cover. Nevertheless, taking advantage of this redundancy in image classification is by itself an important asset to explore.

With the rapid evolution of deep learning classifiers, and increased availability of computing power (e.g., GPU and cloud computing), deep learning classifiers have become one of the most active topics for the sUAS image classification field. This is not only motivated by its successful performance in computer vision, but also due to its operational advantages in comparison with traditional classifiers. For example, deep learning classifiers do not require manual extraction of features, while manually selecting appropriate features are important to achieve good performance for traditional classifiers. Deep learning classifiers, however, are not without shortcomings. Generally, they require large amount of training data accompanied by a computationally intensive training process. Conducting comparison studies to examine the performance of deep learning and traditional classification using sUAS images is necessary. In addition, it is also interesting to investigate whether the multi-view data extracted from sUAS is useful in triggering the power of deep learning classifiers.

Traditional sUAS image processing methods usually treat individual classification targets separately without considering its surrounding information. However, in natural environment, plant functional groups often predictably co-occur (Carranza et al. 2011; Chytrý et al. 2008; Frouz 1997), providing context information for each group. It's interesting to study how this context information can be exploited to improve land cover classification accuracy.

This chapter introduces several approaches for land cover classification that involve (1) multi-view information, (2) two deep learning classification methods, and (3) contextual information modeling. The results of implementing these approaches on sUAS images acquired for a wetland area in Central Florida are summarized and synthesized. The chapter examines how each of these three components can be utilized to improve the land cover classification and how combinations of these approaches can be exploited together to improve classification accuracy.

7.2 Background

7.2.1 Small Unmanned Aircraft System Image Acquisition and Processing

sUAS is a valuable platform for assisting wetland management efforts (Pande-Chhetri et al. 2017), fueled by their temporal flexibility, relative ease of use, and high spatial resolution. This is specifically true considering the high-frequency changes and small community size of wetland land cover. Compared to space-borne remote sensing, sUAS can fly at a much lower altitude, and thus is able to generate remote sensing images with sub-decimeter resolution. This feature is very important because even though civilian remote sensing satellites can collect images with resolution as high as 25 cm (e.g., WorldView-3), this resolution is still insufficient for some natural resource management missions conducted at the species level (Lu and He 2017a). Even though piloted aircraft can collect images with a resolution comparable to sUAS images (e.g., 5–6 cm), operational expense and safety for pilots favor sUAS implementation (Rango et al. 2006). In addition, flight route and time can be flexibly controlled by the sUAS operator. These characteristics make sUAS one of the most favorable remote sensing platforms for small sites and timely and repetitive natural resource management and monitoring practices such as invasive plant species control efforts.

Both fixed-wing and multi-rotor sUAS are used in remote sensing applications. Heavier payload such as LiDAR and pushbroom hyperspectral sensors are normally mounted on multi-rotor systems. Fixed-wing systems, however, enjoy longer flight time, which makes them suitable for imaging extended areas. Multispectral cameras composed of several individual single-band cameras, each recording specific visible to near-infrared (Vis-NIR) part of the electromagnetic spectrum have been used for wetland and aquatic land cover classification (Samiappan et al. 2017). Hyperspectral and LiDAR sensors mounted on sUAS were recently used in wetland and coastal land cover classification motivated by consistent sensor price drop, reduced weight, and power consumption, and the availability of high-quality and light-weight navigation sensors (Cao et al. 2018; Li et al. 2017; Zhu et al. 2019). In the meantime, the use of multi-spectral cameras (e.g., Micasense™ and Tetracam™) and low-cost visual-band (RGB) consumer-grade cameras still dominate sUAS use in wetland management, primary due to their low cost and the high spatial resolution of the consumer grade RGB cameras.

Many studies highlighted the effect of using Structure from Motion (SfM) (Westoby et al. 2012) to produce orthoimages and Digital Surface Models (DSM) and utilize them in wetland and coastal land cover characterization (Chiabrando et al. 2015; Vázquez-Tarrío et al. 2017). Images captured by multispectral cameras and consumer grade RGB cameras do not require expensive onboard Real Time Kinematic (RTK) Global Navigation Satellite Systems (GNSS) or high-quality Inertial Measurement Unit (IMU) for direct georeferencing in order to produce orthoimages with acceptable qualities for most wetland classification purposes. They only need a few ground control points to be used in the photogrammetric solution to produce the orthoimages and digital surface models. The orthoimage is generated from hundreds or thousands of overlapped sUAS images (Wolf et al. 2013), where pixel values usually come from a single UAS image view or a mashup (e.g., using the mean value) of multi-view UAS images.

7.2.2 Object-based Image Classification

Pixel-based image classification is the main approach for processing when medium and coarse resolution imagery (e.g., Landsat, SPOT, ASTER, and MODIS) are the primary source of remote sensing images, where pixel size is usually similar or coarser than the objects to be detected (Blaschke 2010). OBIA techniques were introduced with the emergence of high-resolution satellite images provided by high spatial resolution civilian satellite platforms such as the IKONOS and QuickBird satellites back in the late 1990 and early 2000s. The increase in spatial resolution leads to in-class variability and reduced cross-class separability, which negatively affects classification results (Hsieh et al. 2001; Yu et al. 2006). Pixel-based classification can show salt and pepper effects and ignores important contextual, topological, and semantic information in the images (Whiteside et al. 2011; Zhang et al. 2013). On the other hand, OBIA usually not only creates a more appealing appearance but also tends to obtain comparable if not higher classification accuracy (Cleve et al. 2008; Fu et al. 2017; Gao and Mas 2008). OBIA is commonly used for classifying high-resolution sUAS imagery of natural areas whether the classification is conducted at the vegetation community (Kaneko and Nohara 2014; Nagol et al. 2015; O'Brien 2016; Zweig et al. 2015), or species (Laliberte and Rango 2011; Lu and He 2017b) levels. Most previous OBIA classification studies were based on the orthoimages produced after conducting a bundle-block adjustment process.

OBIA clusters similar adjacent pixels into relatively homogeneous and meaningful objects (Blaschke 2010) and treats each object as a superpixel enabling the use of many spectral, geometric, and textural features extracted from the individual pixel information within the object. Another benefit to using image objects is facilitating contextual modeling of adjacent objects at the same segmentation level or across different levels. Image segmentation raises several questions, however, as object definition is scale dependent and the right parameters must be used to avoid having mixed class objects or fragmented classification results. Optimal segmentation parameters are often determined empirically through visual and quantitative inspection or quantitatively (Drăguţ et al. 2010; Flanders et al. 2003; Kim et al. 2014; Radoux and Defourny 2007), depending on the size and the complexity of the target objects (Whiteside et al. 2011).

Many algorithms have been developed for image segmentation (Blaschke 2010), with region-based being one of the most widely implemented algorithms in the remote sensing field. Region-based image segmentation can be conducted using region growing and merging techniques, where adjacent regions are aggregated into a single object until a certain heterogeneity threshold is reached. This threshold often determines the scale of the segmentation. The process starts from certain seed points until the whole image is segmented (Baatz and Schäpe 2000; Blaschke et al. 2014). This region merging technique is used in the eCognition software, which utilizes the multi-scale fractal net evolution object segmentation iterative algorithm (Baatz and Schäpe 2000). The algorithm starts with individual pixels, where neighboring pixels (objects in subsequent iterations) continue to be merged until a pre-set degree-of-fitting cost (scale parameter based on spectral and spatial similarity) is reached. The scale parameter controls object size and leads to the multi-scale characteristic of the algorithm (Laliberte and Rango 2011).

OBIA enables the use of hundreds of spectral, textural, and geometrical object features computed from the individual pixels within each object. Additionally, contextual features based on the topological relationship among features in the same segmentation level and across parent and child levels are also facilitated. The large number of often highly correlated features highlights the typical high dimensionality problem associated with OBIA. Laliberte et al. (2012) and Georganos et al. (2018) compared commonly used feature selection methods and stressed the importance of using few optimal features in object-based image classification (Laliberte et al. 2012; Ma et al. 2017). Some of the feature selection methods quantify the statistical separability among the classes using different feature combinations or identify important variables through a machine learning classifier such as the Random Forests (RF) (Breiman 2001) that has built-in mechanisms to identify important features.

OBIA is commonly used for classifying high-resolution UAS imagery of natural areas whether the classification is conducted at the vegetation community (Kaneko and Nohara 2014; Nagol et al. 2015; O'Brien 2016; Shin et al. 2015; Torres-Sánchez et al. 2014; Zweig et al. 2015), or species (Laliberte and Rango 2011; Lu and He 2017b) levels. Most previous OBIA classification studies were based on the orthoimages produced after conducting a bundle-block adjustment process using the sUAS images (Wolf et al. 2013).

Traditional machine learning algorithms, such as Support Vector Machine (SVM) (Cortes and Vapnik 1995; Veenman et al. 2002), Random Forests (RF) (Breiman 2001), and Artificial Neural Networks (ANN) (Paola and Schowengerdt 1995), as well as other supervised classification algorithms (e.g., Maximum Likelihood), have been implemented in OBIA. The SVM algorithm separates the classes with a decision surface, known as optimal hyperplane, such that it maximizes the margin between the classes (Cortes and Vapnik 1995; Veenman et al. 2002). Random Forests is an ensemble-based classifier that uses subsets of the data and features (bagging) to build classification trees and use rule (e.g., maximum voting, summation, or Bayesian) to assign final classes (Amit and Geman 1997; Breiman 2001; Rodriguez-Galiano et al. 2012). This RF algorithm can handle a large number of features and provide measures for feature importance. The ANN algorithm was invented out of inspiration of biological neural network. The algorithm differs from SVM, which tries to solve a convex optimization problem. One of the most popular ANN implementations aims to solve a non-convex optimization problem using a technique called error back-propagation (Paola and Schowengerdt 1995). Shadowed by SVM, it gradually lost favor by the machine learning community towards the end of the 1990s and early 2000s. Now, as the deep learning technique becomes popular, ANN has gained significant attention in the machine learning community (Deng and Yu 2014; LeCun et al. 2015).

7.2.3 Multi-view Data Extraction

Studies utilizing multi-view data from sUAS images are difficult to find. However, as piloted aircraft mounted with a digital camera may collect data in a similar way to sUAS, examining the publications based on imagery collected from piloted aircraft may cast some light onto the potential benefits of involving multi-view image information. One example is the study conducted by Hu et al. (2007). The authors designed a special camera called Frequent Image Frames Enhanced Digital Orthorectified Mapping (FIFEDOM) mounted on a piloted aircraft for acquiring multi-view images. The authors also developed software to perform bundle adjustment and space intersection (Wolf et al. 2013) associated with this camera. Their data product contains all the rectified images over the area of interest and the view geometries of each pixel in these images for deriving multi-view datasets. Another example for studying multi-view data extracted from piloted aircraft is the study by Koukal and Atzberger (2012), who manually found the overlapping areas in sUAS images corresponding to given forest stands with areas ranging from 232 m^2 to 2835 m^2 and then derived the multi-view data based on the view geometry using coordinates of camera and center of the forest stands. While this method works for research purpose that only needs a small number of patches of large area, it is insufficient for large scale application that requires accurate multi-view information for thousands of objects with areas as low as sub square meters. The information needed to relate positions on the orthoimage to corresponding locations on individual sUAS imagery is already available as a result of the bundle-block adjustment of the sUAS images. This information can be utilized to extract multi-view image information with a relatively moderate amount of efforts. Considering object-based image analysis (OBIA), rather than pixel-based approach, it may be unnecessary to process the view geometries for all the pixels in an orthoimage. Liu et al. 2018c employed collinearity equations, bundle-block adjustment data, and optimization method to project any point on the orthoimage to sUAS images, enabling of producing multi-view data for the orthoimage object, as demonstrated in Figure 7.1.

7.2.4 Deep Convolutional Neural Network Classifier

Deep Convolutional Neural Network (DCNN) is the workhorse behind the deep learning algorithm. Different architectures of DCNN may contain different types and number of layers, such as the convolutional, pooling, skip connection, batch normalization, and activation layers, etc. (Goodfellow et al. 2016). Developing new architecture to enable better performance of DCNN is still an active research area, which is beyond the scope of our discussion. Since it made a breakthrough by almost halving the error rate of

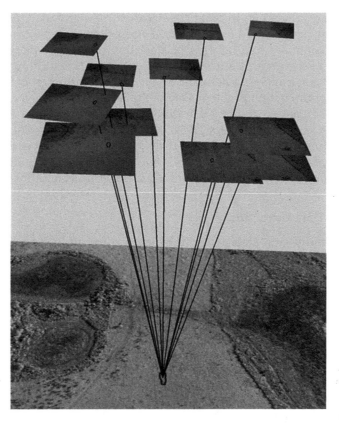

FIGURE 7.1
Multi-view objects on UAS images corresponding to a ground object.

the 2010 Large Scale Visual Recognition Challenge (ILSVRC2010), DCNN was rapidly utilized in many industrial applications and other academic areas in recent years as it continues to advance technologies in areas like speech recognition, medical diagnosis (Suk et al. 2014), autonomous driving (Huval et al. 2015), or even the gaming world (Silver et al. 2016). The success of DCNN in these fields has motivated researchers in the remote sensing community to investigate its usefulness for remote sensing image analysis (Ma et al. 2016; Makantasis et al. 2015; Vetrivel et al. 2017; Zhang et al. 2016; Zhao and Du 2016; Zhong et al. 2017).

A large amount of training datasets is necessary for DCNN training, but collecting training data for a remote sensing application is often expensive and time-consuming. Different techniques have been tested to overcome this shortcoming. For example, to augment training dataset, Zhao and Du (2016) transformed the original very high-resolution images into Laplacian pyramid images as multiscale datasets, and their results showed a significant increase in classification accuracy. To utilize unlabeled samples, Ma et al. (2016) pre-labeled each unlabeled sample with local and global decisions and added some pre-labeled ones with high confidence to expand the training samples. Their results indicated that this was an effective way to apply deep learning for hyperspectral image analysis. Transfer-learning is a technique used to train the classifier in one domain in order to enhance its generalization in another domain (Pan and Yang 2010; Torrey and Shavlik 2009). It was exploited in trying to overcome the difficulty caused by limited training samples. Xie et al. (2015) aimed to map poverty using high-resolution images. Due to the limited training samples, they first trained the convolutional neural network to predict the nighttime light with a rich training dataset. Then the model trained with the nighttime light dataset was further trained with the insufficient poverty dataset. This research demonstrated that features learned with transfer learning methods were informative for poverty mapping.

None of these approaches have tried to address the issue of limited samples by taking advantage of multi-view data from UAS images to enrich the training dataset for DCNN applications, even though it feels natural to do so. Multi-view information inherently improves the diversity and quantity of the training dataset as a result of the different object-light-sensor geometries in these views. Integrating the OBIA, multi-view, and DCNN gives rise to a challenging but promising classification framework for processing UAS data, not only by smoothly incorporating the power of DCNN into general OBIA framework, but also extending existing options to cope with the scarce training samples issue when applying DCNN. Integrating DCNN into OBIA and investigating how changing the size of training dataset would impact the performance of DCNN in comparison of RF and SVM was conducted in Liu et al. (2018b). The impact of multi-view dataset on the performance of DCNN within the OBIA framework was studied in Liu and Abd-Elrahman (2018b).

7.2.5 Fully Convolutional Neural Network Classifier

The DCNN can only tell whether an image patch contains some type of classes or not and it cannot tell where the classes are located at the pixel level. Alternatively, Fully Convolutional Networks (FCN) (Long et al. 2015) attempted to solve this problem. Given an image with unknown label for each pixel, the final objective or output of FCN is to label each pixel within the frame. The main approach used to convert DCNN to fully convolutional networks is to conduct an up-sampling operation on the end layers of the DCNN to extend its end layer to exactly the same dimensions of the input frame (i.e., having the same number of rows and columns). This way, training can be carried out pixel-to-pixel. It is natural to consider applying this type of network for invasive vegetation detection since we do not only want to know whether one area is impacted by invasive vegetation but also want to delineate the area occupied by the invasive vegetation within an object or frame.

FCN has been used to deal with various computer vision related problems successfully in recent years since it was developed in 2015 (Long et al. 2015). For example, FCN has been used in liver cancer diagnosis via analysis of cancerous tissue pathological image (Li et al. 2017), diagnosis of smaller bowel disease through automatically marking cross-sectional diameters on small bowel images (Pei et al. 2017), osteosarcoma tumor segmentation on Computed Tomography (CT) images (Huang et al. 2017), traffic sign detection (Zhu et al. 2016), etc. Applications using FCN in remote sensing can also be seen, even though the number of them is small, but growing. Several studies were conducted using the ISPRS Vaihingen dataset achieves. This dataset contains 8-cm resolution aerial orthoimage of near IR, R, and G, point-cloud (4 points/m^2) and Digital Surface Model (DSM). This dataset was collected for urban object detection and the classes to be detected includes building, road, tree, low vegetation/grass, and artificial ground. Piramanayagam et al. (2016) applied FCN using orthoimage and DSM, obtaining 88% overall accuracy, which is higher than the 86.3% achieved by random forest classification. Sherrah (2016) compared patch-based DCNN and FCN using only orthoimage and found FCN outperformed patch-based

DCNN with 87.17% and 83.46% accuracy, respectively. Marmanis et al. (2016) achieved overall accuracy larger than 90%, by combining edge detection result with orthoimage and DSM as input to the FCN and employing an ensemble classification strategy. Liu and Abd-Elrahman (2018a) and Liu et al. (2018b) applied FCN within OBIA framework for the first time and explored two options of preparing training image patches for FCN: one of them labels as background all the pixels surrounding object (denoted as FCN-I-OBIA) and the other one assigns ground truth class labels to all the pixels within image patch (denoted as FCN-II-OBIA). Liu and Abd-Elrahman (2018a) extended the application of FCN to multi-view datasets within the OBIA framework.

7.2.6 Bidirectional Reflectance Distribution Modeling

Bidirectional Reflectance Distribution Function (BRDF) should describe the intrinsic directional reflectance of a vegetation land cover that can be attributed to the combined effect of energy interaction with the leaf, crown, wood, and background objects, as well as the vegetation gaps at the stand or landscape scales (Ross 2012), which can be a significant contributor to the classification process. However, retrieving bidirectional reflectance defined as the ratio between received radiance at a certain direction to the incident irradiance from an energy source at a certain direction, from remotely sensed imagery does not satisfy the definition and is often conducted through modeling. In practice, remote sensing information involves hemispherical sky irradiance illuminating the surface, finite angular measurements, path radiance, and absorption of the atmosphere (Liang and Strahler 1994).

Several researchers explored the use of multi-view spectral reflectance data to model semi-empirical BRDF models (Koukal et al. 2014; Pacifici et al. 2014; Su et al. 2007) and used the parameters of the modeled BRDF in image classification. Using multi-view spectral data of automatically segmented objects to model the BRDF and extracting BRDF parameters from high-resolution images captured by sUAS for classification purposes have rarely been studied. sUAS provides a unique opportunity to model BRDF since a typical flight mission involves many overlapping pictures, where the same object can be seen from different viewing angles within a relatively short acquisition time (small variation in sun angle). Both physical and statistical models (Liang and Strahler 1994) have been used to model BRDF. Physical modeling requires information about the land cover such as leaf biophysical parameters and canopy structure, while statistical modeling tends to characterize the shape of the BRDF. Statistical models can be divided into empirical models that mathematically fit the BRDF shape and semi-empirical models that integrate some of the physical characteristics when describing the BRDF form.

Semi-empirical models, such as the Rahman-Pinty-Verstraete (RPV) model (Rahman et al. 1993) and the Ross-Thick-LiSparse (RTLS) model (Lucht et al. 2000; Roujean et al. 1992) have been used to model BRDF from multi-view remote sensing images (Koukal et al. 2014; Pacifici et al. 2014). Each of these models requires a certain number of parameters that can be estimated by fitting the BRDF model using multi-view remote sensing observations (Roberts 2001). Once the BRDF model parameters are estimated they can be used as additional features in image classification (Koukal et al. 2014; Liu et al. 2018a) and the model can be used in many applications including albedo estimation and vegetation structure detection (Gao et al. 2003; Schaaf et al. 2002). Rather than relying on BRDF modeling to utilize multi-view data, Liu and Abd-Elrahman (2018c) proposed a novel approach to conduct OBIA using multi-view data and compared its performance with BRDF-based methods.

7.2.7 Contextual Modeling

Conditional Random Field (CRF) (Sutton and McCallum 2012) provides a framework to encode context information for classification. Together with Markov Random Field (MRF) (Wang et al. 2013), a generative model version corresponding to CRF has become a ubiquitous tool to crack numerous problems in the computer vision field using a variety of inference and learning methods developed in the past decade (Blake et al. 2011; Wang et al. 2013). These probabilistic graphical models also attracted attention from the remote sensing community. Kasetkasem et al. (2005) employed Maximum Likelihood Estimate (MLE) to first generate initial super-pixel map from IKONOS and Landsat satellite images and then applied a pair-wise potential function to encourage the smoothness of labeling between neighboring pixels for classification optimization. The results showed a significant increase in classification accuracy compared to linear optimization approaches.

Liu et al. (2018c) incorporated both spatial and temporal context information to help classification for forest change detection. Their results showed significantly improved accuracy of identifying the change from forest to non-forest land cover compared with traditional approaches. Li et al. (2016) applied CRF to map landslides from aerial orthoimage. In their study, unary term of the CRF model (Prince 2012) was derived from Gaussian mixture models and pair-wise potential function using spectral difference of neighboring pixels to impose classification smoothness. Their result outperformed region-based level set evolution method. Zhong et al. (2014) applied SVM to generate label probability for each pixel of high spatial resolution satellite image to create unary energy with two types of potential functions, respectively. Together with the difference of neighboring pixel values for pair-wise potential function, this produced two classification results, respectively. Then, the final classification was created by combining the results using a connected-component labeling algorithm. Their approach showed a competitive quantitative and qualitative performance for high spatial resolution image classification when compared with other state-of-the-art classification algorithms.

Even though MRF and CRF have shown potential to improve classification accuracy to varying degrees in previous studies, the gains are generally not impressive and have been disproportionate to the greater model sophistication, thus decreasing model utility and accessibility. Liu et al. (2018c) attributed the limited improvement to the fact that previous models usually required the users to empirically determine and input model parameters, which was the main motivation to develop a fully learnable CRF model. This model is superior in two aspects. First, all the parameters in the CRF can be automatically learned from the training data, without requiring the user to manually select any parameters. Secondly, multi-view information is effectively utilized by the CRF model to allow the context and multi-view information simultaneously to improve land cover mapping accuracy.

7.3 Case Study

This case study compares different approaches to classify sUAS images. The OBIA framework is used in all classification approaches. A multi-view approach for image classification is implemented using traditional and deep learning classifiers. The results of using the multi-view approach are compared with the results of using the orthoimage objects. The results of using the DCNN and FCN deep learning classifiers were contrasted with the SVM and RF classification results. Finally, the results of implementing a CRF contextual modeling approach is presented.

7.3.1 Study Area and Data Preparation

The study area is a part of a 31,000-acre ranch, located in Southern Florida, within Lake Arcadia City (Figure 7.2). The ranch is comprised of diverse tropical forage grass pastures, palmetto wet and dry prairies, pine flatwoods, and large interconnecting marsh native

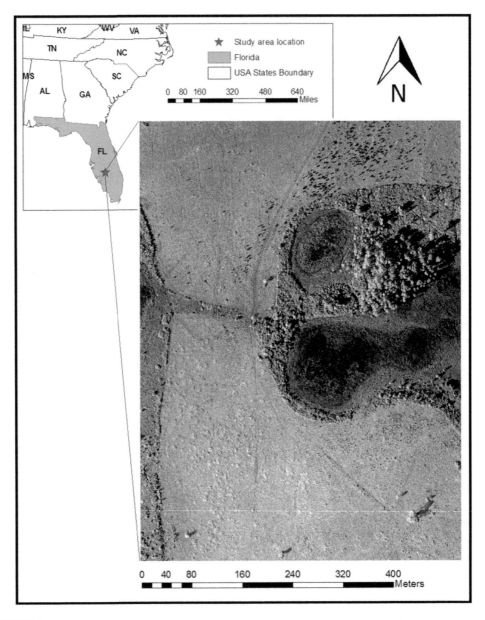

FIGURE 7.2
Study area (600 m × 400 m) in a Ranch in Lake Acardia City, Fl (Liu and Abd-Elrahman 2018c).

grass wetlands (Grasslands). The land also hosts cabbage palm and live oak hammocks scattering the lengths of copious creeks, gullies, and wetlands. The subset area (600 m × 400 m) is representative of the whole ranch that includes all the land cover types appearing in the ranch. All classes, except the Shadow class, were assigned according to the standard of vegetation classification for South Florida natural areas (Rutchey et al. 2006), as listed in Table 7.1.

A total of 1,397 images used in this study were captured by the U.S. Army Corps of Engineers -Jacksonville District (ACE) using the NOVA 2.1 fixed-wing small UAS, which weights around 6.4 kg and flies for up to 50 minutes in a single flight. A flight mission was planned and implemented with about 83% forward and 50% side overlaps. A consumer grade 3456 × 5184-pixel Canon EOS REBEL SL1 digital camera mounted on the NOVA 2.1 UAS was used. The images were synchronized with an onboard navigation grade GPS receiver to provide image locations. Four ground control points established near the four corners and a control point close to the center of the study area were used in the bundle-block adjustment solution. Details of the camera and flight mission parameters are listed in Table 7.2. The images were normalized for the changing light condition caused by differences in sun angle position during the approximately one-hour flight mission. During the one-hour UAS flight mission, the solar zenith changed from 70 to 80 degrees, leading to the change of the pixel values for the same target/land cover. Such a pixel value change resulting from solar angle variation instead of the land cover characteristics could impair the land cover classification using features extracted from the orthoimage data. Considering solar zenith change within a narrow range (≤10 degrees) during the flight operation, given an original UAS image i with zenith angle θ_i, the pixel values of the original UAS images were corrected using ImgCorrected$_i$ = ImgOrginal$_i$ * (cos(θ_i)/cos(75 degrees)) following previous study by Koukal et al. (2014), so that they share a common sun illumination; where the 75 degrees normalization angle is the solar zenith angle mid-time into the flight mission, facilitating the BRDF simulation.

The Agisoft Photoscan software (Agisoft 2016) was used to implement the bundle-block adjustment using the 1,397 UAS images of the study area. The software was used to produce

TABLE 7.1

Land Cover Classes in the Study Area

Class ID	Class Name	Description
CG	Cogon grass	Cogongrass (*Imperata cylindrica*) is a non-native invasive grass which occurs in Florida and several other Southeastern states.
IP	Improved Pasture	A sown pasture that includes introduced pasture species, usually dominated by Bahiagrass (*Paspalum notatum*). These are generally more productive than the local native pastures, have higher protein and metabolizable energy, and are typically more digestible. In our case, we also assume it is not infested by Cogongrass.
SUs	Saw Palmetto Shrubland	Saw Palmetto (*Serenoa repens*) dominant shrubland.
MFB	Broadleaf Emergent Marsh	Broadleaf emergent dominated freshwater marsh.
MFG	Graminoid Freshwater Marsh	Graminoid dominated freshwater marsh.
FHp	Hardwood Hammock - Pine Forest	A co-dominate mix (40/60 to 60/40) of Slash Pine (*Pinus elliottii*) with Laural Oak (*Quercus laurifolia*), Live Oak (*Quercus virginiana*), and/or Cabbage Palm (*Sabal palmetto*).
Shadow	Shadow	Shadow of all kinds of objects in the study area.

TABLE 7.2

Summary of Sensor and Flight Procedure

Items	Description
Sensor Name	Canon EOS REBEL SL1
Length of Focus	20 mm
Sensor Size	14.9×22.3 mm
Channels	RGB
Takeoff Time	29-Oct-2015 16:54:51 Eastern Daylight Time
Landing Time	29-Oct-2015 17:49:33 Eastern Daylight Time
Takeoff Latitude	27.22736549
Takeoff Longitude	−81.51152802
Average Wind Speed	5.1 m/s
Average Altitude	302.7 m

and export a 3-band (Red, Green, and Blue) 6-cm ground pixel-size orthoimage, a 27-cm Digital Surface Model (DSM), and the camera interior and exterior orientation parameters. The orthoimage was segmented using the Multi-Resolution Segmentation (MRS) algorithm implemented in eCognition based on the red, green, and blue bands of orthoimage and the DSM layer. Scale is the most important segmentation parameter controlling the final segmentation results. Even though there seems to be a consensus among the GEOBIA community to the need of developing a standard approach to generate best segmentation, the simplest but most commonly used method is still visual inspection of segmentation results among 76 papers in OBIA reviewed by Im et al. (2014). We tested and inspected the segmentation results using scale 10, 20, 30, 40, 50, 60, and 70, and finally chose the scale (50), shape (0.20), and compactness (0.50) parameters for generating the segmentation results for our study, because these parameters avoided the under-segmentation, and at the same time alleviated the over-segmentation as much as possible based on visual inspection. A total of 40,239 objects were generated from the segmentation process. The centroids of these objects were extracted and used as proxy for the objects to select the training and testing samples used in the classification experiments in Sections 7.3.2, 7.3.3, and 7.3.4. In all experiments, the number of samples used were equal across all land cover types. After a certain number of samples were selected, 90% of them were used to train the classifier and the remaining 10% were reserved for assessing the performance of the classification.

7.3.2 Orthoimage Classification

Image objects generated by eCognition were exported into MATLAB® for classification experiments. Deep learning classifiers (i.e., DCNN and FCN) were implemented using the MatConvNet (Vedaldi and Lenc 2015) package, a third-party deep learning package specifically developed for Matlab users, while the RF and SVM classifiers were provided natively by Matlab. Readers are referred to Liu et al. (2018b) for detailed description of the classification procedure. Figure 7.3 shows the overall accuracies obtained from classifying the orthoimage using five classification methods (DCNN, SVM, RF, FCN-I, and FCN-II), trained with three different numbers of training samples. When 700 training sample were used, DCNN achieved lower accuracy compared with RF and SVM (58.9% for RF, 62.4% for SVM, versus 51.8% for DCNN). However, as more training samples were added, DCNN quickly overtook RF and SVM (62.8% for RF, 66.1% for SVM, versus 67.6% for DCNN), and continued to improve the accuracy with more training samples used in in the training

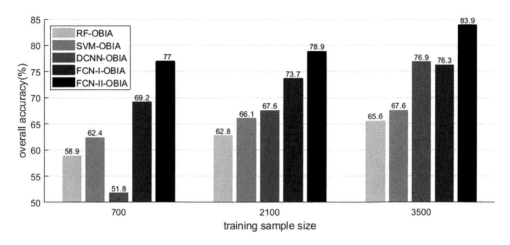

FIGURE 7.3
Orthoimage classification accuracies obtained with a different number of training samples (Liu et al. 2018b).

process (65.6% for RF, 67.6% for SVM, versus 76.9% for DCNN). FCN always shows superior performance than DCNN, RF, and SVM in our experiments under all the settings no matter whether FCN training samples contain class information for the pixels surrounding the object pixels within the object bounding box (FCN-II-OBIA) or not (FCN-I-OBIA).

7.3.3 Multi-View Classification

Each object in the orthoimage was projected and reconstructed on the original individual sUAS images to form the multi-view dataset using the method proposed by Liu et al. (2018a). In this dataset, each object in the orthoimage had 10–14 multi-view object instances in the individual sUAS images. Multi-view objects, rather than orthoimage objects were used to train the classifier in order to enhance its robustness. Given the trained classifier and multi-view objects corresponding to a testing orthoimage object, the classifier was applied to the multi-view objects, followed by a majority voting of multi-view object classification results to obtain the classification label for the testing orthoimage object. This procedure is named Multi-View Object-Based Image Analysis (MV-OBIA), and readers are referred to Liu and Abd-Elrahman (2018c) for a more detailed explanation. The results obtained from using multi-view information are shown in Figure 7.4, together with results obtained from direct orthoimage object classification. When multi-view objects were used in the classification much higher accuracy was achieved compared to the orthoimage classification results for the RF, SVM, and DCNN classifiers (63.6% vs. 77.2% for RF, 65.3% vs. 78.5% for SVM, and 65.3% vs. 82.1% for DCNN). Testing whether the multi-view training samples can boost accuracy on the pre-voting classification results shows that pre-voting accuracy for the multi-view classification is also higher than the accuracy that is achieved using the orthoimage only (63.6% vs. 68.1% for RF, 65.3% vs. 70.0%, and 65.3% vs. 75.9%). While DCNN produced accuracy similar to the RF and SVM for orthoimage data (63.6% for RF, 65.3% for SVM, versus 65.3% for DCNN), it performed better than RF and SVM when multi-view data were used (77.2% for RF, 78.5% for SVM, versus 82.1% for DCNN).

Ortho-OBIA-12st denotes the classification using 12 features in total including three spectral and nine texture features extracted from the orthoimage. MV-RPV-3P-12st represents the classification using a combination of the 12 spectral and textural orthoimage

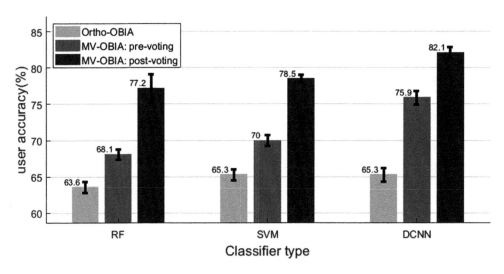

FIGURE 7.4

Classification accuracy improvements due to the use of multi-view classification for classifier RF, SVM, and DCNN (Liu and Abd-Elrahman 2018b).

features and the 3-parameters of the RPV model, and MV-OBIA-12st is the MV-OBIA using 12 spectral and textural features extracted from multi-view objects.

A series of classification experiments was conducted using the parameters of two semi-empirical BRDF models (RPV and RTLS). These semi-empirical models were fit to Multi-view sUAS data and model parameters (3- and 4-parameter RPV models; 4- and 5-parameter RTLS models) which were used in the classification in addition to 12 other spectral and textural parameters. The RF classifier in MATLAB was used in these experiments. Details of the RF classification implementation can be found in Liu and Abd-Elrahman (2018c). Figure 7.5 compares the performance of the classification experiments employing 12 spectral and textural orthoimage features (Ortho-OBIA-12st), BRDF parameters, and 12 spectral and textural features (MV-RPV-3P-12st; MV-RPV-4P-12st, MV-RTLS-3P-12st, and MV-RTLS-5P-12st) and 12 spectral and textural features extracted from multi-view images (MV-OBIA-12st). It should be mentioned here that the same spectral and textural parameters were used in all the classification experiments. Figure 7.5 shows the BRDF classification outperforms the orthoimage features (65.6% for MV-RPV-3P-12st, 65.7% for MV-RPV-4P-12st, 66.3% for MV-RTLS-3P-12st, 64.5% for MV-RTLS-5P-12st, versus 63.6% for Orhto-OBIA-12st). In contrast, multi-view classification achieved substantially higher accuracy than all of the BRDF-based methods (65.6% for MV-RPV-3P-12st, 65.7% for MV-RPV-4P-12st, 66.3% for MV-RTLS-3P-12st, 64.5% for MV-RTLS-5P-12st, versus 77.2% for MV-OBIA). These results indicate the accuracy improvement brought from using BRDF parameters is minimal compared with multi-view classification accuracy.

7.3.4 Post-Classification Contextual Modeling

The CRF model used in this study was designed to have a form belonging to the exponential function family, which guarantees the finding of global minimum during the model learning procedure using gradient descent method (Liu et al. 2018c). Class membership probability vector derived from summation pooling of direct classification

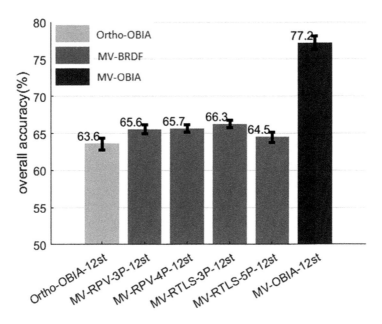

FIGURE 7.5
Classification accuracy obtained based on orthoimage, BRDF, and multi-view classification (Liu and Abd-Elrahman 2018c).

results of multi-view objects served as features for a unary term of CRF, while compatibility matrix of seven land-cover types and object feature (e.g., mean spectral values) distance of neighboring objects constitute the pairwise term of CRF model. In this way, both multi-view and context information are embedded in the model. All the weight parameters associated with the features were automatically learned by optimizing Negative Log Likelihood (NLL) of training samples using the gradient descent method. Belief propagation was used for the inference phase of the CRF model to generate the results in Figure 7.6, since it performs better than alpha expansion as demonstrated by Liu et al. (2018c). A third party MATLAB package developed by Schmidt (2012) was used in this study to model and implement the CRF. Readers are referred to Liu et al. (2018c) for details of implementation.

Figure 7.6 compares the classification accuracies obtained using orthoimage, multi-view classification, and contextual modeling. The multi-view classification results in Figure 7.6 are based on summation pooling, instead of majority voting that was used to produce the results shown in Figure 7.4. Figure 7.6 shows multi-view classifications based on summation pooling also lead to substantial accuracy improvements compared with classifications using orthoimage for all the classifiers (75.8% vs. 62.4% for RF, 75.0% vs. 64.9% for SVM, and 80.3% vs. 65.4% for DCNN). When multi-view information and contextual information are combined in the CRF model, accuracy was further increased considerably (87.3% vs. 75.8% for RF, 84.4% vs. 75.0% for SVM, and 86.3% vs. 80.3% for DCNN).

Figure 7.7 presents the maps generated by orthoimage classification, multi-view classification, and context modeling methods, alongside the orthoimage and ground truth image. The map quality becomes progressively better visually as multi-view information and context information was added to the classification procedure for all the classifiers, which is consistent with the quantitative evaluation of these classification methods shown in Figure 7.6.

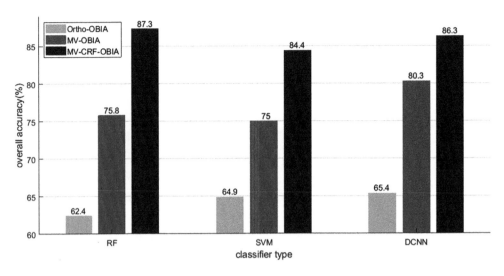

FIGURE 7.6
Performance improvements by CRF modeling (Liu et al. 2018c).

7.4 Discussion

This chapter presented several techniques that can be used to improve the classification accuracy of sUAS imagery under the OBIA framework. We believe that as the sUAS images become increasingly ubiquitous, more effective classification techniques are needed. In this chapter, we explored the potential of the *multi-view information* embedded in sUAS image acquisition missions for improving classification accuracy. We examined the direct use of the different individual representations of an object in multi-view images, where each representation is classified using spectral and textural information and a final class type is assigned through majority voting. Alternatively, we examined using object views to model BRDF for each object and use the BRDF parameters as well as other spectral and textural features of the object in the classification. Both techniques outperformed the widely used technique of orthoimage object segmentation and classification. Classification of the multi-view instances of the object and voting for a final class assignment proved to be a simple and effective technique that outperformed BRDF and orthoimage classification.

The case study results presented in this chapter show very limited accuracy improvement based on the BRDF modeling, making a sharp contrast with improvement achieved using the direct classification of multi-view instances and majority voting. Besides, BRDF modeling requires selecting the appropriate BRDF model and fitting the model to each object in order to get the object BRDF parameters, which can be computationally intensive and sensitive to outliers, and requires laborious quality control efforts. In contrast, the multi-view classification method applies to the multi-view data directly and only requires training a single model during the whole procedure, simplifying the usage of multi-view to a large extent. We believe that this technique can be easily implemented in commercial and widely used sUAS image analysis software such as the Pix4D (www.pix4d.com/) and the AgiSoft (www.agisoft.com/) software, especially with the current move to provide add-on application-driven capabilities beyond the orthoimage and 3D models typically produced by these software packages.

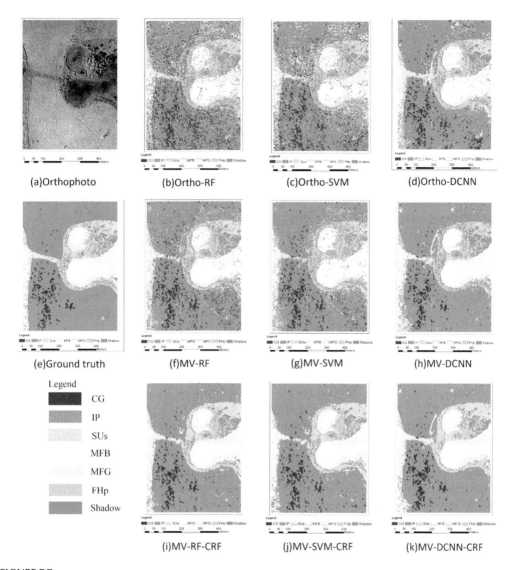

FIGURE 7.7
Classification maps [(a) through (k)] generated by different classification methods (Liu et al. 2018c).

This chapter presented *two deep learning approaches (DCNN and FCN)* to classify multi-view sUAS images under the OBIA framework. The massive training samples used to evaluate the performance of deep learning classifiers in computer vision usually cast doubts on their value for remote sensing applications for which training sample collection is expensive and the sample dataset is limited. Multi-view data cater to the needs of massive training samples for DCNN, and enable the DCNN of producing higher accuracy while its advantage over RF and SVM is not obvious when only orthoimage objects are used. Figure 7.4 indicates the number of training samples indeed affects the performance when comparing DCNN with other traditional machine learning classifiers. When the training set is relatively small (e.g., 700 samples for seven classes in our study), the DCNN deep learning classifier presented inferior performance compared to the RF and SVM traditional classifiers. This indicates that deep learning is better at learning patterns and is inherently more responsive to the increase in training set size compared to traditional classifiers.

FCN obtained better performance than DCNN in our study regardless of training set size, indicating that the end-to-end training scheme employed in FCN leads to better utilization of the training samples compared to DCNN. In addition, due to this end-to-end training scheme, FCN can utilize all the label information within the image patch bounding each object, as evidenced in Figure 7.3. The figure shows that FCN performed better when all the pixels including the ones surrounding and within the object in the training patch were labeled (FCN-II-OBIA) compared to when only the pixels within the object are labeled (FCN-I-OBIA). This constitutes one important advantage of FCN over DCNN. However, the better classification accuracy of FCN comes with a cost, since manually labeling all pixels within each training image patch requires more time and energy than just assigning one label to all the pixels within the object. It should be noted that unlike the RF and SVM that have been established as reliable and matured classifiers for years, deep learning classifiers are still under active development. RF and SVM are widely available with abundant support in several open source and commercial libraries. On the other hand, deep learning classifiers require a relatively steep learning curve to understand, develop, and optimize their network architecture and to master the implementation in a few, but growing number of, software packages such as Tensorflow, Keras, or Pytorch.

The last classification improvement technique presented in this chapter is *Context Modelling*. Object context information provides clues to help infer its class type and has been explored using various models in the remote sensing community. Combining context and multi-view information in a single CRF model presents a novel approach to make the most of the available information for better classification. To retain the multi-view information in the CRF model, the classification results in the form of class label possibility vector on the multi-view objects are pooled with summation operation (Liu et al. 2018c). These pooled multi-view results were also used to generate the multi-view classification results shown in Figure 7.6, which is different from the multi-view results in Figure 7.4 obtained by majority voting. Figure 7.6 demonstrates that summation pooling can not only improve the classification considerably for multi-view classification but also be effectively utilized in a CRF model to improve the classification together with context information. Figure 7.7 shows the primary contribution of context information is to remove noise, making the map smoother and more accurate.

7.5 Conclusions

This chapter discusses several cutting-edge approaches as promising techniques for land cover mapping using sUAS images under the OBIA framework. First, we presented the multi-view object-based image analysis approach as an alternative to conventional ortho-image analysis classification. Second, we demonstrated the potential of two deep learning classifiers DCNN and FCN to improve land cover classification and compared their performance with the RF and SVM traditional machine learning classifiers. Finally, a CRF model that assimilates both multi-view and object context information is presented. The CRF model was also evaluated using RF, SVM, and DCNN classifiers and compared with multi-view classification and conventional classification methods. The case study analysis and discussions presented in this chapter reveal the following conclusions:

1. DCNN outperforms RF and SVM when a relatively larger set of training samples is available.
2. FCN performs better than DCNN, especially when including label information for the pixels surrounding the object in the FCN image patch of the training dataset.
3. Multi-view OBIA can substantially improve classification accuracy of traditional machine learning classifiers (RF and SVM) as well as the DCNN deep learning classifier. This is true when applying both majority voting and summation pooling method to assign final object label from object multi-view instance classification results.
4. Direct classification of multi-view objects is more efficient in utilizing multi-view information than the BRDF modeling approaches.
5. Contextual modelling using CRF can effectively exploit multi-view and context information to generate map with higher accuracy and cartographic quality.

Acknowledgment

This manuscript has been authored by UT-Battelle, LLC under Contract No.DE-AC05-00OR22725 with the U.S. Department of Energy. The United States Government retains and the publisher, by accepting the article for publication, acknowledges that the United States Government retains a non-exclusive, paid up, irrevocable, world-wide license to publish or reproduce the published form of this manuscript, or allow others to do so, for United States Government purposes. The Department of Energy will provide public access to these results of federally sponsored research in accordance with the DOE Public Access Plan.

References

Agisoft. (2016). Professional Edition, Version 1.2. Photoscan Manual.

Amit, Y., & Geman, D. (1997). Shape quantization and recognition with randomized trees. *Neural Computation, 9*, 1545–1588.

Baatz, M., & Schäpe, A. (2000). Multiresolution segmentation: An optimization approach for high quality multi-scale image segmentation. *Angewandte Geographische Informationsverarbeitung, XII*, 12–23.

Blake, A., Kohli, P., & Rother, C. (2011). *Markov Random Fields for Vision and Image Processing*. MIT Press.

Blaschke, T. (2010). Object based image analysis for remote sensing. *ISPRS Journal of Photogrammetry and Remote Sensing, 65*, 2–16.

Blaschke, T., Hay, G.J., Kelly, M., Lang, S., Hofmann, P., Addink, E., Feitosa, R.Q., van der Meer, F., van der Werff, H., & van Coillie, F. (2014). Geographic object-based image analysis–towards a new paradigm. *ISPRS Journal of Photogrammetry and Remote Sensing, 87*, 180–191.

Breiman, L. (2001). Random forests. *Machine Learning, 45*, 5–32.

Cao, J., Leng, W., Liu, K., Liu, L., He, Z., & Zhu, Y. (2018). Object-based mangrove species classification using unmanned aerial vehicle hyperspectral images and digital surface models. *Remote Sensing, 10*(1), 89.

Carranza, M.L., Ricotta, C., Carboni, M., & Acosta, A.T. (2011). Habitat selection by invasive alien plants: A bootstrap approach. *Preslia, 83*, 529–536.

Chiabrando, F., Donadio, E., & Rinaudo, F. (2015). SfM for orthophoto to generation: A winning approach for cultural heritage knowledge. *The International Archives of Photogrammetry, Remote Sensing and Spatial Information Sciences, 40*, 91.

Chytrý, M., Maskell, L.C., Pino, J., Pyšek, P., Vilà, M., Font, X., & Smart, S.M. (2008). Habitat invasions by alien plants: A quantitative comparison among Mediterranean, subcontinental and oceanic regions of Europe. *Journal of Applied Ecology, 45*, 448–458.

Cleve, C., Kelly, M., Kearns, F.R., & Moritz, M. (2008). Classification of the wildland–urban interface: A comparison of pixel-and object-based classifications using high-resolution aerial photography. *Computers, Environment and Urban Systems, 32*, 317–326.

Cortes, C., & Vapnik, V. (1995). Support-vector networks. *Machine Learning, 20*, 273–297.

Deng, L., & Yu, D. (2014). Deep learning: Methods and applications. *Foundations and Trends in Signal Processing, 7*, 197–387.

Drăguţ, L., Tiede, D., & Levick, S.R. (2010). ESP: A tool to estimate scale parameter for multiresolution image segmentation of remotely sensed data. *International Journal of Geographical Information Science, 24*, 859–871.

Flanders, D., Hall-Beyer, M., & Pereverzoff, J. (2003). Preliminary evaluation of eCognition object-based software for cut block delineation and feature extraction. *Canadian Journal of Remote Sensing, 29*, 441–452.

Frouz, J. (1997). The effect of vegetation patterns on oviposition habitat preference: A driving mechanism in terrestrial chironomid (Diptera: Chironomidae) succession? *Researches on Population Ecology, 39*, 207–213.

Fu, B., Wang, Y., Campbell, A., Li, Y., Zhang, B., Yin, S., Xing, Z., & Jin, X. (2017). Comparison of object-based and pixel-based Random Forest algorithm for wetland vegetation mapping using high spatial resolution GF-1 and SAR data. *Ecological Indicators, 73*, 105–117.

Gao, F., Schaaf, C., Strahler, A., Jin, Y., & Li, X. (2003). Detecting vegetation structure using a kernel-based BRDF model. *Remote Sensing of Environment, 86*, 198–205.

Gao, Y., & Mas, J.F. (2008). A comparison of the performance of pixel-based and object-based classifications over images with various spatial resolutions. *Online Journal of Earth Sciences, 2*, 27–35.

Georganos, S., Grippa, T., Vanhuysse, S., Lennert, M., Shimoni, M., Kalogirou, S., & Wolff, E. (2018). Less is more: Optimizing classification performance through feature selection in a very-high-resolution remote sensing object-based urban application. *GIScience & Remote Sensing, 55*, 221–242.

Goodfellow, I., Bengio, Y., & Courville, A. (2016). *Deep Learning*. MIT Press.

Hsieh, P.-F., Lee, L.C., & Chen, N.-Y. (2001). Effect of spatial resolution on classification errors of pure and mixed pixels in remote sensing. *IEEE Transactions on Geoscience and Remote Sensing, 39*, 2657–2663.

Hu, B., Zhang, K.F., Gray, L., Miller, J.R., & Zwick, H. (2007). The frequent image frames enhanced digital orthorectified mapping (FIFEDOM) camera for acquiring multiangular reflectance from the land surface. *IEEE Transactions on Geoscience and Remote Sensing, 45*(10), 3110–3118.

Huang, L., Xia, W., Zhang, B., Qiu, B., & Gao, X. (2017). MSFCN-multiple supervised fully convolutional networks for the osteosarcoma segmentation of CT images. *Computer Methods and Programs in Biomedicine, 143*, 67–74.

Huval, B., Wang, T., Tandon, S., Kiske, J., Song, W., Pazhayampallil, J., Andriluka, M., Rajpurkar, P., Migimatsu, T., & Cheng-Yue, R. (2015). An empirical evaluation of deep learning on highway driving. arXiv preprint arXiv:1504.01716.

Im, J., Quackenbush, L.J., Li, M., & Fang, F. (2014). Optimum scale in object-based image analysis. *Scale Issues in Remote Sensing*, 197–214.

Kaneko, K., & Nohara, S. (2014). Review of effective vegetation mapping using the UAV (unmanned aerial vehicle) method. *Journal of Geographic Information System, 6*, 733.

Kasetkasem, T., Arora, M.K., & Varshney, P.K. (2005). Super-resolution land cover mapping using a Markov random field based approach. *Remote Sensing of Environment, 96*, 302–314.

Kim, Y.H., Im, J., Ha, H.K., Choi, J.-K., & Ha, S. (2014). Machine learning approaches to coastal water quality monitoring using GOCI satellite data. *GIScience & Remote Sensing*, *51*, 158–174.

Koukal, T., & Atzberger, C. (2012). Potential of multi-angular data derived from a digital aerial frame camera for forest classification. *IEEE Journal of Selected Topics in Applied Earth Observations and Remote Sensing*, *5*, 30–43.

Koukal, T., Atzberger, C., & Schneider, W. (2014). Evaluation of semi-empirical BRDF models inverted against multi-angle data from a digital airborne frame camera for enhancing forest type classification. *Remote Sensing of Environment*, *151*, 27–43.

Laliberte, A.S., Browning, D., & Rango, A. (2012). A comparison of three feature selection methods for object-based classification of sub-decimeter resolution UltraCam-L imagery. *International Journal of Applied Earth Observation and Geoinformation*, *15*, 70–78.

Laliberte, A.S., & Rango, A. (2011). Image processing and classification procedures for analysis of sub-decimeter imagery acquired with an unmanned aircraft over arid rangelands. *GIScience & Remote Sensing*, *48*, 4–23.

LeCun, Y., Bengio, Y., & Hinton, G. (2015). Deep learning. *Nature*, *521*, 436–444.

Li, S., Jiang, H., & Pang, W. (2017). Joint multiple fully connected convolutional neural network with extreme learning machine for hepatocellular carcinoma nuclei grading. *Computers in Biology and Medicine*, *84*, 156–167.

Li, Z., Shi, W., Lu, P., Yan, L., Wang, Q., & Miao, Z. (2016). Landslide mapping from aerial photographs using change detection-based Markov random field. *Remote Sensing of Environment*, *187*, 76–90.

Liang, S., & Strahler, A.H. (1994). Retrieval of surface BRDF from multiangle remotely sensed data. *Remote Sensing of Environment*, *50*, 18–30.

Liu, D., Song, K., Townshend, J.R., & Gong, P. (2008). Using local transition probability models in Markov random fields for forest change detection. *Remote Sensing of Environment*, *112*, 2222–2231.

Liu, T., & Abd-Elrahman, A. (2018a). An object-based image analysis method for enhancing classification of land covers using fully convolutional networks and multi-view images of small unmanned aerial system. *Remote Sensing*, *10*, 457.

Liu, T., & Abd-Elrahman, A. (2018b). Deep convolutional neural network training enrichment using multi-view object-based analysis of unmanned aerial systems imagery for wetlands classification. *ISPRS Journal of Photogrammetry and Remote Sensing*, *139*, 154–170.

Liu, T., & Abd-Elrahman, A. (2018c). Multi-view object-based classification of wetland land covers using unmanned aircraft system images. *Remote Sensing of Environment*, *216*, 122–138.

Liu, T., Abd-Elrahman, A., Dewitt, B., Smith, S., Morton, J., & Wilhelm, V.L. (2018a). Evaluating the potential of multi-view data extraction from small unmanned aerial systems (UASs) for object-based classification for wetland land covers. *GIScience & Remote Sensing*, *56*(1), 130–159.

Liu, T., Abd-Elrahman, A., Jon, M., & Wilhelm, V.L. (2018b). Comparing fully convolutional networks, random forest, support vector machine, and patch-based deep convolutional neural networks for object-based wetland mapping using images from small unmanned aircraft system. *GIScience & Remote Sensing*, *55*(2), 243–264.

Liu, T., Abd-Elrahman, A., Zare, A., Dewitt, B.A., Flory, L., & Smith, S.E. (2018c). A fully learnable context-driven object-based model for mapping land cover using multi-view data from unmanned aircraft systems. *Remote Sensing of Environment*, *216*, 328–344.

Long, J., Shelhamer, E., & Darrell, T. (2015). Fully convolutional networks for semantic segmentation. In *Proceedings of the IEEE Conference on Computer Vision and Pattern Recognition* (pp. 3431–3440).

Lu, B., & He, Y. (2017a). Optimal spatial resolution of UAV imagery for species classification in a heterogeneous grassland ecosystem. *GIScience & Remote Sensing*.

Lu, B., & He, Y. (2017b). Species classification using unmanned aerial vehicle (UAV)-acquired high spatial resolution imagery in a heterogeneous grassland. *ISPRS Journal of Photogrammetry and Remote Sensing*, *128*, 73–85.

Lucht, W., Schaaf, C.B., & Strahler, A.H. (2000). An algorithm for the retrieval of albedo from space using semiempirical BRDF models. *IEEE Transactions on Geoscience and Remote Sensing*, *38*, 977–998.

Ma, L., Li, M., Gao, Y., Chen, T., Ma, X., & Qu, L. (2017). A novel wrapper approach for feature selection in object-based image classification using polygon-based cross-validation. *IEEE Geoscience and Remote Sensing Letters, 14*(3), 409–413.

Ma, X., Wang, H., & Wang, J. (2016). Semisupervised classification for hyperspectral image based on multi-decision labeling and deep feature learning. *ISPRS Journal of Photogrammetry and Remote Sensing, 120*, 99–107.

Makantasis, K., Karantzalos, K., Doulamis, A., & Doulamis, N. (2015). Deep supervised learning for hyperspectral data classification through convolutional neural networks. In *2015 IEEE International Geoscience and Remote Sensing Symposium (IGARSS)* (pp. 4959–4962): IEEE.

Marmanis, D., Schindler, K., Wegner, J.D., Galliani, S., Datcu, M., & Stilla, U. (2016). Classification with an edge: Improving semantic image segmentation with boundary detection. arXiv preprint arXiv:1612.01337.

Nagol, J.R., Sexton, J.O., Kim, D.-H., Anand, A., Morton, D., Vermote, E., & Townshend, J.R. (2015). Bidirectional effects in Landsat reflectance estimates: Is there a problem to solve? *ISPRS Journal of Photogrammetry and Remote Sensing, 103*, 129–135.

O'Brien, T. (2016). Small Unmanned Aerial Vehicles as Remote Sensors: An Effective Data Gathering Tool for Wetland Mapping. The University of Guelph.

Pacifici, F., Longbotham, N., & Emery, W.J. (2014). The importance of physical quantities for the analysis of multitemporal and multiangular optical very high spatial resolution images. *IEEE Transactions on Geoscience and Remote Sensing, 52*, 6241–6256.

Pan, S.J., & Yang, Q. (2010). A survey on transfer learning. *IEEE Transactions on Knowledge and Data Engineering, 22*, 1345–1359.

Pande-Chhetri, R., Abd-Elrahman, A., Liu, T., Morton, J., & Wilhelm, V.L. (2017). Object-based classification of wetland vegetation using very high-resolution unmanned air system imagery. *European Journal of Remote Sensing, 50*, 564–576.

Paola, J.D., & Schowengerdt, R. (1995). A review and analysis of backpropagation neural networks for classification of remotely-sensed multi-spectral imagery. *International Journal of Remote Sensing, 16*, 3033–3058.

Pei, M., Wu, X., Guo, Y., & Fujita, H. (2017). Small bowel motility assessment based on fully convolutional networks and long short-term memory. *Knowledge-Based Systems, 121*, 163–172.

Piramanayagam, S., Schwartzkopf, W., Koehler, F.W., & Saber, E. (2016, October). Classification of remote sensed images using random forests and deep learning framework. In *Image and Signal Processing for Remote Sensing XXII* (Vol. 10004, p. 100040L). International Society for Optics and Photonics.

Prince, S.J. (2012). *Computer Vision: Models, Learning, and Inference.* Cambridge University Press.

Radoux, J., & Defourny, P. (2007). A quantitative assessment of boundaries in automated forest stand delineation using very high resolution imagery. *Remote Sensing of Environment, 110*, 468–475.

Rahman, H., Verstraete, M.M., & Pinty, B. (1993). Coupled surface-atmosphere reflectance (CSAR) model, 1, Model description and inversion on synthetic data. *Journal of Geophysical Research-All Series, 98*, 20, 779–720,779.

Rango, A., Laliberte, A., Steele, C., Herrick, J.E., Bestelmeyer, B., Schmugge, T., Roanhorse, A., & Jenkins, V. (2006). Using unmanned aerial vehicles for rangelands: Current applications and future potentials. *Environmental Practice, 8*, 159–168.

Roberts, G. (2001). A review of the application of BRDF models to infer land cover parameters at regional and global scales. *Progress in Physical Geography, 25*, 483–511.

Rodriguez-Galiano, V.F., Ghimire, B., Rogan, J., Chica-Olmo, M., & Rigol-Sanchez, J.P. (2012). An assessment of the effectiveness of a random forest classifier for land-cover classification. *ISPRS Journal of Photogrammetry and Remote Sensing, 67*, 93–104.

Ross, J. (2012). *The Radiation Regime and Architecture of Plant Stands.* Springer Science & Business Media.

Roujean, J.L., Leroy, M., & Deschamps, P.Y. (1992). A bidirectional reflectance model of the Earth's surface for the correction of remote sensing data. *Journal of Geophysical Research: Atmospheres, 97*, 20455–20468.

Rutchey, K., Schall, T.N., Doren, R.F., Atkinson, A., Ross, M.S., Jones, D.T., … Burch, J.N. (2006). *Vegetation Classification for South Florida Natural Areas.* St. Petersburg, FL, USA: US Geological Survey.

Samiappan, S., Turnage, G., Hathcock, L.A., & Moorhead, R. (2017). Mapping of invasive phragmites (common reed) in Gulf of Mexico coastal wetlands using multispectral imagery and small unmanned aerial systems. *International Journal of Remote Sensing, 38*(8–10), 2861–2882.

Schaaf, C.B., Gao, F., Strahler, A.H., Lucht, W., Li, X., Tsang, T., Strugnell, N.C., Zhang, X., Jin, Y., & Muller, J.-P. (2002). First operational BRDF, albedo nadir reflectance products from MODIS. *Remote sensing of Environment, 83,* 135–148.

Schmidt, M. (2012). UGM: MATLAB code for undirected graphical models. www.di.ens.fr/mschmidt/Software/UGM.html

Sherrah, J. (2016). Fully convolutional networks for dense semantic labelling of high-resolution aerial imagery. arXiv preprint arXiv:1606.02585.

Shin, J.S., Lee, T.H., Jung, P.M., & Kwon, H.S. (2015). A study on land cover map of UAV imagery using an object-based classification method. *Journal of Korean Society for Geospatial Information System, 23,* 25–33.

Silver, D., Huang, A., Maddison, C.J., Guez, A., Sifre, L., Van Den Driessche, G., Schrittwieser, J., Antonoglou, I., Panneershelvam, V., & Lanctot, M. (2016). Mastering the game of Go with deep neural networks and tree search. *Nature, 529,* 484–489.

Su, L., Chopping, M.J., Rango, A., Martonchik, J.V., & Peters, D.P. (2007). Support vector machines for recognition of semi-arid vegetation types using MISR multi-angle imagery. *Remote Sensing of Environment, 107,* 299–311.

Suk, H.-I., Lee, S.-W., Shen, D., & Initiative, A.s.D.N. (2014). Hierarchical feature representation and multimodal fusion with deep learning for AD/MCI diagnosis. *NeuroImage, 101,* 569–582.

Sutton, C., & McCallum, A. (2012). An introduction to conditional random fields. *Foundations and Trends® in Machine Learning, 4,* 267–373.

Torres-Sánchez, J., Peña, J., De Castro, A., & López-Granados, F. (2014). Multi-temporal mapping of the vegetation fraction in early-season wheat fields using images from UAV. *Computers and Electronics in Agriculture, 103,* 104–113.

Torrey, L., & Shavlik, J. (2009). Transfer learning. *Handbook of Research on Machine Learning Applications and Trends: Algorithms, Methods, and Techniques, 1,* 242.

Vázquez-Tarrío, D., Borgniet, L., Liébault, F., & Recking, A. (2017). Using UAS optical imagery and SfM photogrammetry to characterize the surface grain size of gravel bars in a braided river (Vénéon River, French Alps). *Geomorphology, 285,* 94–105.

Vedaldi, A., & Lenc, K. (2015). Matconvnet: Convolutional neural networks for MATLAB. In *Proceedings of the 23rd ACM International Conference on Multimedia* (pp. 689–692): ACM.

Veenman, C.J., Reinders, M.J., & Backer, E. (2002). A maximum variance cluster algorithm. *IEEE Transactions on Pattern Analysis and Machine Intelligence, 24,* 1273–1280.

Vetrivel, A., Gerke, M., Kerle, N., Nex, F., & Vosselman, G. (2017). Disaster damage detection through synergistic use of deep learning and 3D point cloud features derived from very high resolution oblique aerial images, and multiple-kernel-learning. *ISPRS Journal of Photogrammetry and Remote Sensing, 140,* 45–59.

Wang, C., Komodakis, N., & Paragios, N. (2013). Markov random field modeling, inference & learning in computer vision & image understanding: A survey. *Computer Vision and Image Understanding, 117,* 1610–1627.

Westoby, M.J., Brasington, J., Glasser, N.F., Hambrey, M.J., & Reynolds, J. (2012). 'Structure-from-Motion' photogrammetry: A low-cost, effective tool for geoscience applications. *Geomorphology, 179,* 300–314.

Whiteside, T.G., Boggs, G.S., & Maier, S.W. (2011). Comparing object-based and pixel-based classifications for mapping savannas. *International Journal of Applied Earth Observation and Geoinformation, 13,* 884–893.

Wolf, P.R., DeWitt, B.A., & Wilkinson, B.E. (2013). *Elements of Photogrammetry with Application in GIS* (4th edn). McGraw-Hill Education.

Xie, M., Jean, N., Burke, M., Lobell, D., & Ermon, S. (2015). Transfer learning from deep features for remote sensing and poverty mapping. arXiv preprint arXiv:1510.00098

Yu, Q., Gong, P., Clinton, N., Biging, G., Kelly, M., & Schirokauer, D. (2006). Object-based detailed vegetation classification with airborne high spatial resolution remote sensing imagery. *Photogrammetric Engineering & Remote Sensing, 72*, 799–811.

Zhang, P., Gong, M., Su, L., Liu, J., & Li, Z. (2016). Change detection based on deep feature representation and mapping transformation for multi-spatial-resolution remote sensing images. *ISPRS Journal of Photogrammetry and Remote Sensing, 116*, 24–41.

Zhang, X., Xiao, P., & Feng, X. (2013). Impervious surface extraction from high-resolution satellite image using pixel-and object-based hybrid analysis. *International Journal of Remote Sensing, 34*(12), 4449–4465.

Zhao, W., & Du, S. (2016). Learning multiscale and deep representations for classifying remotely sensed imagery. *ISPRS Journal of Photogrammetry and Remote Sensing, 113*, 155–165.

Zhong, P., Gong, Z., Li, S., & Schönlieb, C.-B. (2017). Learning to diversify deep belief networks for hyperspectral image classification. *IEEE Transactions on Geoscience and Remote Sensing, 55*(6), 3516–3530.

Zhong, Y., Zhao, J., & Zhang, L. (2014). A hybrid object-oriented conditional random field classification framework for high spatial resolution remote sensing imagery. *IEEE Transactions on Geoscience and Remote Sensing, 52*, 7023–7037.

Zhu, Y., Zhang, C., Zhou, D., Wang, X., Bai, X., & Liu, W. (2016). Traffic sign detection and recognition using fully convolutional network guided proposals. *Neurocomputing, 214*, 758–766.

Zweig, C.L., Burgess, M.A., Percival, H.F., & Kitchens, W.M. (2015). Use of unmanned aircraft systems to delineate fine-scale wetland vegetation communities. *Wetlands, 35*, 303–309.

8

UAS for Nature Conservation – Monitoring Invasive Species

Jana Müllerová

CONTENTS

8.1 The Problem of Plant Invasions

Plant invasions are considered among the most threatening processes in current landscapes with wide-ranging consequences for socio-economic concerns, biodiversity, ecosystem services, human health, and traditional landscapes (Leishman et al. 2007; Levine et al. 2003; Vilà and Ibáñez 2011; Pyšek et al. 2012). Despite the worldwide efforts to control and eradicate alien plant species, their impact is increasing (Hulme et al. 2009), and may further grow with climate change and global warming. The success of invasion depends largely on habitat and species characteristics (Richardson and Pyšek 2006). However, because of many external and stochastic factors that can contribute to the process, consequences are case specific and difficult to foresee (Levine et al. 2003). The process of invasion comprises of multiple spatio-temporal scales from seed recruitment to long-distance dispersal. Therefore, the data collected at different spatial and temporal scales needs to be integrated into spatially explicit models to better understand the invasion impact and role of individual factors (Andrew and Ustin 2010; Nehrbass et al. 2007). Exploring the spatio-temporal dynamic of invasions including the habitat selection and distribution dynamics ameliorates our ability to understand the drivers of invasion and increases the efficiency of control measures (c.f. Meier et al. 2014; Holden et al. 2016; Pyšek and Hulme 2005).

The invasion process is fast and dynamic, and once established, invasive species are usually difficult or even impossible to reduce/eradicate (Rejmánek and Pitcairn 2002). Control of highly infested sites is therefore complicated, and targeting primarily recently infested sites was shown to be efficient and economical. Early monitoring of invasion onsets followed by rapid management measures are therefore crucial (Pluess et al. 2012). Traditional ways of monitoring via fieldwork campaigns are laborious and difficult to organize, and new alternatives of cost-effective techniques are therefore necessary for timely, fast, and precise alien species management (Hulme et al. 2009). It is well known that invasive plants are often easy propagating and fast-growing species with high biomass, fast regeneration, and reproduction rate that tend to dominate the space forming large monospecific stands (Leishman et al. 2007). Such characteristics make invasive plants ideal candidates for remote sensing detection. The method can also be used to study spatio-temporal dynamics of the invasion process at the landscape scale, to model invasion spread, and to plan efficient eradication measures (Müllerová et al. 2005; Nehrbass et al. 2007; Laliberte et al. 2004). Still, to be operational, such methods need to be robust and well defined.

8.2 Remote Sensing Monitoring of Invasions – Temporal, Spatial, and Spectral Aspects

The potential of remote sensing to provide a timely means of invasion monitoring depends on the data resolution and processing approach. The choice for type of remote sensing methodology is specific to the species, and must correspond to its characteristics. Generally speaking, alien plants can be detected by remote sensing means when they represent structure or phenology novel to the neighboring native vegetation (Huang and Asner 2009). For data resolution issues, three components must be considered – temporal, spatial, and spectral (Figure 8.1).

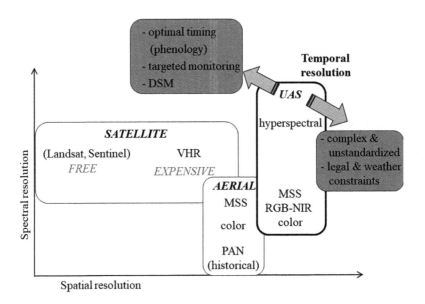

FIGURE 8.1
Different components of resolution related to the remotely sensed data available.

For plant species detection, high temporal resolution covering the plant seasonal variability and/or proper timing of the data acquisition can be very important. Less distinct species might be detected only at particular phenological stage(s) when they differ enough from their surrounding (e.g. in flowering, senescenting or early/late start in the season) (Huang and Asner 2009; Müllerová et al. 2017b; Somodi et al. 2012).

Complementary approaches such as change detection (multi-date assessment) ease the species recognition (Diao and Wang 2016; Martin et al. 2018). Specific timing can even enable detection of forest understorey species (Perroy et al. 2017; Müllerová et al. 2017b; Hernandez-Santin et al. 2019).

Over the last decades, we have seen the rapid development of new accurate classifiers for species recognition, and new sensors and data acquisition systems of high spatial and temporal resolution, such as Unmanned Aerial Systems (UAS). Application of UAS enables flexible data acquisition at very high spatial and temporal scales, potentially bridging the gap between small-scale (point) field surveys and air and space-borne Earth observations (Manfreda et al. 2018). Unmanned platforms are currently being adopted by many fields of environmental sciences (Whitehead et al. 2014; Manfreda et al. 2018) and nature conservation (Jiménez López and Mulero-Pázmány 2019), and bear potential to improve environmental monitoring to assist management decisions. Thanks to its flexibility, UAS is able to collect the imagery at the exact date/phenological stage of the target plant species. They can be used to test influence of data acquisition time on the detection accuracy to establish optimal methodological workflow that can be later adopted for larger scale mapping using satellite imagery (Dvořák et al. 2015).

As fast-growing species, invasive plants (especially herbs) can also be detected using a series of Digital Surface Models (DSM) throughout the season, generated as a by-product of the UAS orthophoto mosaicking process of Structure from Motion algorithm (SfM) (Westoby et al. 2012). However, for these purposes, high spatial accuracy in the UAS imagery is necessary to precisely overlay multi-temporal data (Manfreda et al. 2019). DSMs can serve as valuable input to the analysis of plant invasions since such species tend to be large and fast growing, and can therefore be detected using the change detection approach. This is especially true for herb species that show seasonal cycle, such as knotweeds (Martin et al. 2018). If a detailed Digital Elevation Model (DEM) is available (e.g. LIDAR with high density of points per meter), the DSM can be used to both support classification of optical imagery, and analyze the growth of invasion. The lack of detailed DEM can be substituted by a DSM acquired by a UAS flight out of vegetation season. For change detection and canopy height models, precise matching of layers is crucial, and it is often necessary to coarsen the data to avoid possible mismatches and decrease the rate of false change detection (Müllerová et al. 2017a).

Spatial resolution of the data also plays an important role in species recognition. Coarser resolution can prevent the detection of individual plants and small patches, which are particularly important from a management point of view as they can potentially serve as new dispersal foci in the landscape. Some methods are partially able to overcome this problem, such as sub-pixel classification used to estimate vegetation fraction in mixed pixels (Van De Voorde et al. 2008; Tooke et al. 2009). However, sufficient spatial resolution is still crucial and depends again on the target species and the purpose of the study. On the other hand, very high spatial resolution of UAS imagery tremendously increases the data complexity, and can sometimes be counterproductive, as a millimeter pixel size breaks plant individuals into a complex of branches/stems, leaves, inflorescences, insects, and soil background, making classification extremely difficult. In addition, a large amount of data markedly increases the demand for processing time and data storage.

The third component, spectral resolution, can involve data ranging from historical panchromatic, visible RGB, and multispectral (MSS) imagery to hyperspectral imagery. Here again, the suitable spectral range depends on the species characteristics (Mast et al. 1997; Müllerová et al. 2005). There are some examples of very distinct species, which can be distinguished even on panchromatic imagery. These serve as a valuable source of information on long-term landscape change, and in some cases, historical panchromatic aerial photos have been used to reconstruct the invasion process (Mast et al. 1997; Müllerová et al. 2005). Still, most commonly, higher spectral information is needed, with Near-Infrared (NIR) band being the most important to detect plant photosynthetic activity.

As for UAS, commonly used MSS cameras can be substituted to a large extent by consumer cameras modified to be sensitive in the NIR part of the spectrum (exchanging built-in IR-cut filter by Hoya R72 filter, see e.g. Müllerová et al. (2017a). Such a cheap alternative does not provide perfect spectral information since the filters are not capable of separating the channels completely (Figure 8.2) However, this can still provide valuable information and keep the costs low, making the approach easy to adopt at operational bases. Rasmussen et al. (2016) confirmed the reliability of vegetation indices derived from consumer-grade cameras. On the other hand, hyperspectral sensors represent a comprehensive resource that even enables the separation of similar species and can study plant physiology in detail; yet such sensors are still very costly and not commonly adopted for UAS (Zarco-Tejada et al. 2012; Sankey et al. 2017). In addition, the processing of hyperspectral imagery is very complex and time consuming, and therefore remains reserved for experimental applications; it is not operational in natural conservation efforts.

The large amount of data produced by UAS generates the need for automatic processing that can be complicated especially if the species is not very distinct from the surrounding natural vegetation. To make the detection operational for nature conservation, careful design of the workflow (Figure 8.3) based on testing various data and algorithm settings is extremely important, since the choice of optimal approach is influenced by the data spatial, spectral, and temporal resolution and the target

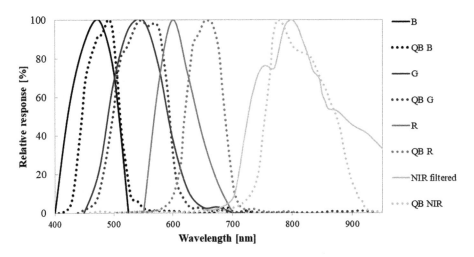

FIGURE 8.2
Non-Gaussian relative spectral response of consumer camera modified to be sensitive in NIR part of the spectrum (exchanging built-in IR-cut filter by Hoya R72 filter) compared with standard RGB camera (adapted from (Müllerová, Bartaloš, et al. 2017).

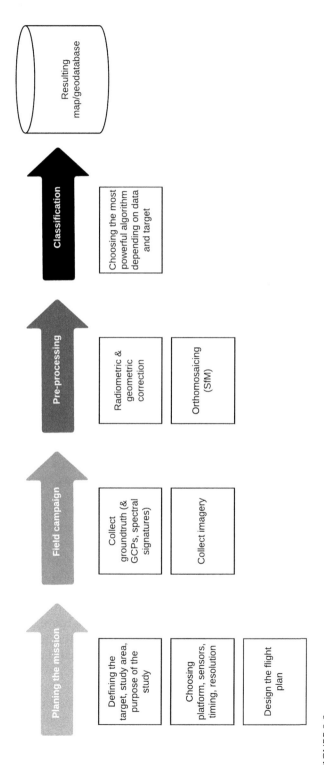

FIGURE 8.3
General workflow of UAS campaign and processing.

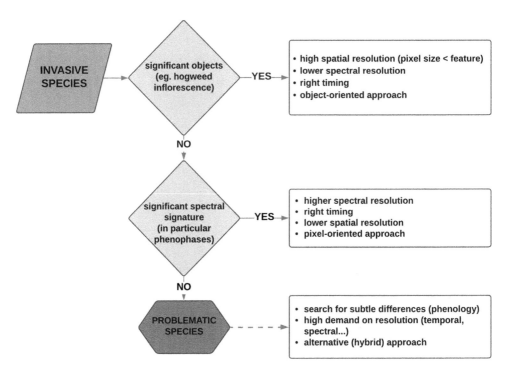

FIGURE 8.4
Decision tree for choosing optimal data and classification approach depending on the target invasive species.

species characteristics. Because different aspects of resolution can compensate for each other to a certain degree, e.g. high spatial resolution can partially overcome the spectral handicap of historical panchromatic imagery (Müllerová et al. 2005). A holistic approach for the UAS workflow design that includes the algorithm characteristics should always be considered. The optimal approach is species specific (Müllerová et al. 2017a), and some general rules still apply (Figure 8.4). Machine learning classifiers are particularly promising, showing high classification accuracy either with a limited number of training samples or upon utilizing highly detailed UAS data (Lary et al. 2016). For such imagery, methods that take into account the spatial structure are often applied instead of the pixel-based approach in order to reduce the effects of shadows, within-class spectral variation, mis-registration, and salt-and-pepper effect caused by increased within-patch variability (Zweig et al. 2015; Chen et al. 2018). The type of the object of interest defines the suitable algorithm; object-based classification is suitable if forming distinct objects larger than the pixel size is required, whereas in case of indistinct/small objects, the pixel-based approach would be more effective (Figure 8.4) (Müllerová et al. 2017b; Mafanya et al. 2017). For distinct species, even low spectral resolution data can be sufficient in case of an adequate spatial resolution and processing algorithm, but a combination of high resolution and sophisticated procedure is necessary for effectively mapping problematic species.

As already mentioned, it is important to keep a balance between the data resolution/complexity of algorithms and the aim of the study since high resolution and/or complex algorithms are extremely demanding on computer processing (complexity, time, data storage, and analysis). Higher resolution is not necessarily always the better choice, and the same applies to classification accuracy. Additionally, we should bear in mind both the extent and aim of the study. Regional or country scale mapping would definitely require a different approach compared to the targeted local monitoring. Acceptable accuracy threshold (especially that of omission error) should be adjusted according to the purpose of study, setting it to high in order to locate invasion hotspots and provide information for modeling, and setting it lower for eradication purposes to make sure none of the possible infestations are omitted (Hamada et al. 2007; Martin et al. 2018).

8.3 UAS for Plant Invasions – Advantages and Challenges

UAS are becoming increasingly popular in many environmental applications mainly because of their easy deployment, operation, and extremely high spatio-temporal resolution. The available number of platforms and sensors is growing, and the choice depends largely on the study target, the purpose, and the budget. Specialized sensors such as LIDAR, thermal, or hyperspectral cameras are not yet common because of the high costs and need for larger and more reliable platforms capable of carrying such a payload. Most commonly, UAS are multi-rotors or fixed-wing platforms loaded with optical sensors, such as MSS, modified near-infrared, simple RGB consumer cameras, and GoPro. Even with the simplest settings, a UAS is still able to provide Digital Surface Models (DSM) (Westoby et al. 2012) and orthoimagery at very high temporal and spatial resolution (Colomina and Molina 2014). Still, challenges remain because of geometrical and radiometric distortions, mainly due to the close land surface proximity, and unstandardized processing workflow (Manfreda et al. 2018). Geometrical precision can be enhanced by carefully designed flight configuration, GCP number and distribution, and navigation system onboard the UAS (James et al. 2017; Manfreda et al. 2019). Radiometric distortions are often caused by changing light conditions during the flight. These can be calibrated using onboard radiometer measuring actual incident solar radiation; however, such sensors are usually not part of the low-cost settings. Extremely high spatial resolution of UAS imagery brings new challenges in the training and validation process (GCP and ground-truth data collection) since the centimeter to millimeter resolution reaches the limits of GPS instrument precision. Expensive differential GPS is meant to provide sub-meter precision in ideal conditions. However, in forest growth or complex environments, it reaches meter precision only, and the data collection is time consuming. New ways of collecting the ground data are yet to be deployed; such software applications would support their own very high resolution basemaps (e.g. UAS imagery) to better locate the ground objects of interest. Such applications (e.g. Collector for ArcGIS, NextGIS Mobile, QField, QGIS Mobile, ODK Collect, and CyberTracker) can be installed on smartphones and tablets, supporting larger scale ground-truth campaigns (Müllerová et al. 2017a).

Such limitations must be borne in mind when planning a flight campaign as there is no reason for collecting data in centimeter resolution if we only need or are able to reach meter precision in the final mosaic. If imagery is acquired in ultra-high spatial resolution, individual images cover a very small area, and finding common tie points for the SfM algorithm can be problematic especially in a complex environment such as a forest. In addition, other already mentioned drawbacks of extremely high spatial resolution exist, such as complexity and amount of giga/terra bytes to be processed and stored.

On the other hand, UAS provide considerable benefits over traditionally used satellite and aerial data (Table 8.1). In addition to ultra-high spatial resolution, UAS supports

TABLE 8.1

Benefits and Disadvantages of UAS Imagery Over Traditionally Used Satellite and Aerial Data

	NASA & ESA Satellites	VHR Satellites	Aerial Data
Advantages of UAS	+ spatial resolution + flexibility (timing) + frequency	+ spatial resolution + flexibility (timing) + frequency + lower costs of acquisition	+ spatial resolution + flexibility (timing) + frequency + lower costs of acquisition
Disadvantages of UAS	– higher costs – acquisition constrains – complicated – complex preprocessing – no archival data	– acquisition constrains – complicated – complex preprocessing	– acquisition constrains – complex preprocessing

the exact timing of the mission, providing flexibility in planning and designing the data acquisition, independent from external providers.

UAS are therefore ideal for targeted monitoring, eradication control, and experimental testing to find the best setup of timing, resolution, and classification workflow since they enable regular acquisition at different seasons to account for environmental and context (background vegetation) variability. They are able to provide very high detail, but they are not suitable for covering large areas. Whereas fixed-wing UAS are able to cover larger areas compared to multi-rotors, it is still no more than several square kilometers per flight. UAS operation is also limited by legal constraints, such as visual contact of the pilot with aircraft and restrictions in urban areas (often highly infested by invasive plant species). For regional or country scale monitoring, other data sources such as satellite (ideally) or manned aircraft imagery become viable (Table 8.1). With the growing availability of free or low-cost higher spatial resolution operational satellites such as Sentinel and CubeSats, larger infestations could be monitored; however, small patches and individual plants are still impossible to detect on coarser resolution data. This is especially true for the highly heterogeneous Central European landscape where the occurrence of invasive plant populations is rather patchy. Small patches and individuals, omitted by satellite campaigns, often serve as dispersal foci and new onsets of invasion. However, in fact such early stages of invasion are crucial for eradication, since after the species becomes widespread, it is often difficult or impossible to eradicate it completely from the landscape (Pluess et al. 2012). This implies that UAS hold great potential to reduce the effects of invasions, identify priorities for management, support targeted monitoring, and control eradication efforts.

8.4 UAS-Based Plant Invasion Case Studies

Examples of invasive studies employing UAS imagery are not very common; still, their number has grown considerably over the last number of years, and early results are promising (Table 8.2). They cover different life forms, not only trees, shrubs, grasses, and herbs, but also wetland species, climbing vines, or cacti, and span from tropics through temperate regions to Mediterranean and dry environments (savannas and deserts). UAS data are typically used for (semi)automatic detection, the most common being by OBIA approach (70% of studies) and machine learning algorithms (70%) such as Random Forests. Many studies are benefiting from structural information provided by UAS, including canopy

TABLE 8.2

Plant Invasion Studies Using UAS

Species	Reference
Acacia dealbata, Ulex europaeus, Pinus radiata	Lopatin et al. (2019), Kattenborn et al. (2019)
Acacia longifolia	de Sá et al. (2018)
Asclepias sp., Bromus inermis, Festuca rubra, Solidago canadensis	Lu and He (2017)
Acacia mangium	Lehmann et al. (2017)
Fallopia sp.	Michez et al. (2016), Müllerová et al. (2017b), Martin et al. (2018)
Hakea sericea	Alvarez-Taboada et al. (2017)
Harrisia pomanensis	Mafanya et al. (2017; 2018)
Heracleum mantegazzianum	Michez et al. (2016), Müllerová et al. (2017b)
Impatiens glandulifera	Michez et al. (2016)
Iris pseudacorus	Hill et al. (2017)
Mikania micrantha	Li et al. (2019), Wu et al. (2019)
Miconia calvescens	Perroy et al. (2017)
Pennisetum ciliare	Elkind et al. (2019)
Phragmites australis	Zaman et al. (2011)
Robinia pseudoaccacia	Müllerová et al. (2017a), Carl et al. (2019)
Spartina alterniflora	Wan et al. (2014), Zhou (2018)

height or surface roughness into the classification algorithm. Despite sophisticated classification procedures, the vast majority of studies (70%) use simple low-cost settings of a consumer camera (RGB), often supplemented by a filtered NIR camera sensor. Only five studies work with hyperspectral sensors, and there is no UAS LiDAR deployed. A majority of studies (70%) combine UAS with very high and medium resolution satellite imagery to explore upscaling to the landscape scale, and authors often employ particular phenology and multi-temporal data to support the detection and to explain invasion dynamics.

To demonstrate applicability of the UAS approach for detection of invasive plants, three plant species (taxons) were selected for being invasive and widely naturalized in a number of European countries. They represent different life forms (one tree and two herbs), span along the gradient of sensitivity to the optimal phenological stage, and either form distinct objects or are difficult to distinguish (Table 8.3). Using these model species of different life forms, phenology, and architecture, it will be shown how different aspects of data quality (spatial, spectral, and temporal), costs, and operational issues affect methodology and can have practical implications (Figure 8.4). These examples indicate how important is it to understand the object of interest we want to monitor, and what benefits the UAS brings to invasive species monitoring; most important being the high spatial resolution and the flexible timing of the flight campaign.

8.4.1 Example 1 – Giant Hogweed

Giant hogweed (*Heracleum mantegazzianum*) is a noxious herb species, native in East Asia, and problematic in colder parts of Europe. It threatens biodiversity and has negative consequences for human health via phototoxicins causing contact dermatitis (Drever and Hunter 1970). Certain characteristics, such as a large size of 2 to 5 meters, a large number of seeds (up to 100,000) (Tiley et al. 1996), and a high rate of spread allows the species to be a successful invader, and have significant implications for its management and potential

TABLE 8.3

Detection of Three Target Species – Description, Data Requirements, Optimal Classification Approach

	giant hogweed	exotic knotweeds	black locust
Species	*Heracleum mantegazzianum*	*Fallopia japonica F. sachalinensis F. × bohemica*	*Robinia pseudoaccacia*
Life form	monocarpic herb	perennial herb	deciduous tree
Size	2 to 5 m	2 to 4 m	12 to 30 m
Leaves	large (up to 2.5 m), well structured	medium (10–30 cm), complicated leaf architecture, forming dense stands	small, compound, folding in the summer heat
Inflorescence	large white round compound umbels (1 m), facing upwards, distinct	small white clustered racemes, not distinct	medium white pendant racemes (10–20 cm), rather distinct
Regeneration	generative – up to 20,000 seeds from single plant	vegetative – extensive fragile rhizomes	mostly vegetative – root suckers
Invaded habitats	fallow land, river banks, margins of forests and arable fields, ruderal stands	fallow land, river banks, open forests and forest margins, ruderal stands, disturbed habitats, urban areas	formerly planted, invading step grasslands, disturbed habitats, road and railway corridors and steep slopes
Native range	Caucasus	Japan	South East United States
Impact	human health (skin allergy), biodiversity	biodiversity, economic losses (extensive root system damages construction work and river banks)	biodiversity (but also an important species for wood and honey production, and erosion control)
Optimal phenophases	peak of flowering	senescence (turns reddish)	peak of flowering, but other periods also good
Optimal data timing	2nd half of June – 1st half of July	end of October – beginning of November	vegetation season
Detection accuracy	high	medium	medium
Tested imagery	UAS (5 cm, RGB + modif. NIR), color ortophotos (25 cm, RGB), Pleiades 1B (0.5 m pan sharpened, RGB, NIR)	UAS (5 cm, RGB + modif. NIR), Pleiades 1B (0.5 m pan sharpened, RGB, NIR)	UAS (5 cm, RGB + modif. NIR), WorldView-2 (0.5 m pan sharpened, RGB, NIR)
Spatial resolution	5–50 cm	5–50 cm	5–50 cm
Spectral resolution	low	medium (NIR necessary)	medium (NIR necessary)
Temporal resolution	high (1 month)	high (1 month)	low
Optimal classification	object-oriented (SVM, RF)	pixel-oriented (ML, RF)	pixel-oriented (SVM, ML)
Reference	Müllerová et al. (2017b)	Müllerová et al. (2017b)	Müllerová et al. (2017a)

Abbreviations: SVM – Support Vector Machine, RF – Random Forest, ML – Maximum Likelihood

Aerial PAN (1962; 0.5 m)   Aerial color (2006; 0.5 m)

UAS (2015; 5 cm)   Pleiades (2013; 2.8 m)   RapidEye (2010; 6.5 m)

UAS; peak of flowering (2016; 5 cm)   UAS; end of flowering (2016; 5 cm)

FIGURE 8.5
Giant hogweed at different phenophases and data sources. Effects of spatial resolution and timing of the campaign are clearly visible.

control (Pyšek et al. 2007). Giant hogweed is an example of herb species forming very distinct phenophase in flowering (Figure 8.5). As a perennial monocarpic species it flowers once, mostly in the third year (Pergl et al. 2006), and its huge white round inflorescences (of up to 1 meter) form distinct well defined objects that are easy to distinguish even from low spectral resolution imagery in case its pixel size exceeds the size of individual plants (i.e. 1 meter). Timing of the data acquisition during flowering is crucial, with a time period of about one month in Central Europe (Perglová et al. 2006). At the beginning and end of flowering, the detection rate decreases markedly, and out of flowering, the species is quite difficult to recognize (Müllerová et al. 2013). If captured at the peak of flowering, it can be detected even from historical panchromatic imagery. In our studies, Müllerová et al. (2013)

and Müllerová et al. (2005), we used historical aerial photo time series to reconstruct the process of invasion at the landscape scale, and model the species spread (Nehrbass et al. 2007; Pergl et al. 2011; Pyšek et al. 2008).

Even though manned aerial imagery provides spatial resolution high enough to support the detection, archives are limited and flight campaigns are costly; the same holds true for VHR satellite imagery. This is not the case for UAS which enable the exact timing of the campaign (Müllerová et al. 2017b). They provide a low-cost flexible solution with potential to be widely used in nature conservation. From UAS imagery, hogweed can be detected with very high classification accuracies (Michez et al. 2016; Müllerová et al. 2017b). Since the hogweed inflorescences are well separable from the background forming distinct objects, the object-oriented approach was shown to be the most effective, using either a combination of multiresolution segmentation and Random Forest (RF) classifier (Michez et al. 2016), or that of contrast split segmentation and rule-based classification. Such an approach successfully separated hogweed inflorescences from other white objects, such as artificial surfaces or harvested fields (Müllerová et al. 2017b). Automatic classification was able to detect even individual flowering plants that could potentially serve as dispersal foci in the landscape. Flowering plants were detected with very high accuracy (Table 8.4); still young (not flowering) and mown/grazed plants were difficult to recognize on low-cost RBG+modified NIR imagery. For such an application implemented outside the flowering period, imagery of higher spectral resolution would be necessary. Compared to manual on-screen digitizing, the automatic approach brought more complexity and detail into the analyses. Since manual processing tends to generalize and omit individual plants, the manual approach was more capable of recognizing non-flowering individuals (Müllerová et al. 2013). It is also important to mention that very fine spatial resolution of UAS imagery is not always beneficial – in the case of hogweed, compound inflorescences are visible as white dots on 0.5 m imagery separated into individual umbels at centimeter resolution, creating artificial gaps in hogweed stands (Figure 8.5) (Müllerová et al. 2017b).

TABLE 8.4

Summary of Classification of Giant Hogweed Using Different Data Sources Capturing Different Parts of the Species Vegetation Season. Lower Spectral Resolution can be Compensated by Timing and Spatial Resolution of the Data

Data	Resolution	Phenology	Method	User's Accuracy	Producer's Accuracy	Reference
RapidEye	6.5 m	early flowering	pixel (MaxLike)/ OBIA	65/44%	76/65%	Müllerová et al. (2013)
Pleiades	2.8 m	mid flowering	pixel/OBIA (RF)	86/70%	94/99%	Müllerová et al. (2017b)
aerial PAN	0.5 m	mid flowering	OBIA	89%	81%	Müllerová et al. (2013)
aerial color	0.5 m	mid flowering	OBIA	57%	94%	Müllerová et al. (2017b)
UAS RGB+NIR	0.05 m	mid flowering	OBIA	99%	100%	Müllerová et al. (2017b)
aerial PAN	0.5 m	ripe fruiting	OBIA	86%	69%	Müllerová et al. (2013)
aerial MSS	0.5 m	ripe fruiting	OBIA/pixel (MaxLike)	51.6/42%	74.2/28.5%	Müllerová et al. (2013)

Hogweed represents a nice example of species for which timing and spatial resolution is crucial, and can successfully compensate for the lower spectral resolution of the data. UAS can clearly bring the benefit of flexibility; and for nature conservation, even low-cost UAS equipped with a simple RGB sensor would be sufficient. Still, because of small area covered, such an approach is ideal for targeted monitoring and checking of eradication efforts, and not for extensive surveys at the landscape scale, where other means such as satellite mapping come into play (Müllerová et al. 2017b).

8.4.2 Example 2 – Black Locust

Black locust (*Robinia pseudoacacia*) is a North American tree, widely planted in Europe in the past mainly for wood production and soil amelioration (Vítková et al. 2017). This pioneer light-demanding species is extremely resistant to disturbance and therefore difficult to eradicate because trunk damage provokes viable regeneration from root suckers. As a nitrogen fixing tree it is able to completely change the soil properties and understorey species composition, having significant consequences for biodiversity, especially for dry grasslands of high natural value that it frequently invades (Vítková et al. 2015; Lazzaro et al. 2018). Locust has small compound leaves and rather distinct white inflorescences (Figure 8.6).

FIGURE 8.6
Black locust (top row) and exotic knotweeds (middle and bottom rows) in the field and on the imagery. NIR-G-B false colour composites of UAS imagery of 0.05 m resolution (top middle), and WV-2 satellite of 2 m resolution (top left). Black locust in bright red. Middle and bottom rows show exotic knotweeds in the spring (UAS imagery of 0.05m), summer (Pleiades satellite, 2.8 m), and autumn (UAS imagery of 0.05m).

According to (Somodi et al. 2012), locust was best separable during flowering. However, its white inflorescences are pendant, and therefore not particularly visible from above. Moreover, the flowering is rather short (one or two weeks), stands usually do not flower exactly simultaneously, and the plant is quite sensitive to the late spring frost that can postpone flowering. This implies that the flowering period is difficult to predict, making this kind of detection rather impractical. Nevertheless, our research showed that locust can also be separated in other parts of the vegetation season based on leaf signature (Figure 8.6) (Müllerová et al. 2017a). It showed good separability in the NIR part of the spectrum, even if assessed using a low-cost modified NIR sensor. Still, precise co-registration of the channels of combined RGB and NIR imagery proved to be an important step in the analysis. Since the tree does not form defined objects but rather heterogeneous stands, the object-based approach was more effective; machine learning algorithms (Support Vector Machine) being less sensitive to the distortion of the orthomosaics, that is an inevitable artefact of UAS data. Deep learning algorithms such as convolutional neural networks represent a powerful option to classify black locust. However, such methods require a high number of labeled training images and are problematic to repeat (Carl et al. 2019).

To improve detection accuracy, ancillary information such as canopy height is often employed in the classification process; however, to create such a canopy height model of forest stand, both detailed Digital Terrain Model (DTM; often unavailable) and accurately located DSM, problematic for ordinary UAS, are needed. To overcome imprecision problems, the vegetation height model can be substituted by surface roughness measure defined as a function of a range of elevation values, i.e. relative elevation differences within a specific subset. In case of black locust, such an approach helped to increase classification accuracy and separate locust from arable fields, which was often misclassified otherwise (Somodi et al. 2012; Müllerová et al. 2017a); still, it did not improve discrimination from other tree species with similar surface heterogeneity values. To detect the species, centimeter resolution of UAS imagery was not necessary and could be substituted by VHR imagery of meter resolution (WV-2 in case of Müllerová et al. 2017a); still, such imagery is very costly, and freely available Landsat imagery seems to be too coarse for the purpose (Somodi et al. 2012).

8.4.3 Example 3 – Invasive Knotweeds

Invasive knotweeds include three taxa expressing subtle morphological differences from the remote sensing perspective. These are *Fallopia* (syn. *Reynoutria*) *japonica, F. sachalinensis,* and their hybrid *F. × bohemica*. Originally from Asia, knotweeds are among the most problematic invasive plants in temperate regions of the world since they are extremely difficult to eradicate or manage due to their vigorous growth and regeneration (Child and Wade 2000). Economic losses related to knotweed invasion are vast, with the annual costs estimated to €2.3 billion in Europe (Kettunen et al. 2008). Improving methods for their detection and monitoring would increase the eradication efficiency, especially along their main dispersal axes, i.e. transportation corridors and rivers (Tiébré et al. 2008).

Knotweeds are herbaceous perennials up to 4 m tall with small flowers and highly heterogeneous canopy architecture, forming dense stands with complicated architecture. Knotweeds represent an example of species that do not form distinct objects; however, they can be discriminated capitalizing on phenological dynamics throughout vegetation season since their spectral signature shows strong intra-annual variability (Figure 8.6).

The NIR part of the spectrum seems to be particularly important for the species discrimination. Because of the lack of distinct objects, pixel-based algorithms are generally more beneficial for detection. Such algorithms are also less computationally demanding and therefore easier to operate on the large data volumes UAS provide. Müllerová et al. (2017b) found that VHR satellite data represent a viable option since even with a coarser resolution of 0.5 m, comparable high accuracies can be achieved depending on phenological stage of the knotweed (Table 8.5). However, at coarser spatial resolution, small emerging patches and those growing under the tree canopy tend to be omitted more frequently.

As already mentioned, phenology plays an important role in species detection. Knotweeds usually start to foliate early, but in some stands, the foliage can be postponed. Therefore, during the spring, knotweed stands form a mixture of old reddish stems from the previous year and new emerging stems and leaves, making the detection particularly challenging. During the summer, they are not particularly distinct from the surroundings, and are separable mainly in the NIR part of the spectrum. Although knotweeds are considered as a light-demanding species (Beerling et al. 1994), they can grow under open forest canopy, and particularly along rivers after flooding disturbances. Remote sensing detection under the canopy is generally difficult, although very high spatial resolution was shown to overcome the problem to a certain extent (Perroy et al. 2017; Hernandez-Santin et al. 2019). During late autumn, knotweeds drop leaves and turn senescenting stems to brown-red (Figure 8.6), making detection the most effective (Müllerová et al. 2017b). Additionally, at this time of year, overstorey deciduous trees drop their leaves, exposing understorey knotweeds. Unlike earlier remote sensing, studies (Dorigo et al. 2012; Michez et al. 2016; Jones et al. 2011; Müllerová et al. 2017b) achieved operational classification accuracies that are meaningful for practical application of the technology. This part of the season therefore seems ideal for the task; however, unstable weather, cloudiness, strong winds, and long shadows can make UAS data acquisition at this period particularly challenging. Martin et al. (2018) showed an alternative option capitalizing on the knotweeds fast growth. They used the summer imagery supported with canopy height models and multi-date band ratios derived from UAS imagery acquired at different parts of the vegetation season that were entering the classification process. Machine learning algorithms (Random Forests) were the most powerful; still this cannot be taken as a rule of thumb as they can in some cases be outcompeted by traditional algorithms such as Maximum Likelihood. For object-based

TABLE 8.5

Classification of Exotic Knotweeds from UAS and Pleiades Satellite Data Acquired in the Summer and Late Autumn During the Senescence. Summary of Accuracies Achieved Using Different Approaches, Data, and Phenological Stages

Data	Resolution	Phenology	Method	User's Accuracy	Producer's Accuracy	Reference
Pleiades 1B	0.5/2 m	green	pixel (RF)	44%	95%	Müllerová et al. (2017b)
UAS RGB/NIR	0.05 m	green	pixel (SVM)	60%	92%	Müllerová et al. (2017b)
UAS RGB/NIR + BTBR/CHM	0.05 m	green	OBIA (RF)	80/78%	83/86%	Martin et al. (2018)
UAS RGB/NIR	0.05 m	senescence	pixel (MaxLike/ SVM)	80/54%	78/95%	Müllerová et al. (2017b)

classification, rule-based algorithms are often more successful compared to machine learning; still, the method is based on expert knowledge and is therefore more subjective and case-specific.

8.5 Implications for Management and Future Research

As demonstrated here by several examples, remote sensing represents an important tool supporting active management of invasions. Remote sensing tools provide precise information on extent of invasion, and supplemented with ancillary data such as environmental variables, land cover/land use, and plant characteristics (means of dispersal, plant traits, population characteristics), it can be used to study the process of invasion. This includes an assessment of the species impact, the susceptibility to invasion of different habitats, and potential of the species to invade the landscape. This is followed by modeling different scenarios of eradication to define the most effective management strategy. This information supports eradication efforts, identifies management priorities, and establishes management protocols (Figure 8.7).

UAS as an easily accessible tool can be used not only for detection of problematic species, but also to study the invasion process and check eradication efficiency. Still, since its areal extent is limited, it is meant for testing the best monitoring approach and targeted monitoring, and to map invasions at landscape scale; other means such as satellites assume importance. UAS can provide unbeatable high spatial resolution and flexibility of the data acquisition at relatively low costs. However, to make UAS mapping operational, there is still a long way to go for standardization of the methods, especially in sense of data acquisition, geometric, and radiometric pre-processing, ground-truthing, and validation.

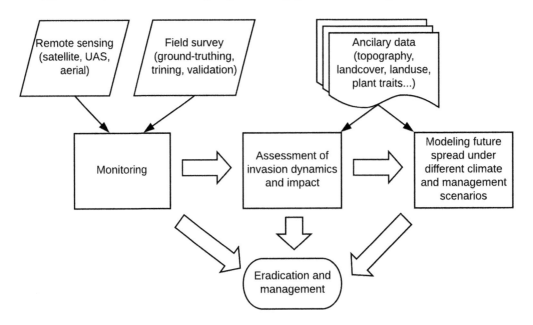

FIGURE 8.7
How can the remote sensing help to tackle the invasions?

8.6 Conclusions

The chapter summarizes the potential and limitations of detection of relevant invasive species in terms of (i) the effect of the spatial/spectral resolution of the RS images, and (ii) the detection rate as a function of the phenological stage of the vegetation. Efficient and cost-effective methods for fast and timely monitoring of plant invasions must be selected based on accuracy of the species detection, data acquisition costs, and labor demands; whereas the selection of methods largely depends on the needed accuracy driven by the purpose of the study, flexibility, and economic efficiency. This holds true for all aspects of the flight mission: (i) selecting the appropriate UAS platform (fixed-wing for mapping larger areas, and multi-rotor for better maneuverability, with size of platform according to the payload), and sensors, ranging from simple GoPro to the most sophisticated hyperspectral cameras and LIDAR; (ii) selecting appropriate spatial resolution according to the study; and (iii) right timing of the mission/temporal coverage (frequency) that often plays a fundamental role in invasive species detection and analysis of invasion dynamics. Still, the trade-offs between detection accuracy and the spatial and spectral resolutions must be borne in mind.

Describing the phenology and species variability at invaded sites at fine spatial and temporal scales not only allows identification of the target alien species, but also the ability to assess species habitat requirements and their impact on soil properties and biodiversity. Evaluation of the negative impact posed to invaded communities and susceptibility of the site to invasion enables the prioritization and focusing of the eradication efforts to the most affected sites and sites important for nature conservation (such as Natura 2000 sites).

Precise detection and monitoring of the invasion process provides knowledge on the mechanisms and spatial patterns of plant invasions and their amelioration. This improves the scientific knowledge on environmental and human-induced factors driving their spread, in order to make the protective measures more effective, and thereby minimize their negative impact on the society, economy, environment, and biodiversity.

Acknowledgments

Preparation of this chapter was supported by the long-term research development project no. RVO 67985939 (The Czech Academy of Sciences) and project LTC18007 (Ministry of Education, Czech Republic). Presented research would not be possible without support from all the colleagues contributing to the cited papers, and technicians in our team.

References

Alvarez-Taboada, Flor, Claudio Paredes, and Julia Julián-Pelaz. 2017. "Mapping of the Invasive Species *Hakea sericea* Using Unmanned Aerial Vehicle (UAV) and Worldview-2 Imagery and an Object-Oriented Approach." *Remote Sensing* 9 (9). doi:10.3390/rs9090913.

Andrew, Margaret E., and Susan L. Ustin. 2010. "The Effects of Temporally Variable Dispersal and Landscape Structure on Invasive Species Spread." *Ecological Applications* 20 (3): 593–608. doi:10.1890/09-0034.1.

Beerling, David J, John P. Bailey, and Ann P. Conolly. 1994. "*Fallopia japonica* (Houtt.) Ronse Decraene." *The Journal of Ecology* 82 (4): 959–79. doi:10.2307/2261459.

Carl, Christin, Jan Lehmann, Dirk Landgraf, Hans Pretzsch, Christin Carl, Jan R. K. Lehmann, Dirk Landgraf, and Hans Pretzsch. 2019. "*Robinia pseudoacacia* L. in Short Rotation Coppice: Seed and Stump Shoot Reproduction as Well as UAS-Based Spreading Analysis." *Forests* 10 (3): 235. doi:10.3390/f10030235.

Chen, Gang, Qihao Weng, Geoffrey J. Hay, and Yinan He. 2018. "Geographic Object-Based Image Analysis (GEOBIA): Emerging Trends and Future Opportunities." *GIScience and Remote Sensing* 55 (2): 159–82. doi:10.1080/15481603.2018.1426092.

Child, Lois, and Max Wade. 2000. "The Japanese Knotweed Manual: The Management and Control of an Invasive Alien Weed". Edited by L. Chlid and Max Wade. Packard Publishing Ltd. www.cabdirect.org/cabdirect/abstract/20013037380.

Colomina, I., and P. Molina. 2014. "Unmanned Aerial Systems for Photogrammetry and Remote Sensing: A Review." *ISPRS Journal of Photogrammetry and Remote Sensing* 92: 79–97. doi:10.1016/j.isprsjprs.2014.02.013.

de Sá, Nuno C., Paula Castro, Sabrina Carvalho, Elizabete Marchante, Francisco A. López-Núñez, and Hélia Marchante. 2018. "Mapping the Flowering of an Invasive Plant Using Unmanned Aerial Vehicles: Is There Potential for Biocontrol Monitoring?" *Frontiers in Plant Science* 9: 293. doi: 10.3389/fpls.2018.00293.

Diao, Chunyuan, and Le Wang. 2016. "Remote Sensing of Environment Incorporating Plant Phenological Trajectory in Exotic Saltcedar Detection with Monthly Time Series of Landsat Imagery." *Remote Sensing of Environment* 182: 60–71. doi:10.1016/j.rse.2016.04.029.

Dorigo, Wouter, Arko Lucieer, Tomaž Podobnikar, and Andraž Čarni. 2012. "Mapping Invasive *Fallopia japonica* by Combined Spectral, Spatial, and Temporal Analysis of Digital Orthophotos." *International Journal of Applied Earth Observation and Geoinformation* 19: 185–95. doi:10.1016/J.JAG.2012.05.004.

Drever, J. C., and J.A.A. Hunter. 1970. "Giant Hogweed Dermatitis." *Scottish Medical Journal* 15 (9): 315–19. doi:10.1177/003693307001500902.

Dvořák, P., Jana Müllerová, Tomáš Bartaloš, and Josef Brůna. 2015. "Unmanned Aerial Vehicles for Alien Plant Species Detection and Monitoring." *International Archives of the Photogrammetry, Remote Sensing and Spatial Information Sciences - ISPRS Archives* 40 (1W4): 83–90. doi:10.5194/isprsarchives-XL-1-W4-83-2015.

Elkind, Kaitlyn, Temuulen T. Sankey, Seth M. Munson, and Clare E. Aslan. 2019. "Invasive Buffelgrass Detection Using High-resolution Satellite and UAV Imagery on Google Earth Engine." *Remote Sensing in Ecology and Conservation* rse2.116. doi:10.1002/rse2.116.

Hamada, Yuki, Douglas A. Stow, Lloyd L. Coulter, Jim C. Jafolla, and Leif W. Hendricks. 2007. "Detecting Tamarisk Species (*Tamarix* spp.) in Riparian Habitats of Southern California Using High Spatial Resolution Hyperspectral Imagery." *Remote Sensing of Environment* 109 (2): 237–48. doi:10.1016/j.rse.2007.01.003.

Hernandez-Santin, Lorna, Mitchel Rudge, Renee Bartolo, and Peter Erskine. 2019. "Identifying Species and Monitoring Understorey from UAS-Derived Data: A Literature Review and Future Directions." *Drones* 3 (1): 9. doi:10.3390/drones3010009.

Hill, David J., Catherine Tarasoff, Garrett E. Whitworth, Jackson Baron, Jacob L. Bradshaw, and John S. Church. 2017. "Utility of Unmanned Aerial Vehicles for Mapping Invasive Plant Species: A Case Study on Yellow Flag Iris (*Iris pseudacorus* L.)." *International Journal of Remote Sensing* 38 (8–10): 2083–105. doi:10.1080/01431161.2016.1264030.

Holden, Matthew H., Jan P. Nyrop, and Stephen P. Ellner. 2016. "The Economic Benefit of Time-Varying Surveillance Effort for Invasive Species Management." Edited by Luke Flory. *Journal of Applied Ecology* 53 (3): 712–21. doi:10.1111/1365-2664.12617.

Huang, Cho-ying, and Gregory P. Asner. 2009. "Applications of Remote Sensing to Alien Invasive Plant Studies." *Sensors* 9 (6): 4869–89. doi:10.3390/s90604869.

Hulme, Philip E., Petr Pyšek, Wolfgang Nentwig, and Montserrat Vilà. 2009. "Will Threat of Biological Invasions Unite the European Union?" *Science* 324 (5923): 40–41. doi:10.1126/science.1171111.

James, M.R., S. Robson, S. d'Oleire-Oltmanns, and U. Niethammer. 2017. "Optimising UAV Topographic Surveys Processed with Structure-from-Motion: Ground Control Quality, Quantity and Bundle Adjustment." *Geomorphology* 280: 51–66. doi:10.1016/j.geomorph. 2016.11.021.

Jiménez López, Jesús, and Margarita Mulero-Pázmány. 2019. "Drones for Conservation in Protected Areas: Present and Future." *Drones* 3 (1): 10. doi:10.3390/drones3010010.

Jones, Daniel, Stephen Pike, Malcolm Thomas, Denis Murphy, Daniel Jones, Stephen Pike, Malcolm Thomas, and Denis Murphy. 2011. "Object-Based Image Analysis for Detection of Japanese Knotweed s.l. Taxa (*Polygonaceae*) in Wales (UK)." *Remote Sensing* 3 (2): 319–42. doi:10.3390/ rs3020319.

Kattenborn, Teja, Javier Lopatin, Michael Förster, Andreas Christian Braun, and Fabian Ewald Fassnacht. 2019. "UAV Data as Alternative to Field Sampling to Map Woody Invasive Species Based on Combined Sentinel-1 and Sentinel-2 Data." *Remote Sensing of Environment* 227: 61–73. doi:10.1016/j.rse.2019.03.025.

Kettunen, M., P. Genovesi, S. Gollasch, Shyama Pagad, U. Starfinger, P. Ten Brink, and C. Shine. 2008. "Technical Support to EU Strategy on Invasive Species (IAS) - Assessment of the Impacts of IAS in Europe and the EU (Final Module Report for the European Commission)." Institute for European Environmental Policy (IEEP). https://researchspace.auckland.ac.nz/ handle/2292/33742.

Laliberte, Andrea S., Albert Rango, Kris M. Havstad, Jack F. Paris, Reldon F. Beck, Rob McNeely, and Amalia L. Gonzalez. 2004. "Object-Oriented Image Analysis for Mapping Shrub Encroachment from 1937 to 2003 in Southern New Mexico." *Remote Sensing of Environment* 93 (1–2): 198–210. doi:10.1016/j.rse.2004.07.011.

Lary, David J., Amir H. Alavi, Amir H. Gandomi, and Annette L. Walker. 2016. "Machine Learning in Geosciences and Remote Sensing." *Geoscience Frontiers* 7 (1): 3–10. doi:10.1016/j. gsf.2015.07.003.

Lazzaro, Lorenzo, Giuseppe Mazza, Giada d'Errico, Arturo Fabiani, Claudia Giuliani, Alberto F. Inghilesi, Alessandra Lagomarsino, et al. 2018. "How Ecosystems Change Following Invasion by *Robinia pseudoacacia*: Insights from Soil Chemical Properties and Soil Microbial, Nematode, Microarthropod and Plant Communities." *Science of the Total Environment* 622: 1509–18. doi:10.1016/j.scitotenv.2017.10.017.

Lehmann, Jan R.K., Torsten Prinz, Silvia R. Ziller, Jan Thiele, Gustavo Heringer, João A.A. Meira-Neto, and Tillmann K. Buttschardt. 2017. "Open-Source Processing and Analysis of Aerial Imagery Acquired with a Low-Cost Unmanned Aerial System to Support Invasive Plant Management." *Frontiers in Environmental Science* 5: 1–16. doi:10.3389/fenvs.2017.00044.

Leishman, Michelle R., Tammy Haslehurst, Adrian Ares, and Zdravko Baruch. 2007. "Leaf Trait Relationships of Native and Invasive Plants: Community- and Global-Scale Comparisons." *New Phytologist* 176 (3): 635–43. doi:10.1111/j.1469-8137.2007.02189.x.

Levine, Jonathan M., Montserrat Vilà, Carla M. D'Antonio, Jeffrey S. Dukes, Karl Grigulis, and Sandra Lavorel. 2003. "Mechanisms Underlying the Impacts of Exotic Plant Invasions." *Proceedings of the Royal Society B: Biological Sciences* 270 (1517): 775–81. doi:10.1098/rspb.2003.2327.

Li, Jianhui, Li Dingquan, Zhang Gui, Xu Haizhou, Zeng Rongliang, Luo Wangjun, and Yu Youliang. 2019. "Study on Extraction of Foreign Invasive Species *Mikania micrantha* Based on Unmanned Aerial Vehicle (UAV) Hyperspectral Remote Sensing." In *Fifth Symposium on Novel Optoelectronic Detection Technology and Application.* Edited by Qifeng Yu, Wei Huang, and You He, 11023:53. *SPIE.* doi:10.1117/12.2520027.

Lopatin, Javier, Klara Dolos, Teja Kattenborn, and Fabian E. Fassnacht. 2019. "How Canopy Shadow Affects Invasive Plant Species Classification in High Spatial Resolution Remote Sensing." Edited by Ned Horning and Dolors Armenteras. *Remote Sensing in Ecology and Conservation.* doi:10.1002/rse2.109.

Lu, Bing, and Yuhong He. 2017. "Species Classification Using Unmanned Aerial Vehicle (UAV)- Acquired High Spatial Resolution Imagery in a Heterogeneous Grassland." *ISPRS Journal of Photogrammetry and Remote Sensing* 128: 73–85. doi:10.1016/j.isprsjprs.2017.03.011.

Mafanya, Madodomzi, Philemon Tsele, Joel Botai, Phetole Manyama, Barend Swart, and Thabang Monate. 2017. "Evaluating Pixel and Object Based Image Classification Techniques for Mapping Plant Invasions from UAV Derived Aerial Imagery: *Harrisia pomanensis* as a Case Study." *ISPRS Journal of Photogrammetry and Remote Sensing* 129: 1–11. doi:10.1016/j.isprsjprs.2017.04.009.

Mafanya, Madodomzi, Philemon Tsele, Joel O. Botai, Phetole Manyama, George J. Chirima, and Thabang Monate. 2018. "Radiometric Calibration Framework for Ultra-High- Resolution UAV-Derived Orthomosaics for Large-Scale Mapping of Invasive Alien Plants in Semi-Arid Woodlands : *Harrisia pomanensis* as a Case Study." *International Journal of Remote Sensing* 39 (15–16): 5119–40. doi:10.1080/01431161.2018.1490503.

Manfreda, Salvatore, Petr Dvorak, Jana Müllerová, Sorin Herban, Pietro Vuono, José Arranz Justel, and Matthew Perks. 2019. "Assessing the Accuracy of Digital Surface Models Derived from Optical Imagery Acquired with Unmanned Aerial Systems." *Drones* 3 (1): 15. doi:10.3390/drones3010015.

Manfreda, Salvatore, Matthew McCabe, Pauline Miller, Richard Lucas, Victor Pajuelo Madrigal, Giorgos Mallinis, Eyal Ben Dor, et al. 2018. "On the Use of Unmanned Aerial Systems for Environmental Monitoring." *Remote Sensing* 10: 641. doi:10.3390/rs10040641.

Martin, François Marie, Jana Müllerová, Laurent Borgniet, Fanny Dommanget, Vincent Breton, and André Evette. 2018. "Using Single- and Multi-Date UAV and Satellite Imagery to Accurately Monitor Invasive Knotweed Species." *Remote Sensing* 10 (10): 1662. doi:10.3390/rs10101662.

Mast, Joy Nystrom, Thomas T. Veblen, and Michael E. Hodgson. 1997. "Tree Invasion within a Pine/Grassland Ecotone: An Approach with Historic Aerial Photography and GIS Modeling." *Forest Ecology and Management* 93 (3): 181–94. doi:10.1016/S0378-1127(96)03954-0.

Meier, Eliane S., Stefan Dullinger, Niklaus E. Zimmermann, Daniel Baumgartner, Andreas Gattringer, and Karl Hülber. 2014. "Space Matters When Defining Effective Management for Invasive Plants." Edited by Ingolf Kühn. *Diversity and Distributions* 20 (9): 1029–43. doi:10.1111/ddi.12201.

Michez, Adrien, Hervé Piégay, Lisein Jonathan, Hugues Claessens, and Philippe Lejeune. 2016. "Mapping of Riparian Invasive Species with Supervised Classification of Unmanned Aerial System (UAS) Imagery." *International Journal of Applied Earth Observation and Geoinformation* 44: 88–94. doi:10.1016/j.jag.2015.06.014.

Müllerová, Jana, Tomáš Bartaloš, Josef Brůna, Petr Dvořák, and Michaela Vítková. 2017a. "Unmanned Aircraft in Nature Conservation: An Example from Plant Invasions." *International Journal of Remote Sensing* 38 (8–10): 2177–98. doi:10.1080/01431161.2016.1275059.

Müllerová, Jana, Josef Brůna, Tomáš Bartaloš, Petr Dvořák, Michaela Vítková, and Petr Pyšek. 2017b. "Timing Is Important: Unmanned Aircraft vs. Satellite Imagery in Plant Invasion Monitoring." *Frontiers in Plant Science* 8: 1–13. doi:10.3389/fpls.2017.00887.

Müllerová, Jana, Jan Pergl, and Petr Pyšek. 2013. "Remote Sensing as a Tool for Monitoring Plant Invasions: Testing the Effects of Data Resolution and Image Classification Approach on the Detection of a Model Plant Species *Heracleum mantegazzianum* (Giant Hogweed)." *International Journal of Applied Earth Observation and Geoinformation* 25 (1): 55–65. doi:10.1016/j.jag.2013.03.004.

Müllerová, Jana, Petr Pyšek, Vojtech Jarosik, and Jan Pergl. 2005. "Aerial Photographys as a Tool for Assessing the Regional Dynamics of the Invasive Plant Species *Heracleum mantaegazzianum*." *Journal of Applied Ecology* 42: 1042–53.

Nehrbass, Nana, Eckart Winkler, Jana Müllerová, Jan Pergl, Petr Pyšek, and Irena Perglová. 2007. "A Simulation Model of Plant Invasion: Long-Distance Dispersal Determines the Pattern of Spread." *Biological Invasions* 9 (4): 383–95. doi:10.1007/s10530-006-9040-6.

Pergl, Jan, Jana Müllerová, Irena Perglova, Tomáš Herben, and Petr Pyšek. 2011. "The Role of Long-Distance Seed Dispersal in the Local Population Dynamics of an Invasive Plant Species." *Diversity and Distributions* 17 (4): 725–38. doi:10.1111/j.1472-4642.2011.00771.x.

Pergl, Jan, Irena Perglová, Petr Pyšek, and Hansjörg Dietz. 2006. "Population Age Structure and Reproductive Behavior of the Monocarpic Perennial *Heracleum mantegazzianum* (Apiaceae) in Its Native and Invaded Distribution Ranges." *American Journal of Botany* 93 (7): 1018–28. doi:10.3732/ajb.93.7.1018.

Perglová, Irena, Jan Pergl, and Petr Pyšek. 2006. "Flowering Phenology and Reproductive Effort of the Invasive Alien Plant *Heracleum mantegazzianum*." *Preslia* 78 (3): 265–85.

Perroy, Ryan L, Timo Sullivan, and Nathan Stephenson. 2017. "Assessing the Impacts of Canopy Openness and Flight Parameters on Detecting a Sub-Canopy Tropical Invasive Plant Using a Small Unmanned Aerial System." *ISPRS Journal of Photogrammetry and Remote Sensing* 125. International Society for Photogrammetry and Remote Sensing: 174–83. doi:10.1016/j.isprsjprs.2017.01.018.

Pluess, Therese, Ray Cannon, Vojtěch Jarošík, Jan Pergl, Petr Pyšek, and Sven Bacher. 2012. "When Are Eradication Campaigns Successful? A Test of Common Assumptions." *Biological Invasions* 14 (7): 1365–78. doi:10.1007/s10530-011-0160-2.

Pyšek, Petr, M.J.W. Cock, W. Nentwig, and H.P. Ravn. 2007. Ecology and Management of Giant Hogweed (Heracleum Mantegazzianum). *CAB International (CABI). CABI.* doi:10.1079/9781845932060.0000.

Pyšek, Petr, and Philip E. Hulme. 2005. "Spatio-Temporal Dynamics of Plant Invasions: Linking Pattern to Process." *Écoscience* 12 (3): 302–15. doi:10.2980/i1195-6860-12-3-302.1.

Pyšek, Petr, Vojtěch Jarošík, Philip E. Hulme, Jan Pergl, Martin Hejda, Urs Schaffner, and Montserrat Vilà. 2012. "A Global Assessment of Invasive Plant Impacts on Resident Species, Communities and Ecosystems: The Interaction of Impact Measures, Invading Species' Traits and Environment." *Global Change Biology* 18 (5): 1725–37. doi:10.1111/j.1365-2486.2011.02636.x.

Pyšek, Petr, Vojtêch Jarošík, Jana Müllerová, Jan Pergl, and Jan Wild. 2008. "Comparing the Rate of Invasion by *Heracleum mantegazzianum* at Continental, Regional, and Local Scales." *Diversity and Distributions* 14 (2): 355–63. doi:10.1111/j.1472-4642.2007.00431.x.

Rasmussen, Jesper, Georgios Ntakos, Jon Nielsen, Jesper Svensgaard, Robert N. Poulsen, and Svend Christensen. 2016. "Are Vegetation Indices Derived from Consumer-Grade Cameras Mounted on UAVs Sufficiently Reliable for Assessing Experimental Plots?" *European Journal of Agronomy* 74: 75–92. doi:10.1016/J.EJA.2015.11.026.

Rejmánek, M., and M.J. Pitcairn. 2002. "When Is Eradication of Exotic Pest Plants a Realistic Goal." In *Turning the Tide: The Eradication of Invasive Species. IUCN SSC Invasive Species Specialist Group*, pp. 249–53.

Richardson, David M., and Petr Pyšek. 2006. "Plant Invasions: Merging the Concepts of Species Invasiveness and Community Invasibility." *Progress in Physical Geography* 30 (3), 409–31. doi:10.1191/0309133306pp490pr.

Sankey, Temuulen, Jonathon Donager, Jason McVay, and Joel B. Sankey. 2017. "UAV Lidar and Hyperspectral Fusion for Forest Monitoring in the Southwestern USA." *Remote Sensing of Environment* 195: 30–43. doi:10.1016/j.rse.2017.04.007.

Somodi, Imelda, Andraž Čarni, Daniela Ribeiro, and Tomaž Podobnikar. 2012. "Recognition of the Invasive Species *Robinia pseudacacia* from Combined Remote Sensing and GIS Sources." *Biological Conservation* 150 (1): 59–67. doi:10.1016/j.biocon.2012.02.014.

Tiébré, Marie-Solange, Layla Saad, and Grégory Mahy. 2008. "Landscape Dynamics and Habitat Selection by the Alien Invasive *Fallopia* (*Polygonaceae*) in Belgium." *Biodiversity and Conservation* 17 (10): 2357–70. doi:10.1007/s10531-008-9386-4.

Tiley, G.E. ., Felicite S. Dodd, and P.M. Wade. 1996. "*Heracleum mantegazzianum* Sommier & Levier." *The Journal of Ecology* 84 (2): 297–319. doi:10.2307/2261365.

Tooke, Thoreau Rory, Nicholas C. Coops, Nicholas R. Goodwin, and James A. Voogt. 2009. "Extracting Urban Vegetation Characteristics Using Spectral Mixture Analysis and Decision Tree Classifications." *Remote Sensing of Environment* 113 (2): 398–407. doi:10.1016/j.rse.2008.10.005.

Vilà, Montserrat, and Inés Ibáñez. 2011. "Plant Invasions in the Landscape." *Landscape Ecology* 26 (4): 461–72. doi:10.1007/s10980-011-9585-3.

Vítková, Michaela, Jana Müllerová, Jiří Sádlo, Jan Pergl, and Petr Pyšek. 2017. "Black Locust (*Robinia pseudoacacia*) Beloved and Despised: A Story of an Invasive Tree in Central Europe." *Forest Ecology and Management* 384: 287–302. doi:10.1016/j.foreco.2016.10.057.

Vítková, Michaela, Jaroslav Tonika, and Jana Müllerová. 2015. "Black Locust—Successful Invader of a Wide Range of Soil Conditions." *Science of the Total Environment* 505: 315–28. doi:10.1016/j. scitotenv.2014.09.104.

Van De Voorde, Tim, Jeroen Vlaeminck, and Frank Canters. 2008. "Comparing Different Approaches for Mapping Urban Vegetation Cover from Landsat ETM+ Data: A Case Study on Brussels." *Sensors* 8 (6): 3880–902. doi:10.3390/s8063880.

Wan, Huawei, Qiao Wang, Dong Jiang, Jingying Fu, Yipeng Yang, and Xiaoman Liu. 2014. "Monitoring the Invasion of *Spartina alterniflora* Using Very High Resolution Unmanned Aerial Vehicle Imagery in Beihai, Guangxi (China)." *Scientific World Journal.* doi:10.1155/2014/638296.

Westoby, M.J., J. Brasington, N.F. Glasser, M.J. Hambrey, and J.M. Reynolds. 2012. "'Structure-from-Motion' Photogrammetry: A Low-Cost, Effective Tool for Geoscience Applications." *Geomorphology* 179: 300–14. doi:10.1016/j.geomorph.2012.08.021.

Whitehead, Ken, Chris H. Hugenholtz, Stephen Myshak, Owen Brown, Adam LeClair, Aaron Tamminga, Thomas E. Barchyn, Brian Moorman, and Brett Eaton. 2014. "Remote Sensing of the Environment with Small Unmanned Aircraft Systems (UASs), Part 2: Scientific and Commercial Applications 1." *Journal of Unmanned Vehicle Systems* 02 (03): 86–102. doi:10.1139/ juvs-2014-0007.

Wu, Zhaocong, Min Ni, Zhongwen Hu, Junjie Wang, Qingquan Li, and Guofeng Wu. 2019. "Mapping Invasive Plant with UAV-Derived 3D Mesh Model in Mountain Area—A Case Study in Shenzhen Coast, China." *International Journal of Applied Earth Observation and Geoinformation* 77: 129–39. doi:10.1016/J.JAG.2018.12.001.

Zaman, Bushra, Austin M. Jensen, and Mac McKee. 2011. "Use of High-Resolution Multispectral Imagery Acquired with an Autonomous Unmanned Aerial Vehicle to Quantify the Spread of an Invasive Wetlands Species." In *2011 IEEE International Geoscience and Remote Sensing Symposium (IGARSS)*, pp. 803–06. doi:10.1109/IGARSS.2011.6049252.

Zarco-Tejada, P.J., V. González-Dugo, and J.A.J. Berni. 2012. "Fluorescence, Temperature and Narrow-Band Indices Acquired from a UAV Platform for Water Stress Detection Using a Micro-Hyperspectral Imager and a Thermal Camera." *Remote Sensing of Environment* 117: 322–37. doi:10.1016/j.rse.2011.10.007.

Zhou, Zaiming, Yanming Yang, and Benqing Chen. 2018. "Estimating *Spartina alterniflora* Fractional Vegetation Cover and Aboveground Biomass in a Coastal Wetland Using SPOT6 Satellite and UAV Data." *Aquatic Botany* 144: 38–45. doi:10.1016/j.aquabot.2017.10.004.

Zweig, Christa L., Matthew A. Burgess, H. Franklin Percival, and Wiley M. Kitchens. 2015. "Use of Unmanned Aircraft Systems to Delineate Fine-Scale Wetland Vegetation Communities." *Wetlands* 35 (2): 303–09. doi:10.1007/s13157-014-0612-4.

9

Small Unmanned Aerial Systems (sUAS) and Structure from Motion for Identifying, Documenting, and Monitoring Cultural and Natural Resources

**Marguerite Madden, Thomas Jordan, Sergio Bernardes,
Cari Goetcheus, Kristen Olson, and David Cotten**

CONTENTS

9.1 Introduction to Geospatial Technologies Supporting Cultural and Natural Resource Management

The increased availability of affordable sUAS, flight planning apps, and software for photogrammetry, image processing, and geospatial analysis has revolutionized the lives of cultural and natural resource managers (Calvo 2018; Wich and Koh 2018). Thirty to forty years ago, there were limited options for managers tasked with identifying, documenting, and monitoring cultural and natural resources at a level of detail that supported management decisions for areas about 0.5 ha in size within conserved lands that ranged from a few hectares to thousands of square kilometers. The staff of most historic sites, natural parks, rangelands, and national forests included resource specialists with field-expertise in domain areas such as history, architecture, forestry, botany, recreation, and outdoor education. To take stock of their physical assets or assess damages caused by disturbances ranging from visitor overuse and decaying structures to exotic insect infestations and hurricane damage, resource managers would call upon groups knowledgeable in the use of geospatial technologies. Government mapping agencies, academic research labs, and private firms often collaborated to fly aerial imagery, perform geometric corrections, interpret features of interest, assess resource condition, and make maps that were classified, symbolized, labeled, and printed at scales large enough to display information upon which management decisions could be made (Bogucki et al. 1980; Welch et al. 1988; Brooks and Johannes 1990; USFS 1990). In the 1990s to 2000s, advancements in computers, geographic information systems (GIS) and the Global Positioning System (GPS), along with national programs for aerial image acquisition and internet access to digital maps, orthophotos, and high spatial and spectral resolution satellite data changed the ways resource managers conducted inventory and monitoring (Welch et al. 1992; Maschner 1996; Lonnqvist and Stefanakis 1999; Welch et al. 1999; 2002; Wheatly and Gillings 2002; Ehlers et al. 2003; Hirano et al. 2003; McCoy and Ladefoged 2009). The next decades saw a shift from 2-Dimensional (2D) to 3-Dimensional (3D) and 4-Dimensional (4D which includes time) for geovisualization, time-series image analysis, and predictive modeling as web portals created free access to large volumes of geospatial data including digital elevation models (DEMs), Light Detection and Ranging (LiDAR), satellite and aerial imagery (Katsianis et al. 2008; Madden et al. 2009; Agapiou et al. 2016; Richards-Rissetto 2017).

In spite of the current abundance of geospatial data and analysis tools, resource managers often find themselves at a loss for the exact aerial imagery and 3D models that fit their needs in terms of spatial detail and timing/frequency of acquisition. Aerial imagery of the National Agricultural Imagery Program (NAIP), for example, has been flown by the U.S. Department of Agriculture annually or biennially since about 2013 in most states during the growing season and at a 1-m spatial resolution. Although the 3-band and sometimes 4-band imagery is an excellent source of repeated, georectified, and relatively high spatial-resolution orthoimagery for general resource monitoring, there are many instances when resource managers find it critical to capture details of cultural heritage monuments and structures, the microtopography of sacred sites or archaeological excavations at particular moments in time or over time periods of days or even hours. Natural resources are often negatively impacted by high winds and flooding, fire, insect/fungus infestation, and herbivory. Responding to natural disasters with rapid damage assessments are essential to recovery and inventory efforts. This chapter will discuss the practical use of sUAS, as a readily accessible methodology for capturing customized aerial images at spatial resolutions of 1 to 2 cm within minutes of arrival at a study site. Coupled with ground-based photographs, the sUAS images can then be processed using SfM photogrammetry to create 3D point clouds, DSMs, and digital orthomosaics at cm-level resolution and accuracy. Researchers at the University of Georgia's (UGA) Center for Geospatial Research (CGR) in the Department of Geography, UGA Cultural Landscape Laboratory in the College of Environment and Design, and consulting firm, Vertical Access, have used sUAS/SfM techniques in both cultural heritage and natural resource applications (Madden et al. 2015; Cotten et al. 2019). For this chapter, sUAS refers to a system of unmanned aerial vehicle (UAV), navigation system and sensor (e.g., 3-band camera and 5-band multispectral sensor) with first-person view capability weighing less than 24.9 kg (55 lbs.), along with a ground-control system and ground-based pilot in command and spotter.

9.2 Structure from Motion (SfM)

Structure from Motion (SfM) methods for image matching and 3D model reconstruction using overlapping images represent the current state-of-the-art photogrammetry and are steadily growing in utility within a wide variety of applications (Westoby et al. 2012; Fonstad et al. 2013; Madden et al. 2015; Cotten et al. 2019). While the concepts of SfM date to the 1950s, its recent incorporation into mainstream image processing is made possible by improvements in computing power and modern programming methods (Hartley and Zisserman 2003; Theriault et al. 2014; Jackson et al. 2016). Originating in the photogrammetric field of computer vision, SfM employs automatic feature-matching algorithms based on simultaneous, highly redundant, and iterative bundle adjustment procedures to perform image matching enabling the extraction of features from multiple overlapping images (Forstner 1986; Grun 1985; 2000; Grun et al. 2004; Fraser and Cronk 2009). For a comprehensive text on computer vision and 3D reconstruction, see Forstner and Wrobel (2016). After photogrammetric reconstruction, very dense point matching is employed to create a highly accurate 3D point cloud (El-Hakim et al. 2004; Remondino and El-Hakim 2006). The SfM process performs best with sets of highly overlapping (i.e., 80 to 90%) images that capture the full 3D structure of a scene as viewed from a wide array of positions. The name, Structure from Motion, comes from the use of images derived from moving a camera or sensor around an object or through a scene to capture images from many perspectives (Figure 9.1).

Structure from Motion is implemented in a series of steps that is normally performed in an automated process without additional input from the user. The process begins by selecting a series of overlapping images to be used in the solution, the number ranging from as few as five photos to as many as 200 or more. No *a priori* knowledge of the geometry of the images is required, although the geocoding information that is associated with the images with GPS-enabled sensors makes the image matching and geometric reconstruction more efficient. Steps in the SfM process include feature matching, image orientation, and point cloud generation.

Step 1: Feature Matching: The feature matching algorithm typically uses a Scale Invariant Feature Transform (SIFT) by Snavely (2008) and Snavely et al. (2008). This algorithm identifies features in individual images that can be used for image matching key points and descriptors. In this case, a feature is a distinct shape or shadow in an image that can be automatically identified and extracted on multiple

FIGURE 9.1
The Structure from Motion (SfM) concept is used to construct a 3D model of a feature of interest. The camera positions here show the arrangement of images required to create a model of a statue, small building, or other similar object.

images. The feature characteristics resulting from the image matching are stored in a database and then matched to the same feature appearing in additional images. This information is used to derive the relative relationships between images and provides input to the next step, image orientation.

Step 2: Image Orientation: Image orientation and geometric reconstruction uses an analytical photogrammetric method known as Sparse Bundle Adjustment, which was developed for aerotriangulation of photographs (Lourakis and Argyros 2009; McGlone 2013). In this process, the algorithm grabs key points from the frames and SIFT database to reconstruct the geometry of all camera positions from multi-point matches. This process models exterior orientation and corrects for lens distortion. Finally, a low-density, sparse point cloud is extracted and verified. Photogrammetric triangulation is used to estimate the 3D positions and incrementally reconstruct the scene geometry that is fixed into a relative coordinate system.

Step 3: Point Cloud Generation: The final step in the SfM process involves enhanced point matching from overlapping photographs to create a very dense point cloud, where each individual point has a set of XYZ coordinates, as well as Red Green Blue (RGB) values taken from the photographs. Depending upon the size of the object or scene of interest and the scale of the images, the resulting point cloud data can include 1000s of points per square meter and many millions of points overall (Figure 9.2). In general, better point match accuracy and higher point density results from high resolution photographs, but this increased detail and accuracy comes with longer processing times and more memory requirements. In many cases, the process is fully automated - the data are simply uploaded, and the process initiated.

FIGURE 9.2
(a) Image of a small statue in the Wormsloe Garden near Savannah, Georgia; and (b) 3D point cloud model of the statue that is the result of SfM processing.

There are a number of available software packages for SfM applications at a variety of cost points. Some of the major packages include the following.

1. Agisoft Metashape (www.agisoft.com/): Agisoft Metashape (formerly Agisoft Photoscan) is a commercial software package that runs on Windows PC, Mac, or Linux platforms and performs all processing locally. Cost is moderate and educational pricing is available for either the Standard or Professional Editions. The Standard Edition provides most of the functionality one needs to create excellent point clouds from hand-held or drone-based images, at any scale. The photogrammetric solution is strong and accounts for lens distortion and color correction. Output is in the form of a 3D point cloud where each point in the dataset has a unique XYZ coordinate and RGB color values extracted from the images. The coordinates are correct in a relative sense, in that, distances and dimensions are consistent. The scale, however, is relative to the images and not relative to actual ground dimensions. The Professional Edition adds the capability to include ground control points for reference or to use geocoded images (i.e., images tagged with GPS coordinates) to create scaled 3D models that are properly referenced to a ground coordinate system such as Universal Transverse Mercator (UTM). Measurements on these models will be in real-world units. For example, if the dimensions of a window in a 3D model of a building are measured, the results would be in units such as feet or meters, depending on the reference coordinate system. Additional output products include DEMs, DSMs, and orthomosaics (i.e., seamless composite images with the geometry of a map that are constructed by stitching many orthorectified and color-balanced images together).

2. Pix4D (https://pix4d.com/): Pix4D is a high-end photogrammetric package that is frequently the choice for larger projects, and is provided with several UASs (e.g., the SenseFly eBee) as part of a turnkey solution. Cost is higher than Agisoft Metashape and is generally tied to an annual subscription which includes data processing in the cloud. Results are similar to Metashape in that it can produce point clouds, DSMs, DEMs, and orthoimage mosaics from a large number of images acquired from an sUAS.

3. Visual SfM (http://ccwu.me/vsfm/), MeshLab (www.meshlab.net/), and Blender (www.blender.org/): Visual SfM is an open source software package that is

downloadable at no cost, but requires a high level of programming skill to implement. It requires several additional software packages to create the full product line one would expect with this software, including MeshLab and Blender. These downloadable and free software programs may not be easy to use, but they are very flexible and expandable if users are able to program. Products are similar to those produced by Metashape and Pix4D.

4. Quick Terrain Modeler/Quick Terrain Reader (http://appliedimagery.com/): Quick Terrain (QT) Reader is an outstanding point cloud visualization program that is free and available for download from the Applied Imagery website. The more powerful QT Modeler is a commercial program with a free trial license that provides analytical, editing, georegistration, and conversion tools for point clouds and other terrain data types. These are the main data visualization programs we used for projects described later.

5. Rhinoceros (also known as Rhino or Rhino3D) (www.rhino3d.com/6/features): Rhino is a commercial computer-aided design (CAD) and 3D computer graphics software developed by Robert McNeel & Associates. Its 3D geometry is based on the non-uniform rational basis spline (NURBS) mathematical model which mathematically represents curves and irregular surfaces in computer graphics rather than a mesh or DSM. The 3D modeling features of Rhino6 include functions to create, edit, analyze, document, render, animate, and translate NURBS curves, surfaces, and solids, point clouds, and meshes. Operating on Windows, Mac, and iOS platforms, a trial full version is available for 90 days. There are many online resources for learning Rhino and connecting to user communities.

6. Other software packages available for sUAS/SfM applications include such products as Drone2Map for ArcGIS (www.esri.com/en-us/arcgis/products/drone2map/overview/) and SURE/nFrames (www.ifp.uni-stuttgart.de/en/) (www.nframes.com/): Drone2Map provides SfM processing of sUAS-acquired imagery and integration of resulting products across ESRI ArcGIS platforms including ArcGIS Online, ArcGIS Pro, and ArcGIS Enterprise. With free trial and purchasing options, a full suite of products can be created including orthomosaics, DSMs, topographic contours, 3D textured meshes, and 3D point clouds. SURE is a full photogrammetric solution for 3D reconstruction developed by the Institute for Photogrammetry at the University of Stuttgart, Germany and is provided by the company, nFrames. It has multiple interfaces and a modular design to meet a variety of professional requirements. Free trials and extended non-commercial trials for university or research institutions are available.

7. Map Pilot for iOS (www.dronesmadeeasy.com/Articles.asp?ID=254): Map Pilot provides flight planning and flight control capabilities to DJI sUAS rotary copters such as Mavic, Phantom, Inspire, and Matrice. With this software installed on an Apple controller tablet or phone, it is possible to plan photogrammetric flights with specific parameters (overlap, ground pixel size, etc.) and then have the software control the flight of the sUAS to collect very high quality, consistent data for subsequent SfM processing. To initiate the flight, the project area is outlined on an image map (bounding points are entered on a touch screen) and the pixel size is specified (e.g., 2 cm). The program then calculates parallel flight lines to cover the area and upon command will take over control of the sUAS to automatically collect single frames at a set interval and forward overlap. If multiple batteries are required, the sUAS will return to the home point where the battery in the aircraft

TABLE 9.1

sUAS Flying Height, Pixel Size, and Area Covered

sUAS Flying Height		Pixel Size		Area Covered (Single Battery)	
Feet	Meters	Inches	Centimeters	Acres	Hectares
50	15.24	0.25	0.64	1.25	0.51
100	30.48	0.50	1.27	5	2.02
200	60.96	1.0	2.54	20	8.09
400	121.92	2.0	5.01	80	32.37

can be swapped and the flight resumed. With the assistance of this program, it is possible to collect data over an area of about 8 ha (20 ac) with a pixel size of 2.54 cm (1 in) flying at 61 m (200 ft) (Table 9.1). We use Map Pilot for many data collection tasks using Phantom 4 and 4 Pro quadcopters. For further information on Map Pilot documentation and articles about using this software, see Map Pilot (2019).

9.3 Cultural Resource Management in the Eastern United States

Cultural Resource Management (CRM) encompasses the identification, documentation, assessment, and management of historic buildings, structures, cultural landscapes, and archeological sites (Themistocleous and Danezis 2018). The base philosophy of CRM dictates best practice as minimal intervention. If interaction with the historic resource is necessary during CRM processes, the federal agency responsible for defining historic preservation standards and practices in the United States, the National Park Service (NPS), has detailed guidance on how to best intervene while protecting the historic resource. Because best practice advocates for nonintrusive and nondestructive tools to do the work, sometimes it can be a challenge to document and monitor change in historic resources. Typical challenges range from historic sites, cultural landscapes, and archeological sites that are too big or diverse to traverse easily and cannot be understood until seen from several hundred meters above, to a historic structure built in a way that makes it difficult to physically access, understand existing conditions, and determine management options. Observing cultural resources via sUAS is extremely useful for CRM purposes, affording physical access to sites and buildings while being a cost-effective tool for surveying and monitoring. This section provides examples of how sUAS, SfM, and other software and hardware have been used to assist CRM processes on cultural landscapes, historic buildings, and archaeological sites in the Eastern United States.

9.3.1 Stratford Hall Cultural Landscape, Stratford, Virginia

In 2014, the UGA Cultural Landscape Lab, in association with the UGA Center for Geospatial Research, received an NPS grant through the National Center for Preservation Technology and Training (NCPTT). The intent of the grant was two-fold: to identify cost-effective data capture and manipulation tools and assess those tools in terms of applicability to the documentation of cultural landscape characteristics. Some of the tools tested were sUAS and SfM.

The practice of cultural landscape preservation is unique in that landscapes are both historic artifacts and combined ecological and cultural systems that encompass a range of time scales (Alanen et al. 2000). As such, cultural landscape practitioners tend to approach these complex systems through a variety of interdisciplinary lenses - physical and human geography, landscape architecture, historic preservation, ecology, archaeology, botany, and history.

Rooted in the dynamic nature of ecological systems, typically cultural landscapes are overlain by not just one cultural system, but many. As such, documenting short- and long-term human occupations, along with their associated land use and management decisions, is integral to understanding the dynamic cycles of cultural landscapes. Cultural landscape complexity can be summarized as: a) overlapping layers of physical history yet each layer typically contains incomplete fragments; b) dynamic ecological systems that influence, and are influenced by, introduced culturally valued biotic and abiotic factors; and c) variations in scale.

Since the 1980s, the NPS has been an international leader in developing definitions, methods of identification, documentation and analysis, policies and guidelines, along with management strategies for cultural landscapes. In NPS Director's Order 28: Cultural Resource Management, cultural landscapes are defined as 'a geographic area (including both cultural and natural resources and the wildlife or domestic animal therein), associated with a historic event, activity, or person or exhibiting other cultural or aesthetic values' (NPS 1998).

The methodology the NPS crafted for understanding the complexity of cultural landscapes focuses on deconstructing them into their component parts known as landscape characteristics. Landscape characteristics include natural systems and features, spatial organization, land use, cultural traditions, vehicular and pedestrian circulation features, cluster arrangement, topography, vegetation, buildings and structures, water features, archaeological sites, and small-scale features (Page et al. 1998) (Figure 9.3). Landscape characteristics act as a useful framework for assessing cultural landscapes because they provide a system for: 1) gathering, organizing, and understanding information about the site history and existing conditions of a cultural landscape; and 2) documenting the changing appearance of a landscape over time.

9.3.2 History and Significance of Stratford Hall

Stratford Hall is the 7.7-sq. km (1,900-ac) home of the Lee family of Virginia, home of two signers of the U.S. Declaration of Independence and the birthplace of Robert E. Lee, Civil War General of the Confederate Army. Located along the upper Potomac River, Stratford Hall's fairly level upland topography is pierced by three parallel ravines that allow drainage from its upper elevations to the Potomac River over 30.5 m (100 ft) below. This diverse topography has created varied ecological zones with vegetation ranging from wetland species along the river and in the ravines, to oak-hickory and pine forests on the upper plateaus. Four generations of Lees, their slaves, and tenant farmers created and lived on this productive agricultural and forest landscape while simultaneously influencing politics and government at the local, state, and national levels for nearly 100 years.

Dating to 1716, when Colonel Thomas Lee had them built, the H-shaped Georgian two-story brick Great House and its two symmetrically placed rectangular single-story brick outbuildings, are highly remarkable examples of colonial Virginia architecture. Complementary brick and wooden outbuildings, barns, formal gardens, wharves on the Potomac River and a mill were later added by Thomas' son, Phillip Ludwell Lee, who

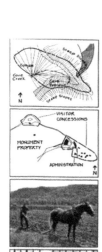

Natural Systems and Features
Natural aspects that often influence the development and resultant form of a landscape.

Spatial Organization
Arrangement of elements creating the ground, vertical, and overhead planes that define and create spaces.

Land Use
Organization, form, and shape of the landscape in response to land use.

Cultural Traditions
Practices that influence land use, patterns of division, building forms, and the use of materials.

Cluster Arrangement
The location of buildings and structures in the landscape.

Circulation
Spaces, features, and materials that constitute systems of movement.

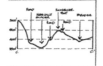

Topography
Three-dimensional configuration of the landscape surface characterized by features and orientation.

Vegetation
Indigenous or introduced trees, shrubs, vines, ground covers, and herbaceous materials.

Buildings and Structures
Three-dimensional constructs such as houses, barns, garages, stables, bridges, and memorials.

Views and Vistas
Features that create or allow a range of vision which can be natural or designed and controlled.

Constructed Water Features
The built features and elements that utilize water for aesthetic or utilitarian functions.

Small-Scale Features
Elements that provide detail and diversity combined with function and aesthetics.

Archeological Sites
Sites containing surface and subsurface remnants related to historic or prehistoric land use.

FIGURE 9.3
Graphic representations of the National Park Service cultural landscape components known as landscape characteristics, from NPS publication, Page et al. (1998) p. 53.

expanded the tobacco plantation to encompass 26.7 sq. km (6,600 ac) at the height of its production in the mid-18th century. Phillip's younger brothers, 'Lighthorse' Harry and later Henry Lee, inherited the property in succession, but due to poor financial management, lost it in the first decade of the 1800s. Following Lee family ownership, the property had just three owners over the next 100 years, continuing the agriculture and forest land uses through the Civil War into the early 20th century. In 1929, the property was acquired by a nonprofit corporation to memorialize the site in honor of Robert E. Lee, its most famous resident. The Robert E. Lee Memorial Association has cared for Stratford Hall as a public historic site ever since.

Throughout its 300-year history, the cultural landscape of Stratford Hall has had numerous additions, deletions, and changes, many of which were captured as part of the NCPTT research. Some of the extant natural and cultural resources documented included the limestone cliffs and Potomac River interface, the Great House with its associated vista cut through the forest to the Potomac River, a three-tiered boxwood parterre garden, two historic spring houses, a late 18th/early 19th century slave quarters archeological site, a mid-20th century garden designed by famed landscape architects, Innocenti and Webel, several large 350-year-old trees, a functioning historic grist mill and associated mill pond, and a tobacco rolling road where hogsheads of tobacco were carefully rolled from higher elevations to the shipping wharfs along the river.

9.3.3 Methods for Data Collection at Stratford Hall

After identifying locations across Stratford Hall that best represented cultural landscape characteristics, this project assessed methods using hardware representing a range of costs and accuracies, to determine how well they captured landscape characteristics (Goetcheus and Jordan 2018). The hardware tested included two types of sUAS, three hand-held digital cameras, two GPS units, and two terrestrial laser scanners. The two sUAS tested were the DJI Phantom 2 Vision+ quadcopter with DJI cameras mounted on 3-axis stabilized gimbals (rotary wing UAS) and the SenseFly eBee (fixed wing UAS). The data used in the study included aerial LiDAR, terrestrial LiDAR, GPS points, and point clouds. The software used was Agisoft PhotoScan (now Metashape), Quick Terrain Modeler, and LAStools.

Data collection and processing procedures were based on photogrammetric methods to create DEMs, 3D point clouds, and orthoimages created from aerial images. Terrestrial and hand-held photographs were used to create architectural maps and point clouds of buildings, dams, and other features best measured from the ground perspective. The photogrammetric methods for this project consisted of ground control data collection, point cloud data acquisition, and data processing leading to result comparison.

9.3.3.1 *Ground Control Data Collection*

Ground control points (GCP) are used as reference points within the photogrammetric model and provide a link between the aerial images and the ground coordinate system. For accuracy and ability to replicate GCP locations on sUAS imagery, 17 white boards (60×60 cm) painted with a 10-cm wide black 'X' were distributed throughout the site for the two-day period of sUAS image acquisition. A 2.5-cm hole was drilled in the center of each board to permit fixing to the ground using a rebar or flag. The center location of the hole was used to record the ground coordinates of the GCP. A professional surveyor was hired to acquire the GCP coordinates using a Topcon HiPer GA differential GPS (DGPS) with base station (XY accuracy of less than .01 m and Z accuracy of .02 m). The GCP coordinate was also measured using a Trimble GeoXH GPS unit with external antenna attached to a 2-m rod using Virginia State Plane North coordinate system with coordinate values in feet (XY accuracy of approximately 1 m). Using the GPSCollect Function in the ESRI ArcPad software on the Trimble unit, 30 observations were collected at each GCP location and averaged to obtain the final point coordinates. The Z values for the GCPs were then extracted from LiDAR that had a vertical accuracy of 15 cm.

9.3.3.2 *sUAS Imagery Acquisition*

Acquiring point cloud data was achieved through a combination of collecting known aerial LiDAR from the Virginia GIS clearinghouse, hiring a private contractor to undertake terrestrial LiDAR scans of sites using a Leica ScanStation P20, as well as employing two different sUAS. Data acquisition from the sUAS is emphasized here to inform the discussion on comparing data processing results from fixed wing and rotor sUAS. The Stratford Hall project employed the DJI Phantom 2 Vision+ quadcopter costing approximately US$1,500 USD and weighing about 1.1 kg. The Phantom quadcopter was equipped with a DJI camera mounted on a 3-axis stabilized gimbal. Although the Phantom 2 Vision+ was the version available at the time of this project, current projects utilize newer generation DJI Phantom 4 and Phantom 4 Pro quadcopters. There have been significant improvements in the camera between the Phantom 2 Vision+ and the Phantom 4, most noticeably in the reduction of lens distortion and increase in image quality in the Phantom 4. The Field of View of the camera is about 140 degrees, but there was a barrel distortion in the images of the DJI Phantom 2 Vision+ that caused the horizon in the images to appear curved. This distortion can be removed using the Photoshop Lens Distortion Filter or modelled in the photogrammetry software; so, for this study the lens distortion was not a problem. Lens distortion is minimized in the Phantom 4 camera. The DJI cameras can record either 14-Megapixel (MP) still photos or High Definition (HD) Video.

Control of the aircraft was conducted using the manual joystick and the DJI Go app operating on an iOS Apple iPad, although the app also works well on a cell phone. The advantage of using a cell phone is that it provides connectivity to the internet. First-Person-View (FPV) images from the camera are transmitted directly to the iPad using a local, long range WiFi connection. The DJI Go App on the iPad controlled the camera angle and still/video selection, as well as manual triggering of the video or still images.

Although there are several useful apps available for planning and programming flight lines of the Phantom quadcopters, these did not become available until the Phantom 3 models. For this reason, we had to manually fly the DJI Phantom 2 Vision+ with the joystick control and iPad FPV controls to capture the images and video. It is not possible to automatically collect still frames using the DJI app, and it is difficult to manually fly the aircraft along precise flight lines while recording individual still frames. For this reason, we have found that collecting HD video and extracting frames from the video work best for photogrammetric analysis when the Phantom 2 Vision+ is used for image data collection. Further, we orient the camera at a slight oblique angle to the ground to permit imaging of the sides of features.

The video from the flight was saved as an MP4-formatted file which is compatible with most video display programs on a PC or Mac. We selected the file that contained East-West flight lines and extracted a total of 324 HD video frames at 1920×1080 pixel-resolution using the (free) VLC Media Player program. The images were saved in PNG format, although they could also have been saved in JPG format. Because the MP4 format is a compressed data format, it is not possible to extract full resolution, uncompressed images from the video file. This does not appear to be a problem in our experience. In the final images, all points and features in the project area were covered by at least nine frames. One significant difference between still images collected with the Phantom vs the video frames is that the still images are tagged with GPS coordinates. Frames grabbed from the video are not geotagged. This lack of GPS reference requires that ground control points be used in order to fully register the point cloud and other products to a known reference coordinate system.

A contractor from Volo Pervidi LLC in Albuquerque, New Mexico was hired to fly the SenseFly eBee sUAS using a small Canon Powershot S110 camera. The 96-cm fixed wing Styrofoam sUAS weighs only 0.70 kg (including the battery and camera), has a radio modem, eMotion software, and autopilot flight planning that can cover up to 1,200 ha in a single flight plan. In this fixed wing sUAS, the camera lens is oriented vertically to the ground plane and not mounted on a stabilizing gimble. In addition, due to the lightweight material that the aircraft is made from, it is not recommended to fly this sUAS in winds over 15 knots (27 kph) to ensure high quality images. The camera images were saved in JPG format and geotags added using Postflight Terra software to prepare for further data processing.

9.3.3.3 Performing SfM and Point Cloud Data Processing

Three software packages were employed for data processing and visualization. Agisoft PhotoScan and SfM methods were used to create point clouds from sets of stereo images collected using the DJI and eBee sUAS. Applied Imagery's Quick Terrain Modeler permitted the georectification of that point cloud data using control points and direct measurement of point locations and distances in the point clouds, while LAStools by rapidlasso GmbH (https://rapidlasso.com/) was used primarily to scale datasets to match measurements of features in the field.

Agisoft PhotoScan Professional was used to perform aerial triangulation on the 324 video frames. This took about 18–20 hours to complete using the 'generic high accuracy' option on an Apple MacBook Pro laptop computer with 8 Gb RAM and an Intel i7-2640M CPU running at 2.8 GHz. The resulting solution found the average flying height to be 63 m above ground level (AGL), the project area was 8.6 ha, the average pixel size was about 3 cm, and the average tie point error was 1.8 cm (0.6 pixel). Initially, a medium resolution dense point cloud (~3.3 million points at 40 pts/sq. m) was created with a corresponding mesh to permit visualization of the 3D model. The primary result of this step was to identify the relationships and orientation between each image using the SfM methods and return the position and altitude of each camera station. This critical information was then used in subsequent steps.

The Ground Control function panel in PhotoScan was enabled and made visible and the GCP file in CSV/txt format was loaded using the Import function. All 17 control points were identified and measured on each image they appeared in using the Marker function. In each case, the image was zoomed until the black 'X' painted on the GCP marker target could be clearly seen. The point was then identified, and its location measured. After identifying the GCP location on at least two images, the locations were automatically transferred to all other images the Point ID appeared on using the orientation information computed in the previous aerotriangulation step. The point locations were then verified on each image in order to be included in the solution. Some points were slightly off location and had to be moved to better fit the GCP marker. Points that were not actually visible, or whose locations were obscured by trees or a building, were not verified on those images and were thus not employed in the solution. In general, each GCP appeared on between 10–29 images. The total time for manually measuring and verifying the GCP locations on the images was about 75 minutes. After all points were measured, the Photoscan 'Adjust' function was used to adjust the initial ungeoreferenced AirTrig solution to the GCPs. The solution was computed with an overall error of 15 cm (2.65 pixels). The Agisoft 'Optimize' function was then used to introduce a modelled correction for lens distortion to the solution. This improved the overall error to 0.03 cm (0.51 pixel).

After georeferencing was accomplished, the medium- and high-density point clouds were recomputed. The medium-density point cloud consisted of 3,411,001 points. This equates to a density of 40 pts/sq. m. Computation time was approximately 30 minutes. The high-density point cloud consisted of 13,107,770 points, which is equivalent to 15.2 pts/sq. m. Computation time was approximately 4 hours. Both point clouds were exported to the standard LiDAR LAS format in Virginia State Plane North Coordinates (Feet). Exported data included XYZ location and RGB photo/color information for each point.

A color-balanced digital orthoimage at 6-cm resolution and a DEM (actually, a DSM) at 15-cm resolution were produced. Both products were exported as TIFF files.

The high-density point cloud was loaded into QT Reader software and markers were placed at the locations of ten control point boards. These were readily visible in the point cloud using the RGB values extracted from the images at each point (point 2 was obscured by the GPS antenna tripod and not used in this evaluation). The XYZ values for the ten marker locations were exported to CSV format and then brought into Excel for comparison with the surveyed values of the GCPs. In mapping, accuracies typically are reported in terms of the root-mean squared error (RMSE) which provides a standardized measure of error that can be compared to real world coordinates in both horizontal (XY) and vertical (Z) positions. It represents a value that says that 68% of all data will fall within an error radius with one RMSE value of its true value. Similarly, 95% of the points will fall within a radius of two RMSE values. For the Stratford Hall Phantom sUAS flight, the $RMSExy = \pm 22.25$ cm while the $RMSEz = \pm 6.1$ cm. These results, measured at the same GCP locations as were used in the photogrammetric solution, demonstrate a very high-quality solution that actually exceeds the estimated accuracy of the uncorrected GPS surveyed GCPs.

The Canon camera in the SenseFly eBee is programmed to record images at specific timings to achieve 80% forward overlap and 80% sidelap between photos. As weather definitely plays a role in the ability of an sUAS to perform, it should be noted that on the primary day of research, the weather was sunny and clear with a very light wind in the morning. However, by the afternoon the wind speed increased to almost 15 knots. This had a negative effect on the SenseFly eBee's data collection, as the lightweight eBee was periodically blown off its flight plan, impacting the normally perpendicular orientation of the camera to the ground. Although the eBee consistently returned to the flight plan, because images were taken at a consistent time interval, if during that interval the sUAS was blown off course it more than likely took an image that was not useful. This led to fewer images for data processing, hence creating a much coarser dataset.

9.3.4 Results and Conclusions for the Stratford Hall Cultural Landscape

Because cultural landscapes are complex, layered, ecological, and cultural systems, they can be challenging to identify, document, assess, and manage. The use of NPS-defined landscape characteristics was a good framework by which to capture a diversity of features on the Stratford Hall cultural landscape. Careful consideration was taken regarding how to properly document each landscape characteristic, to not only reveal as much about the feature as possible, but also permit comprehensive data capture for archiving of current condition and status. Testing a diversity of hardware and software was also useful to compare results of a low cost/limited skill solution versus high cost/more skilled technology to document, assess, and manage cultural landscapes. An orthoimage of high quality was created for Stratford Hall from the images acquired by the Phantom 2 Vision+ quadcopter and a 3D point cloud produced by SfM processing. Figure 9.4a shows details of the Stratford Hall main house and formal gardens and demonstrates the production of a clear and distinct orthoimage that is suitable for documenting the cultural landscape of this historic site. The 3D point cloud can be viewed from any angle to emphasize the cultural features of the historic buildings and grounds (Figure 9.4b). These results show how the combination of sUAS hardware, rigor in applying photogrammetric methods, and SfM software was found to be a very useful approach to data capture for cultural landscapes.

FIGURE 9.4
(a) Orthoimage of Stratford Hall generated from images and point cloud acquired by a Phantom 2 Vision+ sUAS. Details of house and garden are clear, distinct, and suitable for documenting the cultural history of the historic site; (b) the point cloud of Stratford Hall may be viewed from any angle to permit visualization of the house and grounds in a 3D perspective.

9.4 sUAS-Based Data Acquisition in the Investigation of the Soldiers and Sailors Memorial Monument, New York City, New York

The Soldiers and Sailors Memorial Monument in Manhattan's Riverside Park commemorates Union Army soldiers and sailors who served in the Civil War (New York City Landmarks Preservation Commission 1976). First suggested in 1869, but not completed until 1902, investigative work on this monument was initiated in 2016 to assess its condition and level of deterioration. Architectural historians from Vertical Access use rock-climbing ropes to access historic buildings, monuments, bridges, and dams for assessment and restoration (https://vertical-access.com/). They leverage a variety of digital documentation technologies, including nondestructive evaluation and data acquisition by sUAS in order to allow the project team to fully understand existing structural conditions and deterioration mechanisms.

Unknown sub-surface conditions are a major challenge in caring for and designing interventions for older buildings. Frequently, the project team must undertake extensive research and documentation during the discovery phase in order to understand existing conditions or make educated guesses about hidden conditions. The team may have incomplete as-built drawings, design drawings only, or no archival drawings at all. Often adding to the complexity of the investigative work is a lack of personnel access, neglected access systems, or a combination of both. Use of sUAS technology helps architects, engineers, and building investigators access and document existing structures in order to develop appropriate treatments.

In addition to solving access challenges, sUAS can mitigate schedule and financial restraints by allowing an investigation to be carried out in a fraction of the time required for other survey methods. An sUAS survey may supplement a binocular survey or a hands-on inspection using suspended scaffolds, supported scaffolds, or industrial rope access by serving as a first pass or reconnaissance survey, and/or allowing the project team to see inaccessible features such as cupolas, finials, and chimney stacks. Inspection by sUAS may decrease impacts to building users as well as impacts to landscaping or historic materials by eliminating the need to drive heavy equipment onto the site or tie back scaffolds to building façades.

Although current consumer-grade sUAS models leverage sophisticated positioning technology in order to make them easy to operate, the successful use of sUAS in the investigation of buildings requires skill and practice. Because many heritage buildings include structural steel framing, care must be exercised in maintaining enough distance from the subject to avoid interfering with the sUAS compass, yet close enough to capture data in sufficient detail. Precipitation and wind are also a challenge. And, unlike traditional surveys, the project team must develop systems for managing many gigabytes of high resolution video and/or other data captured by a drone. As a recent project documenting a monument interior demonstrates, operating an sUAS indoors presents a different set of challenges, including limited lines of sight, constrained spaces, and other obstacles.

9.4.1 Project Background

The New York City Office of Management and Budget, in conjunction with the Department of Parks and Recreation, retained Perkins Eastman Architects and their consultant team to provide pre-scoping services for a comprehensive existing conditions survey and restoration treatment study of the Soldiers and Sailors Memorial. The conditions survey was completed by Leslie E. Robertson Associates (LERA) and their team of specialty sub-consultants including Vertical Access (VA). It included a detailed conservation and engineering examination and assessment to determine the stability and surface condition of the Monument's architectural and decorative elements and surrounding plazas. Restoration treatment recommendations and a corresponding cost estimate were developed based on this survey. The last major repairs to the Memorial were undertaken in 1962. At the time of the investigation, the exterior stone was in poor to fair condition with widespread spalling and cracking of the masonry, mortar loss, failed Dutchman repairs, and failures of sheet metal roof and flashings.

Rising approximately 28.9 m (95 ft) above the surrounding plaza, the monument consists of a tall cylindrical base, a rusticated drum ringed by Corinthian columns, an ornamental entablature, and a low conical roof. The entablature frieze bears the inscription, 'To the memory of the brave soldiers and sailors who saved the union'. The exterior is of white Vermont marble with a granite water table. A bronze door opens to the marble-clad interior space featuring a mosaic-tiled dome with an oculus opening to a second dome, both constructed by the Guastavino Fireproof Construction Company. During the last major restoration campaign from 1959 to 1962, the marble roof was clad in sheet lead and the marble architrave was replaced in granite.

The Memorial, including the surrounding plazas, was designated a New York City Landmark in 1976, and it was documented as part of the Historic American Engineering Record survey of the Henry Hudson Parkway in 2005–2006 (Martin Reynolds 1988).

9.4.2 Survey Challenges and Solutions

To inspect the roof of the 30-m tall monument, the team's options included pipe scaffolding, a boom lift, or industrial rope access. Pipe scaffolding would have been costly to erect, required tying back to the monument façade and presented a significant aesthetic impact to the monument for the duration of the survey. Due to site constraints, a boom lift would have provided access to view approximately only one quadrant of the monument. Therefore, a combination of industrial rope access (IRA) and lift access was chosen to allow timely hands-on access to all of the exterior materials.

The lack of access at the monument interior presented another challenge during the investigation. The lower portion of the interior is topped with a tile dome featuring a 1.5-m (5-ft) diameter oculus opening into an upper space surmounted by a second dome. No documentation could be found of the as-built conditions of the area above the lower dome, almost none of which is visible from the ground floor. There is currently no means in place for personnel to access the interior space between the two domes, and the project team needed to confirm the construction details at this area and to observe the condition of the materials. In order to avoid the erection of pipe scaffolding at the interior, which would be better suited for a full restoration program in the future, VA's solution was to fly a DJI Phantom 4 Pro sUAS through the oculus in the lower dome, providing video and infrared footage of the interior materials.

Although interior airspace is not subject to FAA regulation or inclement weather, drone pilots face significant challenges flying in interior spaces. Many structures with large interior spaces, such as churches, stadiums, armories, and theaters, contain structural steel that can interfere with the sUAS on-board compass, affecting automatic stabilization and/or increasing the minimum distance from the interior or exterior wall that the pilot must maintain in order to ensure stable flight. Interior spaces often contain other obstacles, such as structural roof trusses or columns, suspended lighting, or audiovisual equipment.

Supplemental lighting is often necessary in order for the sUAS to capture high resolution imagery of interior materials. The VA team used 1000-watt tripod-mounted theatrical lights to illuminate the lower portion of the interior at the Soldiers and Sailors Memorial Monument. Bright daylight through a single window opening was sufficient to illuminate the upper dome and drum. Although the added weight decreases battery life and thus flight time, VA has found drone-mounted LED lights to be useful for indoor applications where the areas of interest are at greater distances from supplemental lighting at the ground.

The interior geometry of the monument posed a significant challenge, even for drone access. Although the obstacle avoidance capability of the sUAS reduces the likelihood of collisions, flying through the very narrow oculus (1.5 m or 5 ft in diameter) would have been impossible with this feature activated. As a result, the pilot had to fly in attitude or 'atti' mode; in other words, manually and with no automatic stabilization. Prior to site work, the pilot practiced flying in atti mode, including spotting the drone from directly below as was necessary within the narrow base of the monument. The team considered constructing a mockup of the oculus, but instead the pilot chose to practice maneuvering the sUAS close to tree canopies in order to simulate conditions within monument while minimizing the likelihood of a crash landing.

The project team scheduled the sUAS investigation immediately following several days of rain for the best opportunity to observe thermal differences in the materials due to water infiltration, in addition to documenting the interior conditions with visual-spectrum imagery. During the flight, a live-feed video downlink allowed multiple project team members to observe the conditions in real time and to direct the sUAS pilot to areas of interest. In order to observe thermal differences in the interior materials, VA's DJI Phantom 4 Pro aircraft was mounted with a Forward Looking Infrared (FLIR) Vue Pro thermal camera designed for sUAS applications.

Once through the oculus, the sUAS recorded visual-spectrum video and infrared video side-by-side simultaneously (https://youtu.be/yfg10jdkH9I). Visual-spectrum video provided information about the materials, construction, and condition at the interior. Crack systems were observed in the brick back-up masonry along with efflorescence and mortar loss. Other observations included the presence of framing pockets in the masonry, used to support interior scaffolding during the monument's construction.

Infrared imagery acquired using a handheld FLIR infrared camera at the ground level of the interior indicated likely water infiltration through open masonry joints. The data gathered by the drone-mounted IR camera showed cooler areas at the marble masonry joints at areas above the lower dome, but did not show any thermal anomalies in the adjacent brick masonry. The data gathered by the sUAS-mounted thermal infrared (TIR) camera was limited due to several factors. The FLIR Vue Pro camera is primarily used for aerial inspections of infrastructure and agriculture. Consequently, the lens focus was not sufficient for the relatively close-range of flight inside the monument and resulted in blurred images. A distance of 4.57 or 6.10 m (15 or 20 ft) would have resulted in sharper imagery, but the diameter of the interior was only about 4.57 m. At the upper dome, the reflective nature of the glazed Guastavino tile prevented any meaningful TIR imagery capture. These lessons learned will better inform future use of sUAS-mounted IR cameras.

9.4.3 Digital Technologies to Process sUAS Monument Data

Several digital solutions were employed throughout the investigation and development of recommendations, including the digital collection and management of survey data, photography and videography, 3D modeling, and structural analysis. Although relatively modest in size, the monument's deteriorated state called for a systematic and efficient means of collecting and analyzing exterior and interior conditions data.

The VA team surveyed the exterior using the Tablet PC Annotation System® (TPAS), a direct-to-digital software developed by VA for recording existing conditions in the field using a predefined library of material conditions and project-specific nomenclature within AutoCAD. It links photographic and video documentation to quantitative condition data (e.g., crack lengths and widths, size of spalls, and areas of soiling) within a single CAD drawing. The data are exportable to a custom web portal and/or other database platforms for further analysis, development of recommendations, and cost estimating. The LERA had prepared background CAD plans and elevations of the entire Memorial based on original drawings from both the 1899 construction period and the 1959–1962 restoration that were utilized by VA for the TPAS survey.

Following the site work, interior conditions observed in the visual spectrum sUAS video were recorded on interior elevations in AutoCAD using TPAS, with video stills hyperlinked to condition annotations within the CAD drawing (Figure 9.5). Although somewhat less efficient than recording observations directly during observations *in situ*, post-site documentation in TPAS based on sUAS video enabled all of the conditions data - both interior and exterior - to be analyzed as a whole. For the investigation of the Soldiers and Sailors Memorial Monument, the TPAS data served as part of the recommendation documents forming the basis for cost estimates.

9.4.4 Conclusions and Future Approaches for Monument Evaluation

The Soldiers and Sailors project team leveraged a variety of nondestructive evaluation tools, digital documentation, and 3D modeling, combined with personnel access and remote viewing methods tailored to meet the challenges of investigating the entire

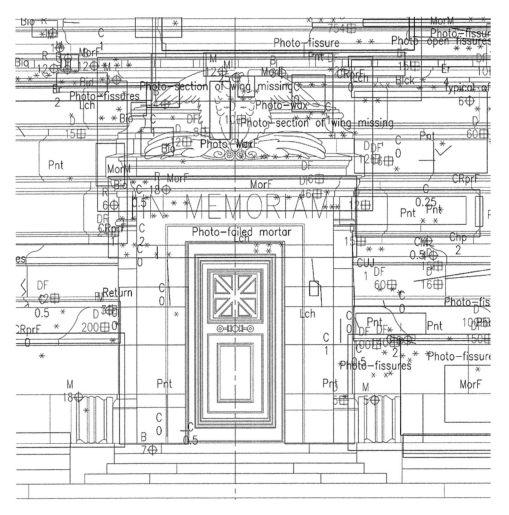

FIGURE 9.5
A portion of the CAD drawing of the Soldiers and Sailors Memorial Monument exterior. The Tablet PC Annotation System® (TPAS), a direct-to-digital software for recording existing conditions in the field using a predefined library of material conditions nomenclature within AutoCAD, links photographic and sUAS video documentation to a CAD drawing.

structure. This multifaceted approach allowed the team to efficiently document existing conditions and analyze the structural performance of the monument, informing the development of recommendations for repairs and cost estimates. Thoughtful deployment of an sUAS provided the project team with much more information than was initially available, and at a much-reduced cost compared to hands-on or close-visual inspection using pipe frame scaffolding.

The data gathered at the exterior and interior of the monument are being used for a full restoration condition assessment with scoping recommendations to determine a comprehensive cost estimate to facilitate fundraising efforts by the New York City Office of Management and Budget and the Department of Parks and Recreation for the eventual restoration of the entire Memorial Monument and surrounding plazas. In addition, Leslie E. Robertson Associates utilized the data obtained to develop a 3D geometry model of the Monument using Rhinoceros to assist with quantity determination and structural analysis. The team is currently investigating future approaches such as the use of virtual reality coupled with sUAS and 360 technology and custom software to better document, analyze, and communicate conditions and proposed remediation.

9.5 Historic Cemeteries in Athens, Georgia

Key requirements in resource management, including condition assessment and resource monitoring, are particularly relevant when involving multi-century objects and often associated rich historical narratives. High resolution images from sUAS can inform the management of cultural resources by providing detailed representations of the landscape, including changes over space and time. The use of sUAS is particularly beneficial for historic cemeteries where multiple strategies for data acquisition and representation can be used for decision making, digital preservation, virtual tours, and exploration of regional or local history with the possibility to further expand diverse narratives.

9.5.1 Considerations about Location and Data Acquisition

Rules on image acquisition in cemeteries from the ground or airborne platforms are usually defined by the administration of the cemetery, with photographs for non-commercial use generally allowed. Beyond the technical aspects of data acquisitions when planning flights over cemeteries, data collectors need to be aware of the environment a cemetery represents and associated sensitivities. Data acquisition efforts need to take into consideration the visitors, family links, and the community where the cemetery is located. In addition, national, state, and local policies and regulations need to be consulted, as those may limit flight operations and may require specific strategies for data acquisition. For instance, when flying an sUAS that weighs less than 24.9 kg (55 lbs.) for work or business in the United States, a waiver should be requested and received from the United States Federal Aviation Administration (FAA). This is needed in order to fly over people who are not participating in the data acquisition operation and who are not protected from falling sUAS by being located under a covered structure or inside a stationary vehicle. Because these are conditions often found when flying over cemeteries and to fully meet requirements, coordination with cemetery administration and the local community may be required. Further, data acquisition may need to be restricted to periods when visitors are not allowed in the area of the cemetery. For information on these and other Part 107 regulations, see www.faa.gov/uas/commercial_operators/.

Additionally, cemeteries can be located under controlled airspace, in built-up areas, and near airports so extra consideration needs to be given to potential limitations to flights due to the proximity to runways and routes used by manned aircrafts. In the United States, the UAS Facility Maps map portal (www.faa.gov/uas/commercial_operators/uas_facility_maps/) provides the location of controlled airspace and indicates maximum allowed flight heights above ground level near airports across the country. Flying in controlled airspace under Part 107 in the United States requires permission that is granted from the FAA. Authorizations to fly over these areas can be requested using the new Low Altitude Authorization and Notification Capability (LAANC) or the DroneZone website (www.faa.gov/uas/commercial_operators/part_107/). Finally, communication with cemetery management is critical for a successful data acquisition campaign, considering that the site may be undergoing periodical maintenance and that historical sites often receive visitors during specific dates or holidays.

Data acquisition may involve flying a planned mission for wall-to-wall mapping of the area occupied by the cemetery and may include one or more flights at lower height in order to collect detail. These lower altitude flights are often conducted manually over specific areas and can include high resolution image acquisition from multiple points of view, circular flights with object of interest located at the center of the circle, and/or video from fly overs.

Depending on the size and location of a cemetery, it is not uncommon that more than one maximum sUAS flight height is assigned to the area by the FAA regulators. When this is the case, a data acquisition mission can be split into multiple flights at different heights. Alternatively, if consistency of ground sample distance is desired, the entire area to be imaged can be flown using the lowest allowed maximum flight height. Individuals planning flights should take into consideration that flying at lower heights may come at a cost, as lower flights will result in a larger number of images collected, increasing storage requirements and image processing times. In addition, data acquisition at lower heights may require longer flights and/or more flights to cover the required area, thus involving more batteries and increasing chances that illumination conditions change during data acquisition.

The timing of data acquisition can significantly affect outputs and, as a result, planning of when collection is conducted should take into consideration the purpose of the collection. Because many historic cemeteries contain numerous old trees, the time of the year and seasonality when considering leaf on/leaf off conditions are key factors as canopies can obstruct the view of surface objects, including graves and monuments. Full canopies may also result in hard shadows that can bring additional challenges to the identification of objects on the ground. Conversely, models or orthomosaics built using images collected under leaf-off conditions may not capture branches in detail and may provide a semi non-obstructed view of the surface. Flight operations are also affected by timing and plant status, particularly when flying under the canopy or between individual trees for detail. These tasks may require fine control over aircraft path and altitude and usually involve rotary copters and flying manually. Because of the required proximity to branches and trunks, a collision avoidance system, when present, may be triggered during flight and may affect acquisition by limiting movement of the sUAS. Depending on flight requirements and conditions, the pilot may choose to disable the collision avoidance system in order to be able to place the sUAS at specific locations.

In addition to time of year, time of the day plays an important role in image acquisition over areas with objects regularly distributed and with similar orientation.

The representation of monuments found in cemeteries, including headstones and obelisks, can be affected by this arrangement. In particular, sun elevation and azimuth directly interact with the spatial distribution of graves and size and orientation of monuments. The geometry of illumination affects shadow distribution and may be particularly important depending on the orientation of objects being imaged. For instance, faces of objects oriented towards east will receive direct sunlight during the morning and data collection should take into consideration that those faces may be poorly illuminated in the afternoon. In addition, interactions of shadows cast by monuments and other objects (e.g., trees, buildings, fences) should be considered when planning flights. The choice of when and how to fly become critical, particularly when the goal of the flight is to capture faces of objects that are oriented vertically. Depending on location, vertical faces may not be directly illuminated during or close to wintertime for the given hemisphere.

In the northern hemisphere, vertical faces of objects at latitudes higher than 23.5 degrees and facing north do not receive direct sunlight. Similarly, in the southern hemisphere, vertical faces of objects at latitudes higher than 23.5 degrees and facing south are not directly illuminated by the sun. Due to the close proximity of objects in cemeteries, hard shadows can be reduced if the area is flown under overcast skies, as clouds work as a light diffuser. In this case, particular attention should be given to camera settings affecting image quality, including shutter speed and exposure. When setting flight parameters, the speed of the sUAS may need to be reduced, if longer exposure times are selected, to avoid the acquisition of blurry images. Advances in cameras and the availability of larger camera sensors have contributed to improved signal to noise ratios and the increase in image quality under reduced illumination.

For SfM and 3D reconstruction of objects, a combination of approaches can be used to achieve high quality results, including a higher altitude flight for general mapping and lower altitude flights for detail. In particular, the goal of the flight and level of detail required by the project need to be identified prior to flight operations, considering that some projects require monument detail and engravings, including lettering, to be represented. For example, a 20 MP camera flying at a height of five meters results in GSD of approximately 1.5 mm/pixel. At this resolution, enough detail of monuments is preserved allowing for the representation of objects and reading of engravings. Additional photos from the ground can be used to enhance the detail of objects.

Multiple strategies can be used when conducting automatic flights. A system based on polygons (multi-side shapes) or grids (four-side shapes) can be used to define flight lines to be followed by the sUAS during image acquisition. Polygons provide more control over the definition of areas to be imaged and may help optimize the acquisition by avoiding flying over areas beyond the area of interest. Thus, the choice between polygons and grids are usually linked to the shape of the area to be flown. Following the definition of a bounding box for an area, the orientation of flight lines and direction of flight can be controlled during flight planning. Although polygons and grids allow for the construction of image mosaics, they may not fully capture architecture or fully represent the 3D nature of monuments, particularly considering vertical faces. To address multiple possibilities for vertical face orientation that are commonly found in cemeteries, a double grid system can be used. This system consists of two flight line grids layered on top of each other and rotated 90 degrees in relation to one another. For instance, a double grid configuration could allow a camera to capture vertical faces oriented north-south when flying the first grid and, later, capture faces oriented east-west, when flying the second grid. This strategy has shown to be particularly effective when combined with tilting a forward-looking camera, for instance at 70 degrees (90 degrees = nadir).

Among other factors, flight planning needs to consider the size of the cemetery and the possible need to use multiple flights to cover the entire area. When defining the area to be imaged, it is good practice to include one or two flight lines (or their area equivalent) around the area of interest, as the extra images will assist with model building and mosaic creation. If the goal is to generate one seamless mosaic and multiple flights missions are used, consider overlapping missions, so individual sets of images can be integrated during processing. In this case, duration of flight and changes in illumination conditions between sets of acquisitions, particularly when shadows are involved, may require special consideration. When these are critical, using multiple sUAS flying at the same time or one sUAS flying during different days (ideally, days should be close in time) may be considered.

Image overlap requirements are dependent on desired final products. In particular, endlap (frontal overlap) and sidelap (side overlap) need to be carefully considered, as they can affect flight and image processing times. For mosaic generation over relatively flat terrain and derived from relatively high-altitude flights, a combination of 75% endlap and 60% sidelap can produce satisfactory results. If 3D reconstruction is involved, higher overlaps (for instance: 85%, 75%) can be used to guarantee the vertical structure of objects is correctly resolved.

Objects located at cemeteries offer multiple possibilities for image georeferencing and the positioning of resulting mosaics. Acquired images should provide enough spatial detail, so monuments or parts of monuments can be easily identified in the images and on the ground. A variety of GPS-based approaches can be used for control point collection and georeferencing, including Real-Time Kinematic (RTK) and Post-Processed Kinematic (PPK), which can reach centimeter-level positioning. Alternatively, existing high resolution images that have already been georeferenced can be used as reference when locating control points.

9.5.2 Athens - Layers of Time with sUAS Content

An example of integration of sUAS-based data and imagery collected on the ground can be found at the Athens - Layers of Time website (https://athenslayersoftime.uga.edu/).

This project occurred as a result of the unearthing of human remains on 17 November 2015, during the construction of an addition to Baldwin Hall on the University of Georgia North Campus, in Athens, Georgia (Parry 2017). Using historic material that captured the Old Athens Cemetery, Oconee Hill Cemetery, the UGA campus, and the city of Athens, from the 1800s to the present, the intent is to provide a virtual walk through the city's changing physical landscape via a story map. For a better understanding of the spatial context and history associated with individuals who had been buried at the cemetery, a team of researchers and students at the UGA Center for Geospatial Research created a GIS database and time-series visualizations. Originally focused on the area surrounding the UGA Baldwin Hall building and Old Athens Cemetery, for contextual reasons the area expanded to include the City of Athens due to the rich source of datasets representing the city over time.

Focused on UGA's original 18th century campus known as North Campus, and extending to the 1800s boundary of the City of Athens, the project collected historic maps, aerial photographs, sUAS and satellite images showing changes in the physical landscape over the last 200 years. With a time-series of maps and images, the work documented those changes in a GIS geodatabase and web-based digital story map that provide online access to geographic, textual, and other visual information that can be used for education and outreach. Figure 9.6 shows part of this dataset, represented by an sUAS-based 3D reconstruction of the Old Athens Cemetery and buildings, as well as a 3D model for one of the monuments in the cemetery.

Flight campaigns for image acquisition were conducted in 2018 (Old Athens Cemetery) and 2019 (Oconee Hill Cemetery) using a DJI Phantom 4 Pro equipped with a 20-MP RGB camera flown at 46 meters. Flight planning included a double grid configuration with camera angle set to 70 degrees (90 degrees = nadir). Overlap was set to 80% for both endlap and sidelap. The area is under controlled airspace and two FAA authorizations were used to conduct the flights. Following acquisitions, images were processed using Agisoft Metashape to generate 3D models and orthomosaics. In addition, images collected on the ground were used with Metashape to create detailed 3D models of monuments in the cemetery (Figure 9.6). Models and technologies are being integrated into the 3D Immersion and Geovisualization (3DIG) system (http://cgr.uga.edu/projects/3dig/) at the UGA Center for Geospatial Research. The system allows for multiple geovisualization strategies, including the use of virtual and augmented reality devices for representation of 3D models derived from sUAS and ground images.

(a)

(b)

FIGURE 9.6
A 3D reconstruction of the Old Athens Cemetery and surrounding buildings (a); and detail of a 3D model of a box tomb created from ground photography and integrated into the model (b).

9.6 Monitoring Hurricane Damage to Vegetation and Shorelines of a Georgia Coastal Barrier Island

Due to their high intensity winds, hurricane impacts on coastal areas often include drastic alterations of physical and chemical environmental conditions that sustain key coastal ecosystems. Strong winds can totally or partially remove vegetation (e.g., uproot trees and/ or break trunks and limbs), change salinity of water available to root systems, increase exposure to salt spray, and change the spatial distribution of sand dunes and associated biota. These impacts are particularly critical considering that climate projections indicate hurricanes, such as those that have affected the United States in recent years, may become more frequent and intense, resulting from the warming of ocean waters and the atmosphere. Thus, understanding vegetation responses to these extreme events is required to understand how extreme events linked to future climate scenarios may impact important ecosystems. This section describes sUAS-based strategies to assess environmental damage caused by extremes events, including damage to vegetation and shoreline erosion.

9.6.1 Considerations about sUAS Location and Data Acquisition

Flight campaigns for hurricane-damage assessment often need to follow specific protocols and restrictions on when flights can be conducted and the location of flights. Unless involved with rescue or other official operations, sUAS should not be flown during and/ or immediately after disasters. In addition, flying over areas affected by hurricanes may require special authorization from agencies or organizations that have jurisdiction over the area. For parks and other conservation areas, flights can be restricted and specific approval may be required.

Flying over coastal areas requires additional caution, as manned aircrafts, for instance the Coast Guard, are frequently involved in a variety of operations during and after hurricanes and may fly at low altitudes. When that is the case, incorporating one or more spotters and radio communication into the data acquisition operation is highly advised. Concerns regarding undesired interactions with wildlife and potential associated impacts also need to be taken into consideration when planning image acquisition campaigns. In addition, precautions need to be taken when there is a possibility that the sUAS will fly over people who are not participating in the data acquisition operation as they are not protected from falling UAS by being located under a covered structure or inside a stationary vehicle. Further, if areas to be flown are located under controlled airspace or near airports, extra consideration needs to be given to potential limitations to flights due to proximity to runways and routes used by manned aircrafts. In the United States, the UAS Facility Maps map portal can be used to identify maximum allowed flight heights above ground level and also the location of controlled airspace and the need for flight permission request (www.faa.gov/uas/commercial_operators/uas_facility_maps/).

9.6.2 Considerations about sUAS Flight Parameters

Multiple possibilities exist for data acquisition in support of vegetation assessment and shoreline erosion, including the acquisition of RGB images, videos, multispectral and hyperspectral remote sensing, use of LiDAR, radar, or other technologies. Here, we emphasize the use of RGB images/video and the collection of multispectral images by sUAS. Images collected in the visible region of the electromagnetic spectrum can be used to create wall-to-wall mosaics of areas affected by the extreme event and are particularly useful when analyzed in combination with pre-event imagery, for change detection and damage assessment. Periodical flights following the extreme event can also be incorporated into operations, in order to assist in the monitoring of recovery efforts. Acquisition may involve flying large areas at maximum allowed flight height using planned missions and automatic flight capabilities embedded into many sUAS missions. Real time monitoring of the photo/video stream from the sUAS during flight operations and FPV capabilities can be used with initial assessment flights at higher altitudes to support the identification of areas to be further detailed. Collection campaigns can then involve lower flights for detailed assessment.

Because vegetation damage may be associated with leaf/limb loss, toppled trees, or leaf damage, time may play an important role in increasing contrast between healthy and damaged plants, as broken limbs and toppled trees stay green for a period of time following initial damage. Similarly, time should be considered when investigating vegetation browning due to salt spray and saltwater reaching the root system as a result of storm surge associated with hurricanes. Due to tidal variations, shoreline erosion may be more easily identified when periods of similar tide level are considered. In addition to the generation of image mosaics, SfM and 3D reconstruction can be used to represent vegetation canopies and other surfaces over coastal areas. Using 3D canopy models, changes in vegetation structure and canopy gaps can then be identified. Similar strategies can be used to generate a 3D model of dunes and shorelines, in support of volumetric analyses of total material removed or added during an extreme event.

Multispectral image collection and analyses support the description of status and processes linked to natural and man-made targets. Descriptions include the association of spectral properties of targets to their physical and chemical properties. Multispectral analyses usually involve the manipulation of images using a numerical or quantitative approach. For that, multispectral cameras should be spectrally calibrated and procedures involving data collection should consider not only radiation reflected by targets, but also the amount of incoming radiation arriving from the illumination source. A multispectral system can then incorporate a downwelling radiation sensor for simultaneous measurement of incoming radiation for each spectral band considered during data collection. Measured incoming radiation is time-stamped and can be associated with individual images, providing required information during system calibration and computation of physical descriptors of spectral behavior, such as percentage of incoming energy that is reflected by the target (reflectance). Synchronization between image acquisition and the measuring of incoming solar radiation allows for target comparisons under different illumination conditions and should be used when available, particularly when illumination conditions may vary during a mission or among missions. When a downwelling radiation sensor is not available, a spectrally characterized calibration panel can be used. Operations involving calibration panels require one or more images of the panel to be acquired using the same multispectral sensor flown by the sUAS. To account for variability in illumination between the beginning and end of the flight, pre- and post-flight images are acquired.

Following image acquisition and processing, a variety of metrics can be computed and analyzed in conjunction with other data collected in the field. Multispectral systems may have a few or many spectral bands. A variety of works describing vegetation have incorporated systems that are sensitive to the visible part of the electromagnetic spectrum (red, green, and blue spectral bands) and include bands involving longer wavelengths, such as red-edge and near-infrared (Jensen 2007). Observations in the red-edge and near-infrared regions of the electromagnetic spectrum provide insights into plant stress and structure and a variety of indices can be derived from these bands, including those descriptive of vegetation status, such as the Normalized Difference Vegetation Index (NDVI) = (near infrared band – red band)/(near infrared band + red band). Values of NDVI range from –1 to +1 and green, healthy vegetation tends to show high NDVI values. Individual bands or derived indices can be used in the creation of mosaics that show the distribution of vegetation quantities and/or status over an area.

9.6.3 Monitoring Vegetation and Shorelines at Sapelo Island, Georgia

This work used remote sensing tools and techniques to assess impacts resulting from recent hurricanes at Sapelo Island, a barrier island off the coast of Georgia. Analyses included a field campaign conducted in September 2017, which involved flying three sUAS over the island and collecting high-overlap 20-MP RGB images at two spatial resolutions (2.5- and 5-cm pixels) (Figure 9.7a and b).

A 5-band MicaSense RedEdge camera, a downwelling radiation sensor, and calibration panel was used to collect calibrated multispectral images of multiple vegetation types, including healthy vegetation and vegetation affected by browning due to storm surge and salt spray (Figure 9.7c). Drone images covering over 242.8 ha (600 ac) were then analyzed for vegetation status and damage. The work captured drastic alterations in shorelines and spatial distribution of sand dunes as a result of Hurricane Irma, particularly the separation of a new island from Blackbeard Island adjacent to Sapelo Island as a result of the storm surge and a tidal creek (Figure 9.7d). A video time series of satellite imagery created by the UGA Center for Geospatial Research and posted on YouTube shows the dramatic changes in the shoreline following Hurricane Irma's passing on 11 September 2017 (www.youtube.com/watch?v=Q7swZz2MMTo&feature=youtu.be).

9.7 Future Directions in sUAS and SfM Photogrammetry for Cultural and Natural Resources

There is no question that sUAS technology has changed the lives of cultural and natural resource managers. The leap from typically available digital imagery of 1-m spatial resolution to sUAS images of mm- to cm-level pixel size provides details never before seen in standard aerial imagery. Flying only 20 to 100 m above the surface truly allows a bird's eye view and permanent documentation of the current status of structures and landscapes from monument surface cracks to tree leaf shapes and branch arrangements. No longer dependent upon the schedules of national imaging programs, the sUAS is available at a moment's notice and only requires a few minutes of set up and flight planning preparation before the touch of a button sends it on its mission. The total cost of an sUAS is an order of magnitude less than contracting aerial imagery with a photogrammetric firm; a budget of less than US$2,000 is sufficient for a fully functional sUAS that will perform feature identification, documentation, and monitoring.

This chapter has described real-world projects using sUAS and SfM to create 3D models and ultra-high resolution orthomosaics of cultural and natural resources. Although challenges also were included to give the reader an honest picture of the current limitations of these techniques, we fully expect the future of sUAS and SfM to solve these issues and give us even more capabilities beyond our current imagination.

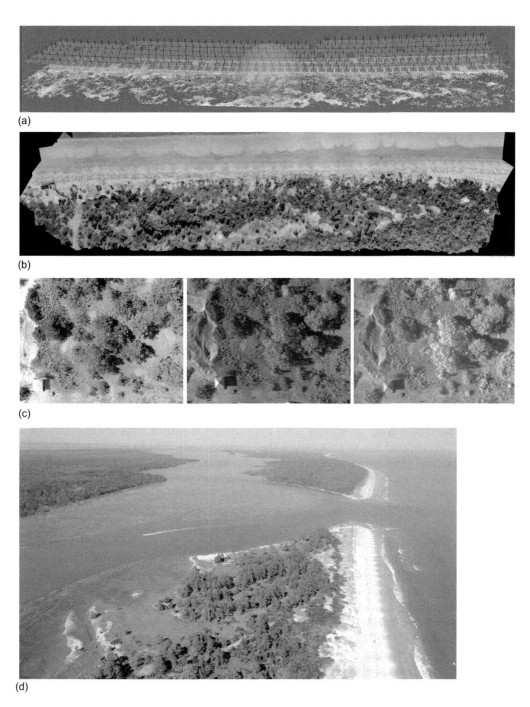

FIGURE 9.7
Vegetation browning and dune removal in Sapelo Island, Georgia. Point cloud and location of photo collection (a) and orthomosaic (b) generated using 367 images. The area affected by vegetation browning in Sapelo Island as depicted by an RGB image (c-left), by a red-edge band (c-center), and by a near-infrared band (c-right). Images were acquired by a 20-MP RGB camera and by a 5-band multispectral system mounted on a small UAS. Vegetation browning along the shoreline of Blackbeard Island imaged with an sUAS (d) shows beach discontinuity, shoreline removal, and new island formation as a result of Hurricane Irma on 11 September 2017.

As the value and application of sUAS in all aspects of daily life become more evident and the demands for expanded flight configurations increase, the FAA has responded with more efficient request reviews and considerations for authorizations and waivers of Part 107 restrictions. On January 14, 2019, U.S. Department of Transportation Secretary, Elaine L. Chao, announced proposed new rules and a pilot project to allow sUAS to fly overnight and under certain conditions even over people without waivers. The proposed Part 107 changes would '...attempt to balance the need to mitigate safety risks without inhibiting technological and operational advances' (FAA 2019). Since 2015, scientists at NASA Ames Research Center have worked on the Unmanned Aircraft Systems Traffic Management or UTM project aimed to manage sUAS flying at low altitude in multi-user airspace (www.nasa.gov/ames/utm). According to NASA, by 2020, 400,000 commercial and two million recreational sUAS are estimated to be flying in U.S. airspace (NASA 2019). In May 2019, the last phase of this project launched test flights of multiple sUAS in cities including Reno, Nevada, and Corpus Christi, Texas to assess enhanced maneuvering among complex obstacles, collision avoidance with moving objects, making safe landings in unfavorable conditions, and two-way communication among multiple sUAS, other aircraft, and ground communications (Tabor 2019).

Recent articles by Cohn et al. (2017) and NASA (2017) agree the future of UAS is commercial use that will open up new markets, promote economic growth, and create new jobs once they are fully integrated with the national airspace system. The UAS technology that is routinely used in military operations will be passed on for civilian use allowing flights of longer duration, heavier payloads, multiple sensors combining optical hyperspectral, thermal and LiDAR data collection, and always, improved reliability and safety. The computing power on-board the sUAS also will transform our ideas about image acquisition and processing. If an application requires a 3D surface model of an historic building, there will be no need to record images by the sUAS sensor, transmit large image data files to the ground, and spend hours processing the imagery with SfM to create the desired 3D model. If the SfM photogrammetry can be performed on-board the sUAS, then only the rendered mesh needs to be transmitted to the user and the data product would be available in near real time. Researchers at the UGA Small Satellite Research Laboratory (SSRL) (www.smallsat.uga.edu/) are designing and constructing a CubeSat nanosatellite to do just this. The Multi-View Onboard Computational Imager (MOCI) mission will acquire imagery of the Earth and perform near real time SfM on-board the CubeSat using custom algorithms and off the shelf, high performance computational units. This proof-of-concept mission will perform data compression, feature detection, and matching, and SfM to produce 3D models (www.smallsat.uga.edu/research). The resulting DSMs are the end products that will be downlinked to the ground receiving station on the roof of the UGA Geography-Geology building. Currently in the construction phase, the CubeSat satellite will be delivered for launch in the second quarter of 2020.

Augmented and virtual reality using content derived from sUAS imagery and 3D models have been and will continue to be used for interpretation of natural and cultural resources. Painstakingly collected ground-based and aerial image data are used to create multi-dimensional, digital maps that look back in time and reveal how ecosystems and cultural systems interacted centuries ago. Improved understanding of human-environment interactions allows trends to be modeled and impacts projected including urban development, storm events, and climate change occurring over time steps of hours, years, decades, or centuries. Integrated advanced technologies provide managers with deeper levels of understanding for natural and cultural resources and inspire creative thinking about resource conservation and preservation. Advancements such as on-board processing, lighter and more efficient batteries, advanced sensors, increased data storage, obstacle avoidance, enhanced communication, and secure data transfer are expected to improve sUAS platforms in the very near future. Continued technology transfer from sUAS military operations to civilian applications, an expanding user base, and constantly improving hardware, software, and workflows guarantee operational use of sUAS to identify, document, and monitor cultural and natural resources.

References

Agapiou, A., V. Lysandrou, K. Themistocleous and D.G. Hadjimitsis, 2016. Risk assessment of cultural heritage sites clusters using satellite imagery and GIS: The case study of Paphos District, Cypress, *Journal of International Society for the Preservation and Mitigation of Natural Hazards*, 1–15, doi:10.1007/s11069-016-2211-6.

Alanen, A.R., R.Z. Melnick and D. Hayden, 2000. *Preserving Cultural Landscapes in America*, John Hopkins University Press, Baltimore, MD, 264 pp.

Bogucki, D.J., G.K. Gruendling and M. Madden, 1980. Remote sensing to monitor water chestnut growth in Lake Champlain. *Journal of Soil and Water Conservation*, 35(2): 79–81.

Brooks, R.R. and D. Johannes, 1990. Photoarchaeology. In T.R. Dudley (Ed.) *Historical, Ethno- and Economy Botany Series Vol. 3*, Dioscorides Press, Portland, OR, 224 pp.

Calvo, K., 2018. *Drones for Conservation: Field Guide for Photographers, Researchers, Conservationists and Archaeologists*, Blurb Publishing, San Francisco, CA, 92 pp.

Cohn, P., A. Green, M. Langstaff and M. Roller, 2017. Commercial drones are here: The future of unmanned aerial systems, McKinsey & Company, Last accessed 5 May 2019. www.mckinsey.com/industries/capital-projects-and-infrastructure/our-insights/commercial-drones-are-here-the-future-of-unmanned-aerial-systems

Cotten, D., T. Jordan, M. Madden and S. Bernardes, 2019. Structure from motion and 3D reconstruction, In Chapter 7, *Image Processing and Analysis Methods*, Section 7, Bernardes, S. and M. Madden (Chapter Eds.), In Morain, S., A. Budge and M. Renslow (Eds.) *Manual of Remote Sensing*, 4th edition, 12 Sections, American Society for Photogrammetry and Remote Sensing, Bethesda, MD, In Press.

Ehlers, M., M. Gahler and R. Janowsky, 2003. Automated analysis of ultra-high resolution remote sensing data for biotope type mapping: New possibilities and challenges, *ISPRS Journal of Photogrammetry and Remote Sensing*, 57(3): 315–326.

El-Hakim, S.F., J.A. Beraldin, M. Picard and G. Godin, 2004. Detailed 3D reconstruction of large-scale heritage sites with integrated techniques, *IEEE Computer Graphics and Applications*, 24(3): 21–29.

Federal Aviation Administration (FAA), 2019. Recent UAS initiatives, Last accessed 9 May 2019. www.faa.gov/uas/programs_partnerships/DOT_initiatives/

Fonstad, M.A., J.T. Dietrich, B.C. Courville, J.L. Jensen and P.E. Carbonneau, 2013. Topographic structure from motion: A new development in photogrammetric measurement, *Earth Surface Processes and Landforms*, 38: 421–430, doi:10.1002/esp.3366.

Forstner, W., 1986. A feature-based correspondence algorithm for image matching, *International Archives of Photogrammetry and Remote Sensing*, 26: 150–166.

Forstner, W. and B.P. Wrobel, 2016. *Photogrammetric Computer Vision: Statistics, Geometry, Orientation and Reconstruction*, Springer International Publishing, Cham, Switzerland, 816 pp.

Fraser, C.C. and S. Cronk, 2009. A hybrid measurement approach for close-range photogrammetry, *ISPRS Journal of Photogrammetry and Remote Sensing*, 64: 328–333.

Goetcheus, C. and T.R. Jordan, 2018. *Comparison of 3D Technologies for Cultural Landscape Documentation and Visualization*, NCPTT Grants 2013, Grant Number P14A00141.

Grun, A., 1985. Adaptive least square correlation: A powerful image matching technique. *South African Journal of Photogrammetry, Remote Sensing and Cartography*, 14(3): 175–187.

Grun, A., 2000. Semi-automated approaches to site recording and modeling. *International Archives of Photogrammetry and Remote Sensing*, 33(5/1): 309–318.

Grun, A., F. Remondino and L. Zhang, 2004. Photogrammetric reconstruction of the Great Buddha of Bamiyan, Afghanistan. *The Photogrammetric Record*, 19(107): 177–199.

Hartley, R. and A. Zisserman, 2003. *Multiple View Geometry in Computer Vision*, Cambridge University Press, Cambridge, UK.

Hirano, A., M. Madden and R. Welch, 2003. Hyperspectral image data for mapping wetland vegetation. *Wetlands*, 23(2): 436–448.

Jackson, B.E., D.J. Evangelista, D.D. Ray and T.L. Hedrick, 2016. 3D for the people: Multi-camera motion capture in the field with consumer-grade cameras and open-source software, *Biology Open*, 5: 1334–1342, doi:10.1242/bio.018713 (The Company of Biologists Ltd).

Jensen, J.R., 2007. *Remote Sensing of the Environment: An Earth resource perspective*, 2nd edition, Pearson Prentice Hall, 592 p.

Katsianis, M., S. Tsipidis, K. Kotsakis and A. Kousoulakou, 2008. A 3D digital workflow for archaeological intra-site research using GIS, *Journal of Archaeological Science*, 35: 655–667, doi:10.1016/j.jas.2007.06.002.

Lourakis, M.I.A. and A.A. Argyros, 2009. SBA: A software package for generic sparse bundle adjustment. *ACM Transactions on Mathematical Software*, 36(1): 1–30, doi:10.1145/1486525.1486527.

Lonnqvist, M.A. and E. Stefanakis, 1999. GIScience in archeology: Ancient human traces in automated space, In M. Madden (Ed.) *Manual of Geographic Information Systems*, American Society for Photogrammetry and Remote Sensing, Bethesda, MD, pp. 1221–1259.

Madden, M., T. Jordan, S. Bernardes, D. Cotten, N. O'Hare and A. Pasqua, 2015. Unmanned aerial systems (UAS) and structure from motion (SfM) revolutionize wetlands mapping, In R. Tiner, M. Lang and V. Klemas (Eds) *Remote Sensing of Wetlands: Applications and Advances*, CRC Press Taylor & Francis Group, Boca Raton, FL, Vol. 10, pp. 195–222.

Madden, M., T. Jordan, M. Kim, H. Allen and B. Xu, 2009. Integrating remote sensing and GIS: From overlays to GEOBIA and geo-visualization, In M. Madden (Ed.) *The Manual of Geographic Information Systems*, American Society for Photogrammetry and Remote Sensing, Bethesda, MD, pp. 701–720.

Map Pilot, 2019. Map Pilot for DJI: Documentation and articles regarding the use of Map Pilot app for iOS devices, Drones Made Easy Support Center, Last accessed 28 May 2019. https://support.dronesmadeeasy.com/hc/en-us/categories/200739936-Map-Pilot-for-DJI

Martin Reynolds, D., 1988. *Monuments and Masterpieces: Histories and Views of Public Sculpture in New York City*, Macmillan Publishing Company, New York.

Maschner, H.D., 1996. Geographic information systems in archaeology, Chapter 1, In M. Aldenderfer and H.D. Maschner (Eds) *Anthropology, Space, and Geographic Information Systems*, Oxford University Press, Oxford, UK, pp. 1–21.

McCoy, M.D. and T.N. Ladefoged, 2009. New developments in the use of spatial technology in archaeology, *Journal of Archaeological Research*, 17(3): 263–295, doi:10.1007/s10814-009-9030-1.

McGlone, J.C. (Ed. in Chief), 2013. *ASPRS Manual of Photogrammetry*, 6th edition, American Society for Photogrammetry and Remote Sensing, Bethesda, MD.

National Aeronautics and Space Administration (NASA), 2017. A civil future for unmanned aircraft systems, NASA, Ames Research Center, Moffett Field, CA, Last accessed 25 May 2019. www.nasa.gov/aeroresearch/programs/iasp/uas/civil-future-for-uas

National Aeronautics and Space Administration (NASA), 2019. Big city life awaits drones in final year of NASA research, NASA, Ames Research Center, Moffett Field, CA, Last accessed 28 May 2019. www.nasa.gov/feature/ames/big-city-life-awaits-drones-in-final-year-of-nasa-research

National Park Service (NPS), 1998. *NPS Director's Order 28: Cultural Resource Management*, NPS, U.S. Department of the Interior, Washington, DC. www.nps.gov/policy/DOrders/DOrder28.html

New York City Landmarks Preservation Commission, 1976. *Designation Report for Soldiers' and Sailors' Monument*. New York City Landmarks Preservation Commission, September 14, 1976, Last accessed 5 March 2019. http://s-media.nyc.gov/agencies/lpc/lp/0932.pdf

Page, R.R., C. Gilbert and S.A. Dolan, 1998. *A Guide to Cultural Landscape Reports: Contents, Process and Techniques*, Department of the Interior, National Park Service, Cultural Resource Division, Washington, DC.

Parry, M., 2017. Buried History, The Chronicle of Higher Education, May 25, 2017, Last accessed 8 May 2019. www.chronicle.com/article/Buried-History/240164?cid=cp117

Remondino, F. and S. El-Hakim, 2006. Image-based 3D modeling: A review. *The Photogrammetric Record*, 21(115): 269–291.

Richards-Rissetto, H., 2017. What can GIS +3D mean for landscape archaeology? *Journal of Archaeological Science*, 84: 10–21.

Snavely, N., 2008. *Scene Reconstruction and Visualization from Interne Photo Collections*, Dissertation Thesis, Computer Science & Engineering, University of Washington, 210 p.

Snavely, N., S.M. Seitz and R. Szeliski, 2008. Modeling the world from internet photo collections. *International Journal of Computer Vision*, 80(2): 189–210.

Tabor, A., 2019. Big city life awaits drones in final year of NASA research, National Aeronautics and Space Administration, Ames Research Center, Moffett Field, CA, Last accessed 25 May 2019. www.nasa.gov/feature/ames/big-city-life-awaits-drones-in-final-year-of-nasa-research

Themistocleous, K. and C. Danezis, 2018. Monitoring cultural heritage sites affected by geo-hazards, *Proceedings of the SPIE Remote Sensing Conference*, Vol. 10790, 1079009, 14 pp., doi: 10.1117/12.2325455.

Theriault, D.H., N.W. Fuller, B.E. Jackson, E. Bluhm, D. Evangelista, Z. Wu, M. Betke and T.L. Hedrick, 2014. A protocol and calibration method for accurate multi-camera field videography. *Journal of Experimental Biology*, 217: 843–1848.

U.S. Fish and Wildlife Service, 1990. *Photo Interpretation Conventions for the National Wetlands Inventory*, St. Petersburgh, FL, 45 pp.

Welch, R., M. Madden and R. Doren, 1999. Mapping the everglades, *Photogrammetric Engineering and Remote Sensing*, 65(2): 163–170.

Welch, R., M. Madden and T. Jordan, 2002. Photogrammetric and GIS techniques for the development of vegetation databases of mountainous areas: Great Smoky Mountains National Park, *ISPRS Journal of Photogrammetry and Remote Sensing*, 57(1–2): 53–68.

Welch, R., M. Remillard and J. Alberts, 1992. Integration of GPS, remote sensing and GIS techniques for coastal resource management. *Photogrammetric Engineering and Remote Sensing*, 58(11): 1571–1578.

Welch, R., M. Remillard and R. Slack, 1988. Remote sensing and geographic information system techniques for aquatic resource evaluation. *Photogrammetric Engineering and Remote Sensing*, 54(2): 177–185.

Westoby, M.J., J. Brasington, N.F. Glasser, M.J. Hambrey and J.M. Reynolds, 2012. 'Structure-from-Motion' photogrammetry: A low-cost, effective tool for geoscience applications, *Geomorphology*, 179: 300–314, doi:10.1016/j.geomorph.2012.08.021.

Wheatly, D. and M. Gillings, 2002. *Spatial Technology and Archaeology: The Archaeological Applications of GIS*, Taylor and Francis, Inc., London, UK, 231 pp.

Wich, S.A. and L.P. Koh, 2018. *Conservation Drones: Mapping and Monitoring Biodiversity*, Oxford University Press, Oxford, UK, +115 pp., doi:10.1093/oso/9780198787617.001.0001.

10

New Insights Offered by UAS for River Monitoring

Salvatore Manfreda, Silvano Fortunato Dal Sasso, Alonso Pizarro, and Flavia Tauro

CONTENTS

10.1 Introduction

Traditional river monitoring approaches are unlikely to provide the level of detail required to advance our understanding and description of the underlying physical processes and mechanisms due to both technical and economical limitations (Manfreda and McCabe 2019). Indeed, our ability to monitor system processes in the face of recent climate and anthropogenic changes is being increasingly compromised by the significant decline in the number of monitoring installations over the last few decades (Shiklomanov et al. 2002). The dynamic nature and inherent variability of many hydrological processes dictates a need for new monitoring technologies and approaches able to increase spatial and temporal resolution of data.

Field measurements still represent the "gold standard" in observational practice, and it is unlikely that anything will supplant the insights that a quality in situ monitoring network can provide. However, a dense monitoring network is available only in few developed countries, while limited economical resources and high maintenance costs prevent the development of a worldwide monitoring network. Currently, new measuring equipment and novel monitoring approaches are recognized as a cornerstone in promoting advancements in environmental monitoring.

Recent technological advances in both satellite and nearer-to-earth platforms (McCabe et al. 2017b) have redefined our capacity to observe and monitor processes through time, and over large spatial domains, in ways that are not possible via ground-based measurement alone. In particular, new CubeSat satellite platforms (McCabe et al. 2017a), unmanned aerial systems (UAS) (Manfreda et al. 2018), and even high-definition video cameras (McCabe et al. 2017b), offer the possibility to monitor the earth system in ways that existing ground-based infrastructure cannot. These observational advances rely, in large part, on technological developments deriving predominantly from the mobile phone and related consumer electronics industry, which has driven sensor miniaturization and relatively low-cost electronics that have enhanced communication, storage, and power-supplies. More specifically, the proliferation of low-cost digital cameras with high-quality sensors and large on-board storage, has enabled a new range of optically-based hydrological monitoring efforts (Tauro et al. 2018b). Indeed, several authors have exploited these technologies using novel image processing algorithms to investigate snow cover detection (Hinkler et al. 2002), derive rainfall intensity, (Kurihata et al. 2005), and measure stream-flow velocity (Dal Sasso et al. 2018).

Optical techniques provide an efficient and non-invasive method for a variety of hydrologic monitoring tasks. One of the most innovative applications of optical sensing from UAS is the use of computer vision approaches to reconstruct three-dimensional surfaces (Turner et al. 2012). The capacity to map both urban and natural landscapes (Flener et al. 2013) and to respond to dynamically changing surface fields, represents a critical advance in hydraulic assessment (Feng et al. 2015). More advanced applications of image and video capture from UAS include flow visualization methods that can yield a spatially distributed estimation of the surface flow velocity field, based on the similarity of image sequences. Proof-of-concept experiments have demonstrated the feasibility of applying these methods to monitor flood events from crowd-sourced imagery (Le Coz et al. 2016), or even to reconstruct velocity fields of natural stream reaches (Tauro et al. 2015).

The emergence of these new observational platforms presents both opportunities and challenges that will need to be addressed by the broader research community. International projects or initiatives have been put in place to support this task. The COST Action HARMONIOUS (www.costharmonious.eu) and the Measurements and Observations in the XXI century (MOXXI) Working Group (WG) of the International Association of Hydrological Sciences (IAHS) aim to advance, among others, our UAS-based monitoring capabilities (Manfreda et al. 2018; Tauro et al. 2018b). In the following sections, an overview of the available UAS-based methods and algorithms for hydraulics studies is given.

10.2 Monitoring Geomorphological Features of Rivers

UASs equipped with GPS and optical cameras are low-cost alternatives to the classical manned aerial photogrammetry in the short- and close-range domain applications (Nex and Remondino 2014; Manfreda et al. 2019). The introduction of a user-friendly photogrammetric technique, called Structure-from-Motion (SfM), has produced a revolution in the field, where any researcher or technician can afford high-resolution topographic reconstruction (Westoby et al. 2012). SfM and Multi-View Stereo (MVS) algorithms allow the creation of digital surface models (DSMs) and orthomosaics without prior information on camera parameters, such as focal length or radial distortion.

The accuracy of SfM-MVS-derived DSMs may be influenced by flight parameters (e.g. elevation above ground level; AGL), flight speed, direction, orientation of the camera, and the camera's focal length), image quality, processing software, the morphology of the studied area, and the type of vehicle (fixed or rotary wing). In this context, ground control points (GCPs) are commonly used to increase the precision of SfM-MVS products, even though their collection is a laborious and time-intensive part of UAS campaigns.

10.2.1 Photogrammetric 3D Reconstruction

3D models can be obtained by images retrieved by UAS processed using commercial or open source software (e.g. Agisoft PhotoScan, Pix4D, nFramesSURE, VisualSFM, Micmac, PMVS/CMVS, Bundler, Apero/MicMac). A typical workflow generally adopted with the software Photoscan includes the following sequence of commands: (1) photo alignment, (2) optimizing alignment, (3) dense cloud building, (4) mesh building, (5) texture building, (6) tiled model building, (7) DSM building, and (8) orthomosaic generation.

An example of 3D model of the Kolubara stream (Belgrade, Serbia) is reported in the following image (Figure 10.1). From this example, it can be observed that SfM algorithms can reproduce topography above water surface with a good level of detail, but the bathymetry below the water surface requires additional efforts (see section 10.2.2).

Strategies to optimize the DSM accuracy have been investigated in the recent work by Manfreda et al. (2019) which provides the following suggestions:

- UAS-derived orthomosaics can produce a planar accuracy of a few centimeters, whereas the vertical accuracy of DSMs is always lower. This is likely due to the fact that most UASs adopt a camera in a zenithal position that provides a more accurate description of planar features. Vertical measurements are generally more complex, but also critical for the description of river morphology.

- The flight plan and camera configuration may significantly impact the overall quality of the resulting DSM. Therefore, it should be planned thoroughly to produce the best depiction of the area of interest. For instance, a transversal survey with respect to the river system may provide better description and quality of the riverbanks.

- The use of a tilted camera can improve the amount of information (retrieved number of points) for inclined surfaces, providing higher DSM elevation accuracy. Tilted camera images increase the robustness of the geometrical model, providing a possible strategy to reduce the total number of GCPs adopted over a given area.

- The use of a double grid flight path may be extremely beneficial to improve the overall quality of the resulting DSM.

- The planar and vertical accuracies can be improved by increasing the number of GCPs. In particular, the quality of the 3D model tends to increase when both the relative planar and vertical distances of the GCPs increase. It is recommended to evenly spread GCPs in space with a minimum number of four/five (see also James et al. 2017).

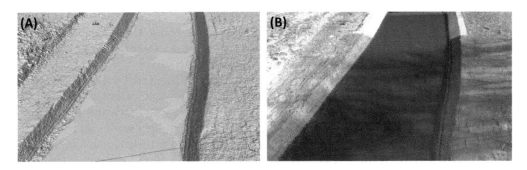

FIGURE 10.1
3D model obtained using an SfM algorithm: mesh model (A) and tiled model (B) derived from a UAS-based survey.

10.2.2 Bathymetric Surveys

Several authors have recently tackled the issue of measuring cross-section bathymetry with UAS. Based on recent works of Woodget et al. (2015) and Detert et al. (2017), SfM-MVS photogrammetry methods can be applied to determine submerged topographies and water level in shallow and clear waters when sufficient texture patches are visible at the bed. Removing refraction and scattering effects in submerged dense point clouds, it is possible to estimate bathymetric data with an accuracy compatible with field measurements. Along the same lines, Langhammer et al. (2017) proposed the use of UAS for monitoring morphological changes of rivers along with characteristics of alluvial depositions using an optical granulometry tool. Finally, Bandini et al. (2018) proposed a methodology to monitor river bathymetry using a tethered single-beam sonar controlled by one UAS platform. This last approach may overcome some of the limitations of optical cameras that can sense only up to a certain water level depth.

10.3 Streamflow Monitoring

Streamflow observations are crucial in water resources management, flood risk assessment, and ecohydrology. Such observations become particularly valuable during extreme flood events, while traditional monitoring approaches (e.g. current meters and acoustic Doppler current profilers) cannot be applied under such adverse conditions. Therefore, non-contact flow measurements approaches, such as image velocimetry, radars, and microwave systems, represent a valuable alternative.

Optical surface flow measurement techniques are becoming particularly popular with the considerable progress of digital recording tools. Images can be acquired from cameras at fixed stations or onboard of UASs, limiting operator risks and allowing remote characterization of river systems and surface flow dynamics. They are preferred instead of other non-contact methods because of their transportability, low costs, and high spatial resolution.

Among optical flow approaches, two main categories can be identified:

- Particle Image Velocimetry (PIV);
- Particle Tracking Velocity (PTV).

These methodologies differ in the observational specifications. Both methods exploit the movement of tracers to derive flow velocities. PIV focuses on the dynamics of a group of particles within a discrete portion of the surface, while PTV adopts a purely Lagrangian approach, whereby the trajectory of each particle detected in the frame is tracked.

The PIV technique estimates the average velocity of particle groups through cross-correlation on two subsequent images. On the other hand, PTV allows the reconstruction of the trajectories of individual particles in the fluid. In both cases, the images show the motion of natural or artificial tracers (or potentially any other pattern or feature), that are transported on the free water surface.

10.3.1 Particle Image Velocimetry (PIV)

PIV has been developed for obtaining the velocity field in laboratory flumes and is based on the analysis of images of a flow, containing tracers, at two different instants in time (t and t′). At each time interval the image is acquired and, based on a cross-correlation algorithm, it is possible to determine the discrete particles displacement and therefore, to estimate the flow velocity field associated to the time interval $\Delta t = t' - t$. The image is discretized using a regular grid with a size set as a function of particle density. In fact, each grid cell, named Interrogation Area (IA), should contain a minimum of three or four particles (Raffel et al. 2018). The pattern of particles within an IA in frame 1 (at time t) is cross-correlated with a Search Area (SA) in frame 2 (at time t+Δt) that may have the same size of IA or may be bigger than IA. The location of the intensity peak in the resulting correlation matrix identifies the new position of the flow pattern in frame 2. This procedure allows the determination of the most probable movement of tracers from frame 1 to frame 2 (see Figure 10.2).

The generally adopted cross-correlation algorithms are: the Direct Cross Correlation (DCC) and the Fast Fourier Transform Correlation (FFT). The main difference between the two is that DCC calculates the correlation matrix in the spatial domain using an interrogation and a search area that can have two different sizes, while FFT calculates the correlation matrix in the frequency domain using areas of identical dimensions.

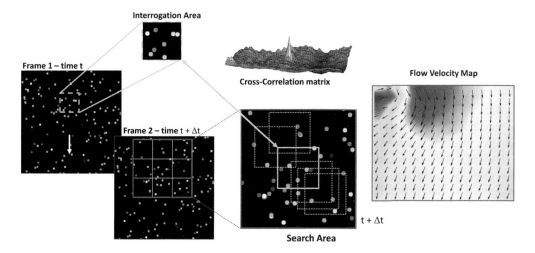

FIGURE 10.2
Schematic description of the PIV algorithm for the estimation of the cross-correlation between the interrogation area and the search area in two consecutive frames separated by time interval Δt. The method is applied on a discrete grid and allows the quantification of the discrete flow velocity map.

Therefore, the FFT reduces computational costs, but it produces some loss of information that may complicate the detection of the correlation peak. Furthermore, the FFT technique allows to identify particles moving at a lower velocity exploiting the displacement measured over multiple frames. Although DCC has been shown to create more accurate results than a standard FFT approach (Huang et al. 1997), it suffers from other disadvantages, such as the fact that due to its linear approach, the method is unable to correlate non-linear particle motion. This problem can be solved by reducing the interrogation window size to a degree that linearizes the particle motion, but this clearly requires a larger computational time.

Regardless of the specific algorithm considered for the analysis, it is worth mentioning that the estimated velocity is recovered from the information of particles and/or features flowing on the water surface. Therefore, seeding density is one of the most relevant parameters in the determination of the velocity field.

Recently, the standard PIV method has been applied at large scales leading to *Large Scale Particle Image Velocimetry (LSPIV)*. LSPIV relies on conventional PIV to remotely monitor surface flow velocities of large-scale hydraulic systems (rivers, wastewater, and irrigation channels), where the field of view is as large as hundreds of squared meters (Fujita et al. 1998). The accuracy of the method is based on the accurate detection and tracking of image patterns and therefore the method assumes that no considerable change occurs in the relative position of particles between the considered consecutive frames. With respect to conventional PIV, LSPIV is able to cover considerably larger areas adopting lower cost cameras. It can be used for the analysis of a wider range of physical phenomena thanks to its adaptability and provides acceptable results, even in the presence of low fluid velocities.

10.3.2 Particle Tracking Velocimetry (PTV)

Particle Tracking Velocimetry (PTV) is a Lagrangian approach that aims at reconstructing the most probable path of every single particle-tracer in a sequence of images. This allows to derive flow velocity measurements even with low tracer density.

A critical step for this method is represented by the particle identification. The phase of particle identification can be addressed through the binarization of the images and identifying a threshold value which allows to separate the background from the particles represented by brighter colors. Thus, the pixels at higher intensity than the threshold will keep their value unaltered and pixels at lower intensities will be assigned a black color.

The procedure just described is called *global threshold*, but other methods also exist, such as i) *Local Threshold* which overcomes the limits of the global approach, varying the value of the threshold within the image depending on the light intensity, and ii) *Otsu's Method* which performs clustering-based image thresholding.

The PTV method identifies the coordinates of the particles from one frame to the next one exploiting the concept of similarity between groups of particles in two consecutive images. It is also possible to exploit a *Multi-Frame* algorithm that uses three or more consecutive frames to solve the problem of correspondences (e.g. *3 Frame in Line Tracking*). Finally, once the particle positions are identified in time, the velocity is estimated by dividing the displacement of particles between consecutive frames by the time interval between the pair of images. Obviously, sampling frequency must be identified properly in order to avoid over or under sampling that may lead to errors.

It is important to highlight that the main characteristics that distinguish PTV from PIV are:

- The possibility of reconstructing the trajectories of the individual particles allowing a Lagrangian description of the velocity field;
- Possibility of determining reliable results, even in the presence of low seeding densities.

Also, it should be noted that in case of LSPIV analysis, velocity vectors are arbitrarily assigned at the center of the analysis grid, where the actual transit of physical objects may not have been observed, while the accuracy of the PTV technique is strongly related to the particle identification, and to the determination of the coordinates of the particle centroid. The velocity vector will be placed at the particle centroid and, therefore, the size of the particles and in turn the particle intensity play a crucial role.

Figure 10.3 shows a schematic diagram of PTV techniques. Velocity is computed by analyzing a pair of frames, in which particles of selected shape and intensity are identified and tracked. The approach aims at reconstructing particle trajectories in the field of view and velocity vectors are computed dividing the displacement by the time interval between consecutive frames.

10.3.3 Optimal Experimental Setup for Surface Flow Velocity

The choice of a proper parameter setting can influence the accuracy of spatial velocity field estimation using both PTV and LSPIV. Critical parameters are: the camera frame rate (that controls the particle displacement between two consecutives images), the seeding density (number of visible tracers per unit of image area), and the number of frames. In this regard, an important source of error is the peak-locking effect, which is a bias error that shifts estimated displacement values towards the closest integer pixel positions, and it is caused by very small particle dimensions (Raffel et al. 2018). Furthermore, the recording frame rate plays a key role, depending on the local flow velocities and physical pixel size (Harpold et al. 2006). Recent studies on PTV performance have shown that frame-by-frame particles displacements should be larger than tracer dimensions in order to allow the motion tracking (Brevis et al. 2011; Tauro et al. 2017).

PTV Approach

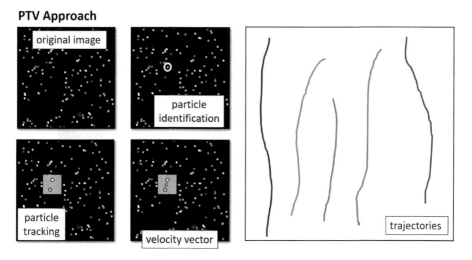

FIGURE 10.3
Schematic description of PTV technique for surface flow velocity estimation based on particle tracking.

To better understand the optimal setup for image-based techniques, Dal Sasso et al. (2018) adopted numerical simulations to reproduce realistic configurations of randomly distributed tracers on a uniform flux. The same approach will be here used to compare the sensitivity of PIV and PTV with respect to particle displacement, seeding density, and number of frames.

A realization of the numerical simulation carried out following the procedure introduced by Dal Sasso et al. (2018) is given in Figure 10.4A. After adjusting image contrast and saturation it becomes the one reported in Figure 10.4B (for more datails, see section 10.5.1). Simulation can reproduce tracers with different properties in terms of particle sizes, brightness, seeding density, and level of aggregation. Analyses are executed on a sequence of 20 frames and results are computed on a smaller square inset of 300×300 pixels in order to avoid border effects.

Numerical simulations allow to test sensitivity of PTV and PIV with respect to the flow velocity that is here described by the dimensionless ratio between the particle frame-by-frame displacement (Dx) and particle diameter (Dxp). Results of several numerical simulations are summarized in Figure 10.5, where errors associated to the use of PTV are described on the first row (Figure 10.5A) and those of PIV are subdivided in three different parametrizations, given in the subsequent rows, used to explore the impact of the Interrogation Area (IA) and Search Area (SA) (see Figure 10.5B, C, and D).

Numerical simulation demonstrated that PTV provides relatively stable results with respect to particle displacement (flow velocity), with errors fluctuating from –2% to 1%. On the other hand, PIV is much more sensitive to this parameter especially when flow velocity tends to extreme values. In particular, the algorithm displays a higher degree of variability of the performances for the lower values of flow velocity and it becomes inapplicable when the displacement of the correlation peak follows outside the SA. Therefore, it may be beneficial to have a preliminary estimate of the frame-by-frame displacement of the patterns to properly set the extent of IA and SA.

Figure 10.5 shows the relative errors of PIV calculated for different displacements using the following configurations: SA = 32×32 px, and IA = 16×16 px (Figure 10.5B), SA = 64×64 px, and IA = 32×32 px (Figure 10.5C), SA = 128×128 px, and IA = 64×64 px (Figure 10.5D). The graphs show that PIV is strongly influenced by the flow velocity and its accuracy tends to increase with the extent of the terms IA and SA. In particular, IA defines the resolution

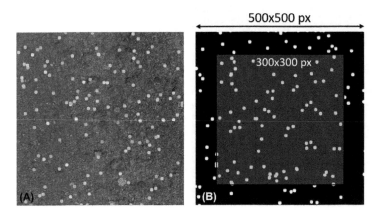

FIGURE 10.4
Numerical simulation of tracers on a realistic background (A) and enhanced image with the area of interest identified by a grey square of 300×300 pixels (B).

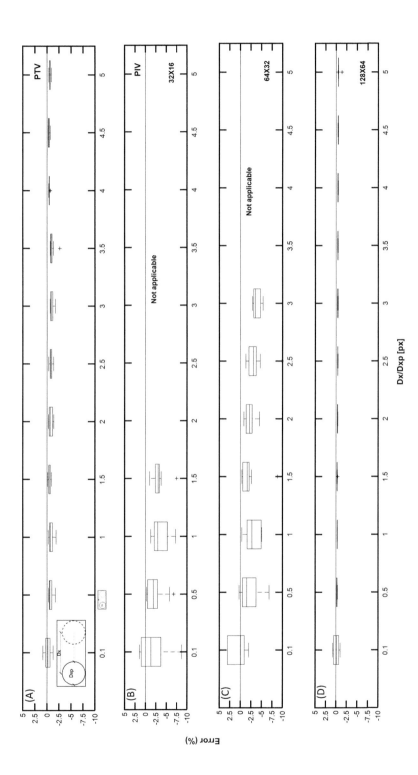

FIGURE 10.5

(A) Box-plot of the relative errors of PTV calculated by numerical simulation with different ratio of particle frame-by-frame displacement to particle diameter. Box-plot of the relative errors of PIV calculated by numerical simulation with different Dx/Dxp on a window size of 32×16 px (B), 64×32 px (C), 128×64 px (D), and Dxp = 10 px.

of the resulting flow velocity map that should be assigned searching for a compromise between accuracy of the estimate and required level of detail.

The influence of seeding densities is described in Figure 10.6 that provides the relative velocity errors computed with the PTV algorithm. The analyses are carried out using 20 images assuming a seeding density ranging from 4×10^{-5} to 1.02×10^{-2} particles per pixel (ppp). Results show that at low and medium seeding densities, i.e. between 4×10^{-5} and 2.2×10^{-4} ppp, there is higher variability in terms of relative velocity errors that tends to be rapidly reduced when tracers density exceeds 2.2×10^{-4} ppp. Nevertheless, some problem may arise with high density of particles in natural conditions due to aggregation of tracers (Kähler et al. 2012a; 2012b).

By contrast, the PIV displays significantly higher sensitivity to particle density with respect to PTV especially under low particle density conditions. Figure 10.6 shows the relative velocity errors for the different densities adopting different window sizes. Results suggest that a possible strategy to improve PIV performances, with low tracers density, is to increase the dimension of the interrogation area (compare lower and upper panel of Figure 10.6).

This result confirms that PIV should be used only with higher tracers density. In fact, it is well known that PIV requires more than three or four particles for interrogation window in order to reach a significant error reduction (Raffel et al. 2018). Of course, larger interrogation windows increase the number of tracers per interrogation area, but this leads also to a less refined flow velocity description not useful when small-scale turbulent flow structures must be detected. Therefore, it is recommended the use of PTV when tracers density becomes a limiting factor for the use of PIV.

The results reported so far tend to emphasize the performances of PTV that is less sensitive to the flux velocity and tracers density, but it must be clarified that it may provide an incomplete characterization of the flow velocity field when there is low seeding density or it is not uniformly distributed. The negative influence of low seeding densities can be reduced by increasing the number of frames (see e.g. Kähler et al. 2006; Dal Sasso et al. 2018). Figure 10.7 describes the minimum number of frames needed to reach a complete characterization of the 2D velocity field as function of the tracers density for different particle displacements. It can be observed that the lack of seeding density can be compensated with an increase of the number of frames that becomes significant for low tracer displacements.

Increasing the number of frames can be beneficial also for the PIV technique. In particular, the influence of the number of frames on the relative velocity error using the PTV and PIV techniques is described in Figure 10.8, adopting a number of frames ranging from 100 to 600. A medium seeding density condition (4.2×10^{-4} ppp) was considered in order to test their capability to reduce the errors. PIV analyses were performed using a search area of 128×128 px. Results show that increasing the number of frames provides beneficial effects to both techniques, but variability of the observed errors of PIV seems to be reduced significantly moving from the use of 100 up to 600 frames, while average errors of PTV seem slightly higher in respect to PIV, with lower variability.

In general, PTV and PIV provide two different strategies for measuring streamflow velocity. The analysis based on numerical simulations displays that the performances of the PTV algorithm are less sensitive to the investigated parameters. Both approaches improve their performances with the increase of tracers density, and number of frames adopted. With low seedings densities, PTV performs better than PIV, but in order to reach a full description of the flow velocity field it requires a larger number of frames. Therefore, PTV can provide a description of the flow velocity field with respect to PIV, but this comes

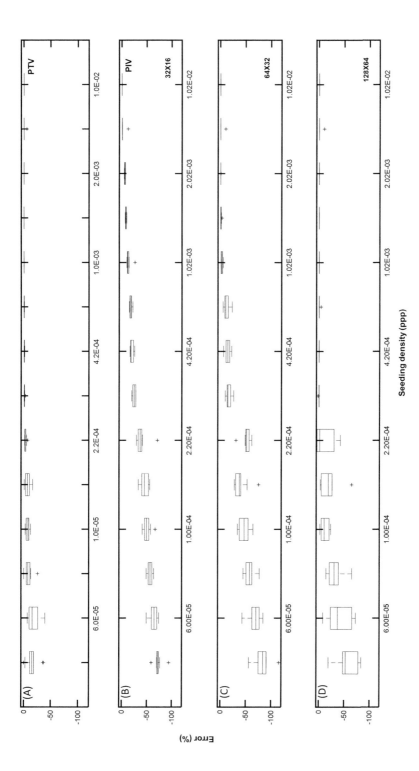

FIGURE 10.6
Relative velocity errors computed by PTV (A) and PIV (B, C, and D) as a function of seeding density using 20 frames. Panel B, C, and D refers to different configurations of PIV in terms of search and interrogation area.

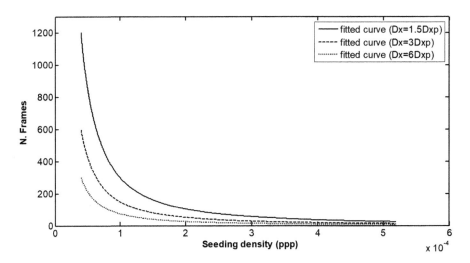

FIGURE 10.7
Fitting curves extrapolated to estimate the number of frames required for optimal image processing in function of particle seeding densities using PTV.

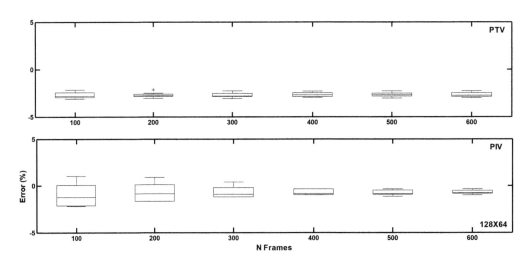

FIGURE 10.8
Box-plot of the relative error obtained for velocity measurements for the different number of frames with PTV and PIV techniques.

at the expense of number of parameters adopted and computational time. On the other hand, it is possible to improve the PIV performances increasing the extent of the IA, but reducing the level of details of the measurements. Therefore, the choice between the two must be done by searching for the right compromise among computational cost, level of details required, and environmental conditions.

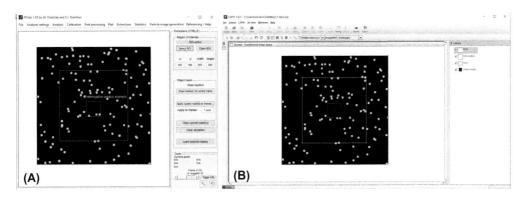

FIGURE 10.9
User interface of two different software for PIV analysis: PIVlab (A) and Fudaa-PIV (B).

10.4 Software Available

Different open source software have been developed during the last number of years for both PIV and PTV analysis, such as OpenPIV (Taylor et al. 2010), PIVlab (Thielicke and Stamhuis 2014), OpenPTV (Liberzon and Meller 2013), and PTVlab (Brevis et al. 2011). Furthermore, practical and user-friendly toolboxes have been developed for flow discharge calculation [Fudaa-LSPIV: Le Coz et al. (2014); RIVeR: Patalano et al. (2017)]. Figure 10.9 presents the user-friendly interface of two software, namely PIVlab and Fudaa-PIV.

At the same time, different smartphone applications have also been developed for measuring surface flow velocity distribution using the sensors embedded in smartphones. For example, Discharge App (Photrack AG Company) is a user-friendly platform dedicated to streamflow measurements for smartphones. The android App uses a non-intrusive optical approach to measure discharge in natural rivers, irrigation, and wastewater channels. The measurements are then synchronized to a remote server for visualization and data management. Measurements rely on five seconds of movie recorded using the smartphone camera. Based on the recording, discharge app detects water levels and estimates surface flow velocities. The streamflow is then computed considering the recovered information and the bathymetry provided by the user.

10.5 Field Campaigns Guidelines

All field campaigns can lead to specific problems that are sometimes unpredictable, but it is possible to trace some guidelines based on our experience to reduce potential errors and improve the final result of our monitoring campaign. We identified potential guidelines to avoid common errors:

1. The choice of the monitoring site can be performed according to the ISO 748/1997 limiting the presence of wind, shadows, light reflections, or flow disturbances on free surface.

2. Adopt UAS digital camera with non-wide-angle lens to avoid image distortion or consider the need for image rectification if wide-angle lens is needed.

3. Define above ground level (AGL) for the drone ensuring that the field of view of the camera includes the entire width of the cross-section or plan the flight in order to cover the cross section in different instants.

4. Define camera resolution such that the tracers occupy at least one pixel.

5. Set camera at nadir position in order to reduce perspective distortion. It may be useful to use a tilted camera to retrieve information on larger rivers.

6. Set a frame rate (e.g. from 24 to 60 fps) proportional to the flow velocity. A sub-sampling may be necessary for slow flows to reduce the computational time.

7. Even if few seconds of video recording may be enough for the image processing, it is wise to record a video of a few minutes (2–4) in order to being able to select the best temporal window in post-processing with good seeding distribution and stable images.

8. Ensure that the field of view of the camera includes a minimum of four fixed GCPs for the image orthorectification and features of known dimensions for geometric calibration of the images.

9. Ensure the presence of visible natural patterns on the water surface in order to track their displacement in time.

10. If necessary, add tracers artificially to overcome the lack of natural flow patterns (usually during low flow conditions) and to improve the performances of image analysis.

11. Use environmentally friendly tracers with diverse color or temperature that increase the contrast against the image background (e.g. wood chips, charcoal).

12. Add tracers on the water surface ensuring that their spatial distribution is uniform across the entire field of view to obtain a full velocity map limiting the effects of particle aggregations.

In some cases, sensing alternative ranges of wavelengths can be highly beneficial to improve image quality. For instance, thermal imaging cameras can considerably help to decrease image noise and highlight the presence of features on the water surface (Tauro and Grimaldi 2017).

10.5.1 Data Processing

Pre-processing is frequently needed to enhance image quality for the subsequent phases. This enhancement is carried out with the orthorectification and stabilization of images as well as with the contrast and brightness correction. Figure 10.10 shows the workflow of surface flow estimation using optical methods and the description of each phase is described as follows:

- *Image recording*: UAS offers the great advantage of reaching remote and inaccessible locations. Video should be recorded under the best possible configuration following the guidelines for data acquisition with UAS discussed in the previous section, but it must be stated that image velocimetry based on fixed installation is also widely utilized.

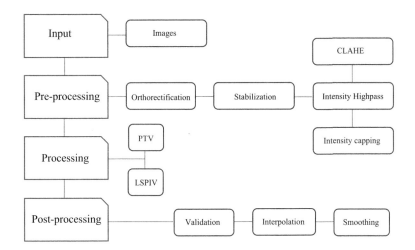

FIGURE 10.10
Workflow of surface flow estimation using optical methods.

- *Pre-Processing*: This requires the following operations: orthorectification, stabilization, and enhancement of images (Contrast Limited Adaptive Histogram Equalization – CLAHE; Intensity Highpass; Intensity Capping). Image orthorectification removes the effects of camera orientation towards the development of planimetrically correct images. Particle displacement should be estimated only on orthorectified images since distances are not distorted from camera perspective and tilt. This phase is crucial and can be accomplished by utilizing previously surveyed ground control points (GCPs) in the camera field of view. Alternatively, orthorectification can be circumvented if the camera optical axis is aligned along the vertical. In this case, distortions are only due to camera lens effects, and they can be accounted for through a correction based on the lens parameters. Sometimes, lens distortion removal software kits are available for selected camera brands (see, for instance, GoPros) or can be implemented after a preliminary camera geometric calibration in laboratory conditions. As the distance among GCPs is dimensional, orthorectification also allows for assigning metric dimensions to pixels (image photometric calibration). If the camera axis is orthogonal to the water surface and orthorectification is not carried out, photometric calibration can be executed by surveying objects visible in the field of view or by creating reference points at known distance through a system of lasers (Tauro et al. 2014). Despite the enhanced performance of platforms currently available in the market, UAS stationary and hovering imaging abilities are limited and it is difficult to obtain experimental image sequences that depict consistent fields of view for several minutes of footage. However, stabilizing images is fundamental for accurate image processing and it is mandatory in case of low video acquisition frequencies. Several commercial software enable video stabilization; alternative custom-built approaches entail the identification of common features in images and the correction of image coordinates through geometric transformation, such as the affine and projective corrections (Tauro et al. 2018a).

  A number of image enhancement operations is also frequently performed on image sequences before velocimetry processing. The objective of this phase is to improve the appearance of features transiting in the field of view with respect to the background, thus facilitating particle detection and/or motion tracking against noise due to reflections or stationary objects. Among such enhancement operations, image histogram equalization increases image contrast by spreading the original histogram to all possible intensity values. High-pass filters help sharpening the image by emphasizing fine details at higher intensity. Intensity capping imposes a maximum threshold on the grayscale intensity of images, thus limiting the presence of bright spots eventually due to noise. Several additional image filtering procedures are possible through convolution with ad hoc developed templates. These should follow preliminary inspection of image sequences by expert users and be tailored to the specific image settings.

- *Processing and Post-Processing*: Only after these preliminary steps, it is possible to proceed with the following steps regarding the application of specific algorithms (PTV, LSPIV, other algorithms) and the Post-Processing that include: 1) validation (Velocity limits, Standard Deviation filter); 2) interpolation; and 3) smoothing of the velocity maps.

10.5.2 Image Velocimetry: Examples of Applications

Several application examples of streamflow estimation using image processing techniques are reported in the literature in different flow conditions (flood, flash flood, debris flow, shallow flows) and configurations (videos captured from fixed camera, UAS, or hand-held camera). The use of these techniques is particularly attractive because people record videos in real-time using their devices and share on the web critical information on hydrological events (e.g. Le Boursicaud et al. 2016).

Figure 10.12 (A and B) show the images extracted by real-time videos recorded by a mobile phone in the Raganello stream at Civita in Calabria (Italy) during a recent flood event. Surface velocity calculation was conducted using the open source software FUDAA-LSPIV (Le Coz et al. 2014). For this purpose, the videos were stabilized using Google tools to limit the camera movements in random directions (Grundmann et al. 2011). The calibration of the images extracted from videos was performed by recognizing relative positioning of the free-surface with stationary elements that are visible in orthophotos. Results show the spatial distribution of velocity vectors along the reach (Figure 10.11B).

10.5.3 Examples of Stream Images and Their Limitations

Image processing techniques in hydrological monitoring may suffer some difficulties due to environmental factors that can deteriorate the image quality during acquisition (Muste et al. 2008; Le Coz et al. 2010). Poor illumination, sunlight reflections, glare, and shadows on the flow surface can introduce some noise resulting in poor results. Image-enhancement methods and trajectory-based image filtering can remove some of these errors. Few examples of such problems are given in Figure 10.12.

Seeding distribution and characteristics also represents a critical factor for LSPIV and PTV. Both techniques require visible patterns on the water surface in order to track their movement and measure velocities. In some cases, natural patterns on free surface like wave crests, vortices, bubbles, natural debris provide sufficient amount of information for image processing (see e.g. Figure 10.13A and B). High seeding densities can be rarely encountered in natural environments and, especially during low flow conditions.

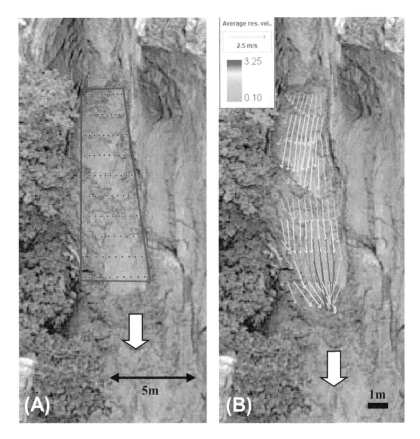

FIGURE 10.11
(A) View of the flow during the flood in the Raganello stream with the sampling grid for the application of cross-correlation between image pairs. (B) Mean surface velocity vectors (m/s) computed with LSPIV technique during the flood event.

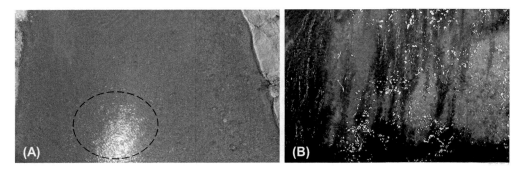

FIGURE 10.12
Examples of environmental factors affecting image quality: (A) reflections and glare (Noce River, Italy); (B) natural shadows on water surface (Brenta River, Italy).

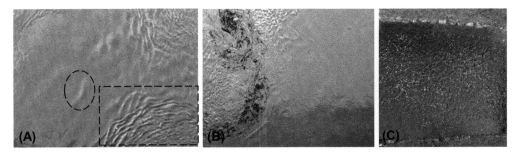

FIGURE 10.13
Examples of moving structures on the water surface: (A) wave crests and vortices; (B) natural seeding during flood (Tiber River, Italy); (C) artificial seeding (Murge River, Switzerland).

Therefore, artificial tracers are often required to increase the surface seeding in the entire field of view (see e.g. Figure 10.13C). Difficulties may arise from the non-uniform distribution of tracers that some time tend to agglomerate on a portion of the investigated domain, thus preventing the possibility to reach a full description of the velocity field (Tauro et al. 2017).

10.6 Recent Advances in Image Velocimetry

More recently, new algorithms have been developed to improve the accuracy of LSPIV and PTV techniques in complex and challenging environmental conditions. Some of these methods and their advantages are introduced in the following.

10.6.1 PTV-Stream

Traditional PTV relies on the identification and tracking of objects of well-known shape (typically circular). This constraint poses several challenges in natural settings, where tracers density and occurrence are difficult to control. While artificial seeding may partially alleviate this issue, it is impractical to manually deploy buoyant objects on the surface of large-scale streams or in areas that are difficult to reach by trained personnel. Also, adverse light conditions may negatively affect the appearance of artificial objects, thus hampering their identification.

PTV-Stream is a versatile alternative to cross-correlation-based PTV that affords the identification and tracking of features of any shape transiting in the field of view (Tauro et al. 2019). Specifically, the scheme is a nearest-neighbor algorithm that allows for reconstructing and filtering the trajectories of features that are more likely to pertain to actual objects transiting in the field of view rather than to water reflections. PTV-Stream revolves around three main phases: particle recognition, particle tracking, and velocity estimation. The particle recognition phase is an appearance-based method that detects objects based on their luminance without searching for specific geometric shapes. Object detection is directly conducted on RGB images by identifying objects whose luminance is higher than a user-defined value and whose area is within an a priori defined range. Once a tracer is identified, the pixel coordinates of its centroid and its area are computed. Then, by setting a simple set of conditions on the distance between tracers identified in successive images, the association between parent and child particle tracers is executed. Once the tracking has been implemented, parent and child tracer coordinates are saved and trajectories can be reconstructed. A filtering procedure is then implemented on the tracer trajectories as in Tauro et al. (2017). Specifically, the filtering scheme aims at identifying trajectories that cover a significant region of the field of view and that minimally deviate from the main flow direction. Finally, the trajectory-based velocity is determined by first computing instantaneous velocities along the same trajectory and then calculating their average.

10.6.2 Feature-Based Velocimetry

The occurrence of visible and evenly distributed tracers on the surface of water streams is rare and related to high flow conditions. However, streamflow measurement techniques should be effective also in case of low and moderate flow regimes. To address this issue, feature-based velocimetry was recently implemented, allowing detection and matching of features rather than physical buoyant objects. Features are defined as scale-invariant key points at regions with high gradient of intensity, such as corners and junctions.

The Feature Tracking Velocimetry (FTV) approach (Cao et al. 2018) combines automatic feature detection through machine learning algorithms (e.g. Scale Invariant Feature Transform, SIFT; Speed Up Robust Feature, SURF; Robust Independent Elementary Features, BRIEF) and matches features by the RANdom SAmple Consensus method (RANSAC).

Perks et al. (2016) implemented the tracking through the sparse first-order differential pyramidal Lucas-Kanade algorithm (Tomasi and Kanade 1991) to investigate high-velocity fields during a flash flood event that occurred in Scotland. The same tracking algoritm is adopted by the recently introduced Optical Tracking Velocimetry (OTV) approach (Tauro et al. 2018c) that adopts many other features detection algorithms (e.g. Features from Accelerated Segment Test, FAST; or Oriented FAST and Rotated Binary robust independent elementary features, ORB; among others) and trajectory-based filtering as in Tauro et al. (2017) to estimate the surface flow velocity field of natural streams. Specifically, only longer trajectories in the average direction of flow are further processed. This guarantees that such reliable trajectories can be related to the transit of physical objects, thus minimizing the probability of fake trajectories and allowing for uncertainty estimation.

Feature-based algorithms are promising in case of diverse seeding densities and non-homogeneous seeding (i.e. in terms of shapes, colour and size), where other techniques produce significant errors. These techniques require less input and pre-processing efforts than alternative PTV and PIV methods since the detection and tracking phases are fully automated. These methods are computationally more efficient with respect to PIV and PTV, and user-involvement is much lower than other techniques such as PTV. Remarkably, these techniques are a valid alternative in complex settings, such as scarcely seeded surfaces and in the presence of severe illumination and shadows. For these reasons these methods are particularly suitable for drone-based velocimetry in real-time.

10.6.3 Space-Time Image Velocimetry (STIV)

Space-time Image Velocimetry (STIV) is an LSPIV-inspired technique that utilizes the time- and space-averaged velocity along a line segment set in the streamwise direction (Fujita et al. 2007). The brightness distribution along the search line is used to characterize the image patterns with time and derive the averaged velocity along the specified direction.

This method can be more efficiently applied than LSPIV when there is a less visible texture at the free surface. For example, floods or high-velocity flows involve naturally occurring small-scale waves and ripples at the free surface. At the same time, this technique is computationally efficient and can cover a measurement area larger with respect to other image processing techniques. However, STIV requires predetermined streamwise direction for setting a search line, which is sometimes difficult to perform when the flow direction changes in time, i.e. for recirculation flows.

10.7 The Use of Image Processing Techniques for Hydrodynamic Applications

Flow measurements are of great significance in many engineering applications such as sediment transport modelling, flooding analyses, and dispersion of pollutants in different environmental contexts (rivers, lakes, coastal lines, and reservoirs). Investigations of such applications include monitoring mean flow characteristics that are costly, time consuming, and subject to many practical difficulties especially during extreme events. Image processing techniques represent a real low-cost alternative that can successfully be employed for the complete analysis of large-scale river systems with high spatial and temporal resolutions.

Surface velocities, river bathymetry, and water depth data acquired from UASs can be used to estimate river stream flows, adopting some assumptions (vertical velocity profile, relationship between surface and mean velocities. These technologies can supply hydrometric data needed for understanding floodplain inundation processes (e.g. Wright et al. 2008), constructing reliable flow rating curves (Manfreda 2018), investigating hydraulic properties such as bed roughness (Simeonov et al. 2013), and for calibration and validation purposes in terms of hydrodynamic models (Refice et al. 2014).

Image-based techniques coupled with UASs have the potential of monitoring river systems in real-time conditions during different flow levels and in consequence, they represent a significant method for developing a modern flood forecasting and warning system. In recent years, non-contact flow measurement has increased the ability to monitor extreme events (e.g. debris flows and flash floods) (Theule et al. 2018; Perks et al. 2016).

Furthermore, UAS platforms have the potential to generate remote and distributed hydrometric data over any river system, also under difficult-to-access environments. Accurate topographic data from the riverbed and floodplain areas are essential for simulating flow dynamics and forecasting flood hazards, predicting sediment transport and streambed morphological evolution, and monitoring instream habitats (Bandini et al. 2018). From this point of view, UASs can fill the gap between satellite/aerial imagery and ground-based measurements. Whereas exposed floodplain areas can be directly monitored from aerial surveys, riverbed topography is not directly observable. At the same time, satellite methods have been successfully applied in monitoring water surface elevation in large rivers, but are ineffective for smaller streams due to low spatial resolution.

The development of an accurate and cost-effective image-based system may improve and expand surface water monitoring networks. To date, data availability from in situ monitoring stations is declining worldwide for both political and economic reasons (Tauro et al. 2018b). Thus, most river networks are gauged at relatively few locations. In this context, only small streams are often engaged and therefore, low spatial resolution is ensured. The increase in the geographic extent of stream monitoring would greatly enhance our understanding of our water resources allowing the optimization of river maintenance, flood prediction, and management of costs.

Additionally, image-based techniques provide a suitable method for analyzing surface flows around hydraulic structures (e.g. weirs, bridges). The techniques described in this chapter can be used in many other applications, including the investigation of sedimentation in shallow reservoirs, run-of-river hydropower plants, oil spills barriers, groin fields, river confluence, and sediment flushing in reservoirs (Kantoush et al. 2011).

References

Bandini, Filippo, Daniel Olesen, Jakob Jakobsen, Cecile Marie Margaretha Kittel, Sheng Wang, Monica Garcia, and Peter Bauer-Gottwein. 2018. "Technical Note: Bathymetry Observations of Inland Water Bodies Using a Tethered Single-Beam Sonar Controlled by an Unmanned Aerial Vehicle." *Hydrology and Earth System Sciences*. doi:10.5194/hess-22-4165-2018.

Boursicaud, Raphaël Le, Lionel Pénard, Alexandre Hauet, Fabien Thollet, and Jérôme Le Coz. 2016. "Gauging Extreme Floods on YouTube: Application of LSPIV to Home Movies for the Post-event Determination of Stream Discharges." *Hydrological Processes*. doi:10.1002/hyp.10532.

Brevis, W., Y. Niño, and G. H. Jirka. 2011. "Integrating Cross-Correlation and Relaxation Algorithms for Particle Tracking Velocimetry." *Experiments in Fluids* 50(1): 135–47.

Cao, Liekai, Volker Weitbrecht, Danxun Li, and Martin Detert. 2018. "Feature Tracking Velocimetry Applied to Airborne Measurement Data from Murg Creek." *E3S Web of Conferences*. doi:10.1051/e3sconf/20184005030.

Coz, Jérôme Le, A. Hauet, G. Pierrefeu, G. Dramais, and B. Camenen. 2010. "Performance of Image-Based Velocimetry (LSPIV) Applied to Flash-Flood Discharge Measurements in Mediterranean Rivers." *Journal of Hydrology* 394(1): 42–52.

Coz, Jérôme Le, Magali Jodeau, Alexandre Hauet, Bertrand Marchand, and Raphaël Le Boursicaud. 2014. "Image-Based Velocity and Discharge Measurements in Field and Laboratory River Engineering Studies Using the Free Fudaa-LSPIV Software." In *Proceedings of the International Conference on Fluvial Hydraulics, RIVER FLOW*, 1961–67.

Coz, Jérôme Le, Antoine Patalano, Daniel Collins, Nicolás Federico Guillén, Carlos Marcelo García, Graeme M. Smart, Jochen Bind, Antoine Chiaverini, Raphaël Le Boursicaud, and Guillaume Dramais. 2016. "Crowdsourced Data for Flood Hydrology: Feedback from Recent Citizen Science Projects in Argentina, France and New Zealand." *Journal of Hydrology* 541: 766–77.

Dal Sasso, S. F., A. Pizarro, C. Samela, L. Mita, and S. Manfreda. 2018. "Exploring the Optimal Experimental Setup for Surface Flow Velocity Measurements Using PTV." *Environmental Monitoring and Assessment* 190(8). doi:10.1007/s10661-018-6848-3.

Detert, Martin, Erika D. Johnson, and Volker Weitbrecht. 2017. "Proof-of-concept for Low-cost and Non-contact Synoptic Airborne River Flow Measurements." *International Journal of Remote Sensing*. doi:10.1080/01431161.2017.1294782.

Feng, Quanlong, Jiantao Liu, and Jianhua Gong. 2015. "Urban Flood Mapping Based on Unmanned Aerial Vehicle Remote Sensing and Random Forest Classifier—A Case of Yuyao, China." *Water* 7(4): 1437–55.

Flener, Claude, Matti Vaaja, Anttoni Jaakkola, Anssi Krooks, Harri Kaartinen, Antero Kukko, Elina Kasvi, Hannu Hyyppä, Juha Hyyppä, and Petteri Alho. 2013. "Seamless Mapping of River Channels at High Resolution Using Mobile LiDAR and UAV-Photography." *Remote Sensing* 5(12): 6382–6407.

Fujita, Ichiro, Marian Muste, and Anton Kruger. 1998. "Large-Scale Particle Image Velocimetry for Flow Analysis in Hydraulic Engineering Applications." *Journal of Hydraulic Research* 36(3): 397–414.

Fujita, Ichiro, Hideki Watanabe, and Ryota Tsubaki. 2007. "Development of a Non-intrusive and Efficient Flow Monitoring Technique: The Space-Time Image Velocimetry (STIV)." *International Journal of River Basin Management*. doi:10.1080/15715124.2007.9635310.

Grundmann, Matthias, Vivek Kwatra, and Irfan Essa. 2011. "Auto-Directed Video Stabilization with Robust L1 Optimal Camera Paths." In *Proceedings of the IEEE Computer Society Conference on Computer Vision and Pattern Recognition*. doi:10.1109/CVPR.2011.5995525.

Harpold, A. A., Saied Mostaghimi, Pavlos P. Vlachos, K. Brannan, and T. Dillaha. 2006. "Stream Discharge Measurement Using a Large-Scale Particle Image Velocimetry (LSPIV) Prototype." *Transactions of the ASABE* 49(6): 1791–1805.

Hinkler, Jørgen, Steen Birkelund Pedersen, Morten Rasch, and Birger Ulf Hansen. 2002. "Automatic Snow Cover Monitoring at High Temporal and Spatial Resolution, Using Images Taken by a Standard Digital Camera." *International Journal of Remote Sensing* 23(21): 4669–82.

Huang, H., D. Dabiri, and M. Gharib. 1997. "On Errors of Digital Particle Image Velocimetry." *Measurement Science and Technology*. doi:10.1088/0957-0233/8/12/007.

James, M. R., S. Robson, S. d'Oleire-Oltmanns, and U. Niethammer. 2017. "Optimising UAV Topographic Surveys Processed with Structure-from-Motion: Ground Control Quality, Quantity and Bundle Adjustment." *Geomorphology*. doi:10.1016/j.geomorph.2016.11.021.

Kähler, Christian J., Sven Scharnowski, and Christian Cierpka. 2012a. "On the Uncertainty of Digital PIV and PTV near Walls." *Experiments in Fluids*. doi:10.1007/s00348-012-1307-3.

Kähler, Christian J., Sven Scharnowski, and Christian Cierpka. 2012b. "On the Resolution Limit of Digital Particle Image Velocimetry." *Experiments in Fluids* 52(6): 1629–39.

Kähler, Christian J., U. Scholz, and J. Ortmanns. 2006. "Wall-Shear-Stress and Near-Wall Turbulence Measurements up to Single Pixel Resolution by Means of Long-Distance Micro-PIV." *Experiments in Fluids*. doi:10.1007/s00348-006-0167-0.

Kantoush, Sameh A., Anton J. Schleiss, Tetsuya Sumi, and Mitsuhiro Murasaki. 2011. "LSPIV Implementation for Environmental Flow in Various Laboratory and Field Cases." *Journal of Hydro-Environment Research*. doi:10.1016/j.jher.2011.07.002.

Kurihata, Hiroyuki, Tomokazu Takahashi, Ichiro Ide, Yoshito Mekada, Hiroshi Murase, Yukimasa Tamatsu, and Takayuki Miyahara. 2005. "Rainy Weather Recognition from In-vehicle Camera Images for Driver Assistance." In *IEEE Proceedings. Intelligent Vehicles Symposium, 2005*, pp. 205–10. IEEE.

Langhammer, Jakub, Theodora Lendzioch, Jakub Miřijovský, and Filip Hartvich. 2017. "UAV-Based Optical Granulometry as Tool for Detecting Changes in Structure of Flood Depositions." *Remote Sensing*. doi:10.3390/rs9030240.

Liberzon, A., and Y. Meller. 2013. Particle Tracking Velocimetry Software "OpenPTV." https://openresearchsoftware.metajnl.com/articles/10.5334/jors.101/

Manfreda, Salvatore. 2018. "On the Derivation of Flow Rating-Curves in Data-Scarce Environments." *Journal of Hydrology* 562: 151–54.

Manfreda, Salvatore, Petr Dvorak, Jana Mullerova, Sorin Herban, Pietro Vuono, J. José Arranz Justel, and Matthew Perks. 2019. "Assessing the Accuracy of Digital Surface Models Derived from Optical Imagery Acquired with Unmanned Aerial Systems." *Drones*. doi:10.3390/drones3010015.

Manfreda, Salvatore, and Matthew F. McCabe. 2019. "Emerging Earth Observing Platforms Offer New Insights into Hydrological Processes." *Hydrolink* 1(1): 8–9.

Manfreda, Salvatore, Matthew McCabe, Pauline Miller, Richard Lucas, Victor Pajuelo Madrigal, Giorgos Mallinis, Eyal Ben Dor, David Helman, Lyndon Estes, and Giuseppe Ciraolo. 2018. "On the Use of Unmanned Aerial Systems for Environmental Monitoring." *Remote Sensing* 10(4): 641.

McCabe, Matthew F., B. Aragon, Rasmus Houborg, and J. Mascaro. 2017a. "CubeSats in Hydrology: Ultrahigh-Resolution Insights into Vegetation Dynamics and Terrestrial Evaporation." *Water Resources Research* 53(12): 10017–24.

McCabe, Matthew F., Matthew Rodell, Douglas E. Alsdorf, Diego G. Miralles, Remko Uijlenhoet, Wolfgang Wagner, Arko Lucieer, Rasmus Houborg, Niko E. C. Verhoest, and Trenton E. Franz. 2017b. "The Future of Earth Observation in Hydrology." *Hydrology and Earth System Sciences* 21(7): 3879.

Muste, Marian, I. Fujita, and A. Hauet. 2008. "Large-scale Particle Image Velocimetry for Measurements in Riverine Environments." *Water Resources Research* 44(4). doi:10.1029/2008WR006950.

Nex, Francesco, and Fabio Remondino. 2014. "UAV for 3D Mapping Applications: A Review." *Applied Geomatics*. doi:10.1007/s12518-013-0120-x.

Patalano, Antoine, Carlos Marcelo García, and Andrés Rodríguez. 2017. "Rectification of Image Velocity Results (RIVeR): A Simple and User-Friendly Toolbox for Large Scale Water Surface Particle Image Velocimetry (PIV) and Particle Tracking Velocimetry (PTV)." *Computers & Geosciences* 109 (December): 323–30. doi:10.1016/J.CAGEO.2017.07.009.

Perks, Matthew T., Andrew J. Russell, and Andrew R. G. Large. 2016. "Technical Note: Advances in Flash Flood Monitoring Using Unmanned Aerial Vehicles (UAVs)." *Hydrology and Earth System Sciences*. doi:10.5194/hess-20-4005-2016.

Raffel, Markus, Christian E. Willert, Fulvio Scarano, Christian J. Kähler, Steve T. Wereley, and Jürgen Kompenhans. 2018. *Particle Image Velocimetry: A Practical Guide*. Springer.

Refice, Alberto, Domenico Capolongo, Guido Pasquariello, Annarita Daaddabbo, Fabio Bovenga, Raffaele Nutricato, Francesco P. Lovergine, and Luca Pietranera. 2014. "SAR and InSAR for Flood Monitoring: Examples with COSMO-SkyMed Data." *IEEE Journal of Selected Topics in Applied Earth Observations and Remote Sensing*. doi:10.1109/JSTARS.2014.2305165.

Shiklomanov, Alexander I., Richard B. Lammers, and Ch J. Vörösmarty. 2002. "Widespread Decline in Hydrological Monitoring Threatens Pan-Arctic Research." *Eos, Transactions American Geophysical Union* 83(2): 13–17.

Simeonov, Julian A., K. Todd Holland, Joseph Calantoni, and Steven P. Anderson. 2013. "Calibrating Discharge, Bed Friction, and Datum Bias in Hydraulic Models Using Water Level and Surface Current Observations." *Water Resources Research*. doi:10.1002/2013WR014474.

Tauro, Flavia, and Salvatore Grimaldi. 2017. "Ice Dices for Monitoring Stream Surface Velocity." *Journal of Hydro-Environment Research* 14: 143–49.

Tauro, Flavia, Christopher Pagano, Paul Phamduy, Salvatore Grimaldi, and Maurizio Porfiri. 2015. "Large-Scale Particle Image Velocimetry from an Unmanned Aerial Vehicle." *IEEE/ASME Transactions on Mechatronics* 20(6): 3269–75.

Tauro, Flavia, Andrea Petroselli, and Salvatore Grimaldi. 2018a. "Optical Sensing for Stream Flow Observations: A Review." *Journal of Agricultural Engineering*. doi:10.4081/jae.2018.836.

Tauro, Flavia, R. Piscopia, and S. Grimaldi. 2017. "Streamflow Observations from Cameras: Large-Scale Particle Image Velocimetry or Particle Tracking Velocimetry?" *Water Resources Research* 53(12): 10374–94.

Tauro, Flavia, Rodolfo Piscopia, and Salvatore Grimaldi. 2019. "PTV-Stream: A Simplified Particle Tracking Velocimetry Framework for Stream Surface Flow Monitoring." *Catena*. doi:10.1016/j.catena.2018.09.009.

Tauro, Flavia, Maurizio Porfiri, and Salvatore Grimaldi. 2014. "Orienting the Camera and Firing Lasers to Enhance Large Scale Particle Image Velocimetry for Streamflow Monitoring." *Water Resources Research* 50(9): 7470–83.

Tauro, Flavia, John Selker, Nick Van De Giesen, Tommaso Abrate, Remko Uijlenhoet, Maurizio Porfiri, Salvatore Manfreda, et al. 2018b. "Measurements and Observations in the XXI Century (MOXXI): Innovation and Multi-disciplinarity to Sense the Hydrological Cycle." *Hydrological Sciences Journal*. doi:10.1080/02626667.2017.1420191.

Tauro, Flavia, Fabio Tosi, Stefano Mattoccia, Elena Toth, Rodolfo Piscopia, and Salvatore Grimaldi. 2018c. "Optical Tracking Velocimetry (OTV): Leveraging Optical Flow and Trajectory-Based Filtering for Surface Streamflow Observations." *Remote Sensing*. doi: 10.3390/rs10122010.

Taylor, Zachary J., Roi Gurka, Gregory A. Kopp, and Alex Liberzon. 2010. "Long-Duration Time-Resolved PIV to Study Unsteady Aerodynamics." *IEEE Transactions on Instrumentation and Measurement*. doi:10.1109/TIM.2010.2047149.

Theule, Joshua I., Stefano Crema, Lorenzo Marchi, Marco Cavalli, and Francesco Comiti. 2018. "Exploiting LSPIV to Assess Debris-Flow Velocities in the Field." *Natural Hazards and Earth System Sciences*. doi:10.5194/nhess-18-1-2018.

Thielicke, William, and Eize J. Stamhuis. 2014. "PIVlab – Towards User-Friendly, Affordable and Accurate Digital Particle Image Velocimetry in MATLAB." *Journal of Open Research Software*. doi:10.5334/jors.bl.

Tomasi, Carlo and Takeo Kanade. 1991. *Detection and Tracking of Point Features*, Computer Science Department, Carnegie Mellon University, April, 1991.

Turner, Darren, Arko Lucieer, and Christopher Watson. 2012. "An Automated Technique for Generating Georectified Mosaics from Ultra-High Resolution Unmanned Aerial Vehicle (UAV) Imagery, Based on Structure from Motion (SfM) Point Clouds." *Remote Sensing* 4(5): 1392–1410.

Westoby, M. J., J. Brasington, N. F. Glasser, M. J. Hambrey, and J. M. Reynolds. 2012. "'Structure-from-Motion' Photogrammetry: A Low-Cost, Effective Tool for Geoscience Applications." *Geomorphology*. doi:10.1016/j.geomorph.2012.08.021.

Woodget, A. S., P. E. Carbonneau, F. Visser, and I. P. Maddock. 2015. "Quantifying Submerged Fluvial Topography Using Hyperspatial Resolution UAS Imagery and Structure from Motion Photogrammetry." *Earth Surface Processes and Landforms*. https://doi.org/10.1002/esp.3613.

Wright, N. G., I. Villanueva, P. D. Bates, D. C. Mason, M. D. Wilson, G. Pender, and S. Neelz. 2008. "Case Study of the Use of Remotely Sensed Data for Modeling Flood Inundation on the River Severn, U.K." *Journal of Hydraulic Engineering*. doi:10.1061/(asce)0733-9429(2008)134:5(533).

11

The Campus as a High Spatial Resolution Mapping Laboratory – Small Unmanned Aerial Systems (sUAS) Data Acquisition, Analytics, and Educational Issues

J.B. Sharma, J. Zachary Miller, Brian Duran, and Lance Hundt

CONTENTS

11.1 The Emerging sUAS Applications in Higher Education

The maturing of small Unmanned Aerial Systems (sUAS) in recent years has provided a novel and revolutionary platform for the development of remote sensing based mapping applications. The confluence of advances in fast computing, sensor miniaturization, novel algorithms, and simplified regulations have combined to give powerful mapping capabilities to sUAS platforms. The advent of the Federal Aviation Administration (FAA) Part 107 remote pilot examination has opened the skies for commercial, educational, and recreational applications in the United States (FAA 2016). Analogous events elsewhere in the world are giving rise to regulatory frameworks that are nurturing safe sUAS applications that provide social, environmental, and economic enrichment.

According to the Association for Unmanned Vehicle Systems International (AUVSI 2014; Jenkins and Vasigh 2013) UAS applications will create 100,000 new jobs and have an \$82 billion economic impact in the United States by 2025. Extrapolating this impact over the entire planet, the socio-economic impact of sUAS technology will be significant. This heralds a renaissance in aviation technology that has far reaching implications for the mapping sciences as applied to areas including education, infrastructure inspection, disaster response, agriculture, natural resource management (Wallace et al. 2016; Lucieer et al. 2014), historical preservation, and public safety (Wing et al. 2013). This is a great opportunity for higher education to develop teaching, research, and outreach applications (Carrivick et al. 2013). It is necessary to develop the workforce and the academic ecosystem to nurture the talent for innovation in this emergent field. There are many applications of sUAS technology which require highly honed drone piloting skills like search and rescue, cinematography, firefighting, infrastructure inspections, and reconnaissance; not many sUAS pilots have the knowledge to implement sUAS based mapping projects. This opens up a new economic sector in which highly detailed local sUAS based remote sensing mapping projects can be conducted on a worldwide basis. The development of sUAS data based mapping science applications is still an area of active research and development. The ability of an sUAS based sensor to capture imagery with Ground Sampling Distance (GSD) to the order of sub-decimeters opens up a new realm of hyper-resolution spatial data that requires novel approaches for collection, processing, and development of value added geospatial products. These include land cover maps and Digital Surface Models (DSM) that can be utilized for further geo-spatial modeling and engineering applications.

In particular, the advent of the Structure from Motion (SfM) algorithm developed by the computer vision community has revolutionized the sUAS based applications of the mapping sciences (Westoby et al. 2012; Fonstad et al. 2013; Micheletti et al. 2015). With high overlap imagery, the SfM algorithm can generate highly accurate 2D orthoimage mosaics and 3D models (Verhoeven et al. 2011) that can be georeferenced with a few Ground Control Points (GCPs) (James and Robson 2012; Fonstad et al. 2013). The absolute positional accuracy of the sUAS data is still an active area of research and in many instances, the GSD size collected can be smaller than the uncertainty of the Global Navigation Satellite System (GNSS) collected GCPs. In this case, it is problematic to accurately co-register sUAS data collected in different flights such that viable raster algebra calculations can be performed on these data layers. Advanced Real Time Kinematic (RTK) UAS positioning techniques or Post Processing Kinematic (PPK) technology are needed to achieve sub-centimeter accuracy.

The development of sUAS mapping applications requires a foundational literacy in Science, Technology, Engineering, and Mathematics (STEM). Mapping applications span across a range of disciplines and provide an impetus for integrative thinking in the Academy. Specific domain sUAS applications involve interaction of mapping scientists with content specialists. For example, forestry applications may need input from botanists, field technicians, and forest rangers. Urban planning applications need an interaction with architects, civil engineers, public safety and city managers. This presents both a challenge and an opportunity to higher education to develop novel curricula, programs, and outreach activities that promote the deployment of innovative sUAS applications. The issues involved are regulations, positioning technology, computing infrastructure, sensor characteristics, terrain characteristics, weather, sUAS capabilities, limitations, and maintenance. The subject matter involved includes the fundamentals of algebra, trigonometry, statistics, physics, chemistry, biology, programming, computing, sUAS technology, and geospatial science. This builds bridges between the physical and the social sciences. This trans-disciplinary nature of sUAS applications is commensurate with the blurring of boundaries between the traditional sciences and highlights the integrative nature of this emergent new field.

11.2 Integration into Teaching and Research – The Efficacy of using the Campus as a Laboratory

The rapid emergence of robust sUAS technologies is opening a great teaching, research, and outreach opportunity for academic programs ranging from the university level (Mott 2016) to K12 education. The implementation of sUAS platforms as a teaching and research tool requires the development of a workflow that involves several interlinked issues. These range from understanding FAA regulations, obtaining the sUAS remote pilot license, flight-planning considerations, piloting skills, data indexing and warehousing, data processing and developing value added geospatial products like image orthomosaics, three dimensional (3D) terrain models and land cover maps. These products in turn can be used for geospatial modeling applications including non-point erosion, sedimentation, and thermal audits of buildings. In particular, educational institutions have limited resources; each component of this workflow requires detailed knowledge and must be very carefully developed. This is necessary to co-opt legal liability and the risk of damage or injury in an sUAS crash.

A university or school campus is an ideal laboratory to develop sUAS applications that enable research, curriculum, coursework, and outreach. The campus as a study area is readily accessible to students and there is prior knowledge of the physical terrain. Furthermore, a campus can contain all types of small ecosystems that can be studied and modeled at a high spatial, radiometric, spectral, and temporal resolution by utilizing sUAS platforms. The urbanized and built-up areas of campus can require different types of sensors and flight operations as opposed to natural, agricultural, or forested areas. The ability to make ground observations readily in order to validate information collected using an sUAS is crucial for data interpretation and analytics. Developing sUAS based curricula and programs is an interdisciplinary endeavor that involves more than one subject area. This is a technology that enables scholarly opportunities that span the geo-sciences, engineering, urban planning, architecture, archeology, agriculture, and natural resource management; sUAS applications find a natural home in a university environment.

In the United States, the Federal Aviation Administration (FAA) has developed guidelines for the use of sUAS in higher education (Mott 2016; FAA 2019). The FAA stance does not overly restrict educational activity involving sUAS platforms and recognizes the need to educate a workforce capable of developing advanced applications based on sUAS data. In essence, the college or university instructor involved in supervising sUAS flights for a course or a research project involving students must have the Part 107 sUAS license and be the Pilot in Command (PIC) to supervise all aspects of the flight. The students working under the supervision of the PIC may not necessarily have an sUAS pilot's license as long as the activity is a part of an educational course. The most important part for developing productive sUAS operations is assembling a highly trained team of faculty, students, and staff that ensures compliance of university, county, state, and national regulations.

11.2.1 The Emergent sUAS Operations University Policies

The faculty member as a PIC is responsible for following the FAA regulations and ensuring that safety concerns are addressed rigorously. To this end, in addition to the FAA regulations for sUAS flights, there are state-wide, city, and university sUAS policies that are being formulated and need to be adhered to as well. As educational, commercial, and recreational sUAS flights increase into the future, public safety will become increasingly important (Clarke and Moses 2014). Best practices for college and university sUAS operations policies are rapidly emerging and typically involve Web based portals. The regulatory workflow to seek approval for campus based sUAS operations involves alerting all campus stakeholders including public safety, submitting flight plans for risk assessment and approval, going through an extensive pre and post flight checklist, and the recording of post-flight sUAS logs to complete the entire data collection mission. The flight logs, battery logs, the indexing, and warehousing of the data collected are critical pieces of the workflow that ensure data integrity, public safety, and regulatory compliance. Effective and sustainable drone operations require a clear knowledge of airspace types and associated restrictions, proximity to airports, and regulations that range from the Federal to the local level. The development of campus wide sUAS policy is essential for safe and sustained sUAS operations on university property that support teaching, research, and outreach.

11.2.2 Urban and Environmental Mapping on the Distributed Campuses of UNG

The University of North Georgia (UNG) has five campuses distributed over the region of Northeast Georgia. These campuses are located at Dahlonega, Gainesville, Oconee, Cumming, and Blue Ridge. The primary study area for this project is the UNG-Gainesville Campus (Figure 11.1) which has both the urbanized main campus area and the natural

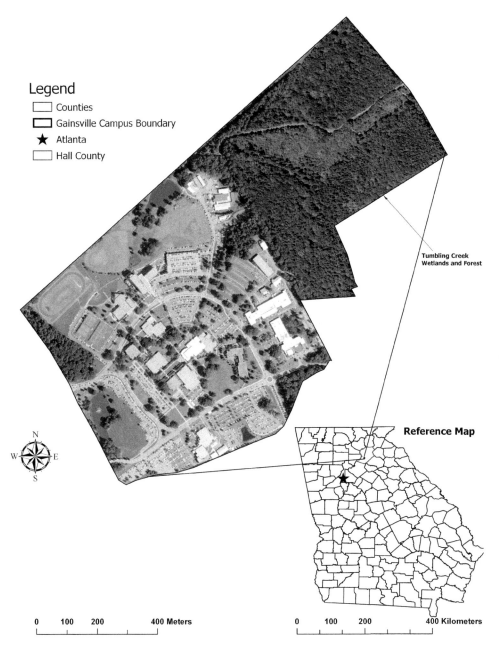

FIGURE 11.1
The University of North Georgia – Gainesville Campus; the Tumbling Creek Forest and Wetland is on the Northern end and the urbanized area is on the Southern end of the campus.

Tumbling Creek Forest and Wetland. The southern portion of the campus has a primarily urban land cover comprising of buildings, roads, parking lots, water bodies, and grass. The forested region north of the main campus is the Tumbling Creek Forest and Wetland. Both of these areas are approximately 0.5 km^2 for a total campus area of about 1.0 km^2. This is the study area for this sUAS based project whose main objective was to incrementally experiment and develop a viable sUAS based 2D and 3D campus mapping project with multispectral and thermal sensors for both urban and natural study areas. The data collected formed the basis of Geographic Object Based Image Analysis (GEOBIA) land cover classification and a thermal modeling application, which are elaborated upon in the Data Analytics section 11.5. This workflow will be scaled up to develop similar sUAS applications for all the UNG campuses in the next phase of this project, beginning with the UNG-Dahlonega campus and the nearby Hurricane Creek Forest Preserve which also provide large study areas that are both urban and natural. These distributed UNG campuses have a variety of land cover types and substantial terrain relief that provide both challenges and opportunities for developing sUAS applications that support teaching, research, outreach, and campus facilities management.

11.2.3 sUAS Flight Planning and Operations Issues

Flight planning involves a confluence of several issues. It is necessary to know exactly what type of FAA airspace designation encompasses the study area and the proximity to airports is particularly important. The UNG-Gainesville Campus is below Class G airspace and is about 3 miles from the Gainesville Airport, which does not have a control tower; since the campus is within 5 miles of this airport, by FAA regulation the Gainesville Airport manager must be informed prior to sUAS operations. This also gives particular importance to line-of-sight observation of sUAS flights and flying higher than 400 feet Above Ground Level (AGL) to facilitate rapid evasion of low flying aircraft, if needed. Campus Public Safety must be informed as they will be the first responders in case of an accident. Emergent best practices are based on university Web based portals that facilitate flight plan logging, approval, pre-flight checks, post-flight checks and log, UAS maintenance, and battery logs. Commercial Web based providers are beginning to provide online capabilities that support all aspects of sUAS operations.

11.2.4 Time Constraints on sUAS Flight Operations

An overcast day with diffused sunlight is perhaps the best for sUAS mapping flights, as shadows are not present. On bright sunlit days, the best flight times for multispectral imaging are from about 10:00 am to 2:00 pm with the shadows being the shortest in this time interval. This is with shadows longer during the winter and shorter during the summer. With limits on the speed and range of the quadcopter sUAS, this constrains the study area to about 0.5 km² for a high spatial resolution 2D and 3D mapping. The flight plan must have a high image overlap in both the forward and sideways directions; this consideration must be balanced with an excessively long flight time required for increasing image overlap. Longer flight times also have implications for battery exchanges and require recharging capability available during flight operations. This will require a gasoline powered generator if an electrical outlet is not available and portability of heavy equipment when needed must be factored into flight operations.

Thermal imaging is best done early right before dawn, at mid-noon, or right after sunset there is a maximum contrast between the ground surface temperature and the ambient temperature; this limits the time window for thermal sensors. It is best to avoid the 'thermal crossover' at dawn and sunset at which time the radiant temperatures of all ground surfaces are indistinguishable. Flight planning is also dependent on the sensor being utilized, its Field of View (FOV) and the height that it is flown; these determine the swath width and GSD of the imagery.

11.2.5 sUAS Flight Control and Staging Area

The staging area for the flights must be chosen to allow for line-of-sight observation of the flight. If terrain considerations do not allow for line-of-sight observation from the flight take-off and landing area, a network of observers in direct telephonic contact must be present in order to give the PIC complete flight awareness in case evasive action is required or if there is a mechanical failure leading to a crash. If needed, the study area can be broken up into a series of flight 'blocks' to ensure that line-of-sight observation is enabled for the entire sUAS operations. The wind speed and direction are important considerations for sUAS return flights for battery exchanges. For study areas as large as 0.5 km^2; if the sUAS has to work against the wind for a battery replacement for about 0.5 km, it is better to interrupt the flight operation at about 30% battery power to ensure a safe return with no less than 15% remaining. There is a greater risk of rapid battery drainage at low power levels leading to a crash, particularly for older batteries. In cold weather battery warmers will be needed for sUAS operations.

A tablet computer is typically needed for flight control. There are several commercial flight software applications or 'Apps' like Pix4DCapture, DJI GO, Atlas-Flight, Drone Deploy, and others that are available for flight-line planning and control. These are critical for effective mapping applications and are available for both iOS and Android based tablets and smartphones. It is important that the tablet computer has a bright screen to allow clarity of observing the software controls and observing the flight telemetry; specialized tablets with bright daylight displays are now available and are invaluable in daylight sUAS operations. Tablets can overheat in hot weather and protective casings must be removed in sUAS flights on warm days. The flight control software applications or 'Apps' must be updated and sUAS and camera firmware must be updated prior to flight.

11.3 sUAS Platforms and Sensors Utilized

There were several sUAS platforms and sensors utilized in this study and incremental experimental flights were conducted to establish the right combination of sUAS platforms and sensors for particular mapping applications. The primary study area was the UNG-Gainesville Campus (Figure 11.1) which has major two-lane highways at its Southern and Eastern boundaries and a railway line running along the entire length of the Northeast boundary. It is a requirement of safety and regulations that the sUAS flights be confined to the limits of the UNG property.

11.3.1 sUAS Platforms Utilized

Both fixed wing and quadcopter sUAS platforms were tested for data acquisition flights in an incremental manner. The eBee Sensefly is a fixed wing sUAS with a wingspan of 96 cm and a mass of 0.69 kg (1.52 lbs). This is a highly capable sUAS and has great safety features. Its foam core construction, a rear-facing propeller, and low mass make for a minimal kinetic energy and mechanical impulse delivered upon collision and is unlikely to incur severe human injury. It can carry a 3-band RGB camera, a 3-band Color IR camera, and a ThermoMap camera for thermal imaging. The eBee is a versatile sUAS that is capable of mapping up to 12 km^2 in a single automated flight. Its disadvantage is that it must overshoot the study area in order to turn and resume the flight lines for mapping. In our specific circumstance, this restricts us to smaller study areas on our campus such that eBee can turn around within the UNG-Gainesville airspace. This constraint of the eBee made it unsuited to mapping the entire UNG-Gainesville Campus, as the turnarounds would take it over private property and over busy highways.

The DJI Inspire 1 quadcopter proved to be the most viable mapping platform given that it can be pre-programmed to flight lines that keep it within the UNG-Gainesville Campus. It is capable of stopping at the campus boundary and changing direction. It has a flight time of about 15 minutes and the flight controlling software must be chosen to allow for pausing the sUAS mission to exchange the batteries. The DJI Inspire 1 has proven to be a very good workhorse and upon incrementally increasing the study area is able to effectively map the entire UNG-Gainesville Campus with high overlap (85%) imagery with mutually orthogonal flight lines from a flying height of 120 m (394 ft) with two days of data collection. With a multispectral 5-band sensor this results in about 60,000 2 MB images and a data volume of about 120 GB. This high overlap image data can then be processed with the SfM algorithm utilizing software like Agisoft Photoscan (renamed to Agisoft Metashape as of May 2019) or Pix4d Mapper. Given the large data volumes involved, there are many other associated issues in data indexing, download, processing, and image classification, which will be elaborated upon in section 11.7.

11.3.2 DJI X3 RGB Camera

The DJI Inspire 1 quadcopter comes with a DJI X3 RGB camera that is capable of taking both still imagery and video (DJI X3 2019). The X3 camera is based on a CMOS Sensor technology and can take up to a 12.4 MB image. It is a good camera to begin practice mapping flights and high overlap RGB imagery can be processed with SfM to produce high quality 2D orthophoto mosaics and realistic 3D models with the imagery draped over them. However, the utility for land cover classification is limited due to the lack of the NIR band, which is needed for clear separation of the different land cover classes. The Digital Surface Models (DSM) and the Digital Elevation Models (DEM) also referred to Digital Terrain Models (DTM) can be subtracted to yield a normalized Digital Surface Model (nDSM) that gives the heights of all objects above the ground. In particular, Agisoft is capable of photogrammetrically generating a good approximation of a DEM if there are sufficient ground class points in the dense cloud. This can be used for biomass and volumetric estimations (Iqbal et al. 2017; Zylka et al. 2014) along with providing surface morphology that greatly complements the spectral information in the RGB bands. However, the absence of the NIR band reduces the utility of this camera for mapping applications. This camera is compatible with several software applications like the DJI Go, Pix4D Capture, DroneDeploy, all of which are available free for iOS and Android platforms.

11.3.3 Micasense RedEdge MX Multispectral Camera

The Micasense RedEdge multispectral camera captures a 5-band image that includes the blue, green red, red-edge, and the Near Infra-Red (NIR) bands (Micasense 2015). It can be mounted on an Inspire 1 or Inspire 2 quadcopter, among other sUAS platforms, with a custom mounting kit. The red-edge band is useful for detecting plant stress and is very useful for vegetation and agricultural mapping. It has a global shutter that minimizes image distortion. An NDVI layer is easily generated and high overlap imagery results in very good 2D orthophoto mosaics that have a spatial resolution of about 8 cm/GSD from a flying height of 120 m. The 3D models, DSM, and DEM created are in excellent registration with the multispectral orthophoto making this data suitable for automated classification using GEOBIA techniques. The RedEdge MX imagery is calibrated using a reflectance panel before and after the flight. It also has an irradiance sensor (downwelling light sensor) mounted on top of the sUAS that is used to normalize the radiance collected by the camera such that the illumination conditions in the imagery collected remain the same throughout the flight operations on a particular day. This type of calibration and normalization yields orthophoto mosaics that are at sub-decimeter GSD and are at a 16-bit radiometric resolution. They are highly suited for comparison and change detection over time. The experiments with GEOBIA rulesets and classification using data from this sensor will be elaborated upon in section 11.5.3.

11.3.4 The Zenmuse XT Thermal Camera

The advent of the solid-state microbolometers for thermal imaging is a recent advance (Nicklaus et al. 2008) and sUAS mountable thermal cameras based on this technology have become commercially available in 2018 from FLIR Systems (DJI Zenmuse XT 2019). This thermal imaging capability has opened a range of applications that were just not possible earlier. The microbolometer Vanadium Oxide (VOx) detector elements on the imaging plane capture the radiant power intensity incident on them leading to a warming and a resulting change in resistivity. This change in resistivity is linearly proportional to the incident power intensity (watts/m^2) in the range of temperatures that exist in our environment ($-10°$ C to $40°$ C)

$$E_\lambda = (2\pi hc^2)(\lambda^5) \cdot 1/(\exp(hc/kT\lambda) - 1) \tag{11.1}$$

This sensor uses Planck's Law (Equation 11.1) to calculate the temperature associated with each GSD on the ground; where $E_\lambda =$ incident power (watts/meter^2/sr/μm), h (Planck's constant) $= 6.626 \times 10^{-34}$ J s, c $= 3.0 \times 10^8$ m/s, $\lambda =$ wavelength, T $=$ temperature in degrees Kelvin, and k (Boltzmann's constant) $= 1.38 \times 10^{-23}$ J/K. It assumes the emissivity to be 1 and as the variation in the incident power intensity is very small as a function of the incident wavelength that the temperatures in our environment (centered about 300 K), the wavelength λ is taken to be a constant (10 μm). Upon acquisition, the emissivity of the different surfaces can be changed using the FLIR Tools $+$ software to infer the correct temperature of the surface. The thermal images are in a radiometric JPEG format in which the incident power intensity is also recorded per GSD. Both Agisoft Photoscan (renamed to Metashape in May 2019) and Pix4D software have the Zenmuse XT thermal camera model and can produce 2D orthophoto mosaics in which temperatures can be queried per GSD and produce 3D thermal models with which the thermal radiative energy emissions can be modeled. This camera can capture up to a resolution of 640$\times$512 pixels, has a focal length of 13 mm, and a narrower Field of View (FOV) of 45°$\times$37°.

11.4 Mapping of the Tumbling Creek Forest and Wetland at the UNG-Gainesville Campus

The forested area of the UNG campuses is an excellent laboratory for sUAS based environmental mapping and modeling applications. The terrain morphology and relief of the

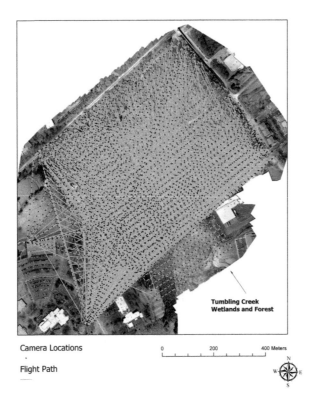

Camera Locations

Flight Path

0   200   400 Meters

FIGURE 11.2
DJI Inspire1 Quadcopter flight lines for mapping the Tumbling Creek Forest and Wetland on March 7 and 11, 2019. Note the return flights needed to go to the staging area for battery exchanges.

Tumbling Creek and Hurricane Creek Forests bring challenges unique to each study area for conducting sUAS mapping flights. These natural areas have mainly vegetative land cover and the NIR and red-edge bands are critical for mapping the vegetation health. The initial trial flights were conducted with an RGB camera on both of these forested areas. This was followed by multispectral mapping of the entire Tumbling Creek Forest and Wetland (Figure 11.2) and a workflow was developed that will be applied to the Hurricane Creek Forest in the future.

11.4.1 sUAS based Multispectral Imaging of the Tumbling Creek Forest

The Tumbling Creek Forest and Wetland has an extent of about 0.5 km^2 and is on the Northern portion of the UNG-Gainesville Campus (Figure 11.1). It has a valley along the middle where there are wetlands and forested ridges on each side that run roughly in the North-South direction. These ridges have substantial terrain height at 384 m altitude with the wetlands in the valley being at 348 m giving a total terrain relief of 36 m (118 ft). Given that the flying height by regulation is restricted to 122 m (400 ft) Above Ground Level (AGL), the GSD in the valley can be about 30% smaller than a ground GSD at the ridges.

The multispectral mapping of the Tumbling Creek Forest and Wetland was conducted for leaf-on conditions in October 2018 (Figure 11.3A) and leaf-off conditions in March 2019 (Figure 11.3B). This took a series of incrementally larger DJI Inspire 1 quadcopter test flights to understand the issues involved in mapping flights over this terrain. Due to lack

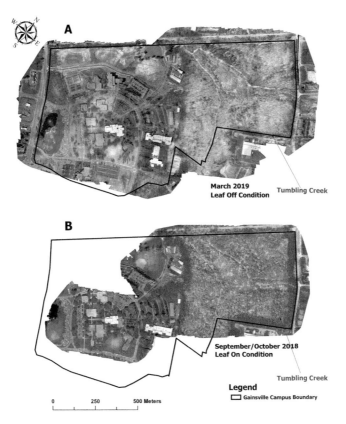

FIGURE 11.3

Multispectral False Color orthophoto mosaics of UNG-Gainesville for March 11, 2019 (leaf-off) and September 21, 2018 (leaf-on) with 9.5 cm GSD. The near-infrared reflectance rendered through the red color represents chlorophyll and is indicative of vegetation health.

of line-of-sight over the entire flight over this study area a team of ground observers were deployed at the Northern end of the Tumbling Creek Forest that were in telephone contact with the flight controllers at the Southern reaches of this Forest. The initial flights were with the X3 RGB camera to establish workflows for data uploads, indexing, and processing. Both Agisoft Photoscan (renamed to Metashape in May 2019) and Pix4D Mapper software were utilized for generating orthophoto mosaics and DSMs both with a 9.5 cm GSD, and an optimal workflow was developed with each of these software packages. These test flights were conducted from February 2018 to May 2018.

11.4.2 Multispectral Mapping of the UNG-Gainesville Main Campus

The multispectral imaging of the urbanized and forested area of the Gainesville Campus was established as a goal upon the acquisition of the Micasense RedEdge MX camera in August 2018. The RedEdge camera was mounted on a DJI Inspire 1 quadcopter using a custom mounting kit provided by Micasense. The aim of the image acquisition for both the urban and forested areas of the campus was not only to acquire a multispectral image mosaic but also to generate a photogrammetrically derived Digital Surface Model (DSM) and a Digital Elevation Model (DEM) also referred to as a Digital Terrain Model (DTM). The difference between the DSM and DEM gives the normalized DSM (nDSM) which gives heights and morphology of all objects imaged, relative to the ground. The leaf-on multispectral imagery for most of the UNG-Gainesville Campus in leaf-on conditions was acquired in September 2018 (Figure 11.3B). The multispectral imagery for the entire UNG-Gainesville Campus with leaf-off conditions was acquired in March 2019 (Figure 11.3A). Both the spectral information from the multispectral imagery and the morphological information from the nDSM was used for developing Geographic Object Based Image Analysis (GEOBIA) mapping applications for mainly the urban portion and also the forested area of the UNG-Gainesville campus. This is elaborated upon in section 11.5.1 through section 11.5.3.

11.4.3 Multispectral Imaging Flight Operations Issues

The generation of a DSM, DEM, and nDSM that is less than 10 cm/GSD requires flight lines that have a substantial side and front overlap. For the urbanized campus study area of about a 0.5 km², an overlap of 85% at flying at 6 m/s required a set of very dense mutually orthogonal flight-line blocks. From the staging area chosen for good line-of-sight, the flight time for each battery was about 12 minutes which was sufficient for about four 0.5 km flight lines and this necessitated frequent battery exchanges. AtlasFlight is the software for flight planning and control of the Micasense camera. Once the flight plan has been finalized, AtlasFlight sets the triggering time for the camera shutter for these particular flight parameters. The flight plan is then loaded onto the DJI Inspire 1 quadcopter using a wifi connection. Images of the reflectance calibration panel provided with the Micasense RedEdge camera are acquired before and after the flight and are essential to obtain calibrated multispectral imagery. The downwelling light sensor adjusts for changes in the irradiance providing uniform illumination over the orthoimage mosaic. Both Agisoft Photoscan and Pix4D Mapper software were utilized for processing the multispectral imagery into orthomosaics and a DSM.

The battery exchanges for the Inspire 1 quadcopter after every 12 minutes required the pausing of the mission on the AtlasFlight software, exchanging the batteries, reconnecting with the Micasense camera via wifi, and then resuming the mission. Initially, for the leaf-on condition flights in Fall 2018, reconnecting with the camera to resume a mission was sometimes problematic and needed a reboot of the UAS. When the leaf-off condition flights began in March 2019, the 'resume mission' feature on AtlasFlight software did not work due to a conflict with the new iOS 12 operating system on the iPad used with the flight controller. The Micasense technical support team was contacted and they were able to replicate the error and provided us with a beta copy with the software bugfix on AtlasFlight that worked well for subsequent mission resumptions after battery exchanges.

The staging areas for these flights were over a roof that had substantial metal reinforcement in it. This caused problems with the UAS compass and we had to recalibrate the compass several times. This issue requires attention as the DJI Inspire 1 manual has a caution about being around excessive amounts of metal in the take-off and landing area and in the area where the compass calibration is performed. The imagery collected has large data volumes and spare SD cards are a necessity, especially during large area (>0.5 km²) 3D mapping, quadcopter sUAS operations where high overlap is necessary. There are several data downloading and handling issues that will be elaborated upon in section 11.7.

The DJI Inspire 1 sUAS crashed due to mechanical failure after about 500 flights and over 100 hours of flight; fortunately, no damage occurred. However, this highlighted the need for greater a-priori due diligence to ensure safety and to hedge against loss of valuable sUAS platforms and sensors. This reinforces the importance of keeping accurate flight logs to monitor total flight times and to monitor battery usage over time. Maintenance procedures need to be articulated based on manufacturer recommendations and post-flight electric motor temperatures need to be recorded to anticipate incipient failure. Handheld point-able thermometers and smartphone attachable thermal imagers will be used for this purpose in the future. Propeller guards can protect humans from injury and prevent propellers from contacting obstacles. This incident has also started a formal process in our University to deploy best practices for sUAS operations that ensure public safety and co-opt liabilities incurred by accidents.

11.4.4 Thermal Mapping of the UNG-Gainesville Central Campus

The FLIR Zenmuse XT thermal camera was utilized on a DJI Inspire 1 quadcopter for thermal mapping of the urban portions of the UNG-Gainesville Campus. By regulation, sUAS flights are not allowed at night and require special permission from the FAA. Therefore, early dawn is the best time for thermal imaging. However, the time window is limited to at best a couple of hours as upon sunrise the land surface temperatures begin rising and the temperature contrast with the ambient temperature decreases. In order to get a good 3D thermal model, at least 80% substantial forward and side overlap is needed which makes for two sets of narrowly spaced mutually orthogonal flight lines and this can easily take almost an hour to image a relatively small urbanized portion of the campus. This limited time window, coupled with weather, and sUAS team logistical concerns early in the morning restricts the thermal imaging opportunities.

Thermal orthophoto mosaic for the Science building area was generated as shown in Figure 11.4. The ambient temperature on April 13th 2018 was 8.0° C at 6 am in the morning. From the temperature histogram in Figure 11.4, it can be seen that some ground surfaces are cooler than the ambient temperature and are absorbing heat from the surroundings; these are mostly the grassy field and the building rooftops. Some other surfaces like pavements and trees are warmer than the ambient temperature and are emitting heat. These temperature orthophotos will be processed into thermal emission 'power-scapes' in which the spatially distributed rate of energy or power absorption or emission can be calculated. These power exchange dynamics change with the rising of the sun as all surfaces heat up to become net thermal power emitters. The 3D thermal model of the Nesbitt Building on campus was created and is shown in Figure 11.5. sUAS based thermal imagery of the faces of the Nesbitt Building was used to conduct a thermal power emission analysis. These thermal emission phenomena will be elaborated upon in the Data Analytics sections 11.5.4 and 11.5.5.

11.5 Developing UAS Data Analytics

Processing the UAS data came with several challenges due to the high radiometric and spatial resolution. The Micasense multispectral imagery was collected at a 16-bit radiometric resolution. ArcMap software could not handle this radiometric resolution in image processing operations; however, ArcPro proved to be proficient in processing high radiometric resolution imagery. The segmentation parameters in eCognition are much larger

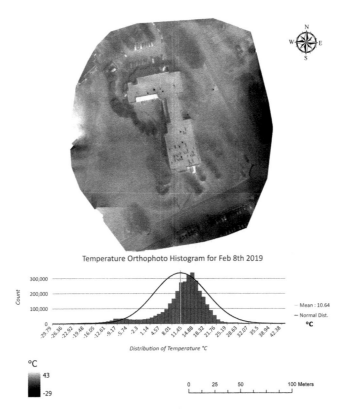

FIGURE 11.4
Temperature orthophoto of the Science Building February 8, 2019 at 6 am in the morning and with an ambient temperature of 8° C. Note the two peaks in the temperature histogram showing thermal absorption and emission.

FIGURE 11.5
3D Temperature model of the Nesbitt Building. The brighter areas are at a higher temperature and identify areas of greater heat loss.

than typically used for aerial and satellite imagery due to the high radiometric resolution of the imagery. The high spatial resolution imagery has a large number of transient artifacts like shadows and cars; these can vary on the study area during the UAS flight operations which can last up to a few hours. Shadows are the bane of high spatial resolution sUAS imagery and are a big source of classification error. The information obtained by analyzing the data collected, and processed into orthophoto mosaics and 3D models, is elaborated upon in the following sections.

11.5.1 Developing Geographic Object Based Image Analysis (GEOBIA) Derived Land Cover

Once the multispectral orthophoto and an nDSM is available, a land cover classification map can be developed using GEOBIA techniques by developing rulesets for automated feature extraction utilizing the eCognition software. Typically, the data is segmented into 'meaningful' objects such that the objects clearly delineate individual land cover classes on the earth surface like buildings, trees, roads, grass, and water. The Quadtree, Multiresolution, and Spectral Difference segmentation algorithms in eCognition were utilized to segment the data, which was then classified into land cover classes utilizing several different types of rulesets.

11.5.2 Developing an nDSM – Heights from the Ground

Several approaches were tried to develop a DEM such that an accurate nDSM could be developed, until a viable methodology was ascertained. The first attempt was to use an aerial LiDAR derived DEM resampled from a 5 m to a 10 cm GSD, such that it could be subtracted from the UAS derived roughly 10 cm/GSD DSM. This was not successful as the spatial uncertainty in the aerial LiDAR derived DEM is much larger than the GSD of the photogrammetrically derived DSM.

The successful approach was to use Agisoft software to generate an nDSM utilizing photogrammetric techniques. This involves an interactive process after sUAS photo alignment and the building of the dense cloud. The dense cloud is then classified for the ground class. A mesh is then generated with just the ground class by interpolation to form a connected DEM; this is done using an algorithm that Agisoft utilizes in which a cell size, a radius, and an angle have to be chosen. The 'ground' surface is then exported as a Digital Terrain Model (DTM) (Stalin and Gnanaprakasam 2017), which is also known as a DEM. The most accurate nDSM was created by utilizing ArcPro software to subtract the DEM from the DSM (Figure 11.6).

11.5.3 Developing a GEOBIA Ruleset

Several different classification rulesets were developed to derive the land cover utilizing the eCognition 9.2 Software. Both the October 2018 leaf-on and the 2019 March leaf-off imagery were utilized in experiments with rulesets to generate land cover classification for the central urbanized portion of the UNG-Gainesville Campus. It was challenging to classify the deciduous trees on campus using the March 2019 leaf-off imagery due to the leaf-off tree canopies and problems in delineating bare tree branches as objects. Further work is needed to develop viable classification approaches for leaf-off trees utilizing sUAS imagery. However, the classification of leaf-on tree canopies using the October 2018 imagery proved to be successful as the leaf-on tree canopies could be easily segmented into viable objects. This 2018 leaf-on imagery was further utilized to develop land cover maps using GEOBIA techniques.

Several different approaches were utilized on eCognition to classify the multispectral imagery and the nDSM leaf-on October 2018 imagery of the UNG-Gainesville urbanized central campus into the following five classes: buildings, trees, grass, roads/pavement, and water

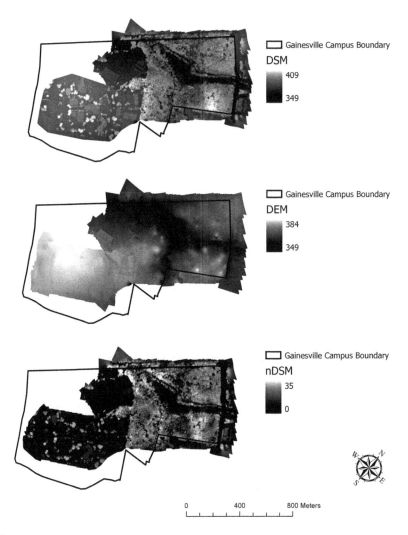

FIGURE 11.6
UNG-Gainesville campus SfM derived Digital Surface Model, Digital Terrain Model, and the Normalized Digital Surface Model for October 2018 with leaf-on vegetation.

(Figure 11.7). The initial efforts to develop a land cover map using a supervised approach involved training samples and a Nearest Neighbor classification; this approach worked but not as well as a ruleset based on thresholding the spectral and spatial data. Due to the spectral variability of the surface, the high spatial resolution sub-decimeter imagery and shadows, a large number of samples would be needed to capture all of the variation. However, thresholding based on the Normalized Difference Vegetation Index (NDVI), the normalized Difference Water Index (NDWI), and nDSM, the standard deviation of the nDSM was sufficient to create a viable land cover map. Since the study area in this case is 2.22×10^5 meters^2 (54.9 acres), manual editing and a close visual inspection was sufficient to establish high viability of the land cover map for the urbanized center of the UNG-Gainesville Campus.

Similarly, rulesets were developed for classification based on machine learning utilizing the following algorithms: both the object based and pixel based Computer Aided Regression Tree (CART), and an object based Support Vector Machine (SVM) Algorithm.

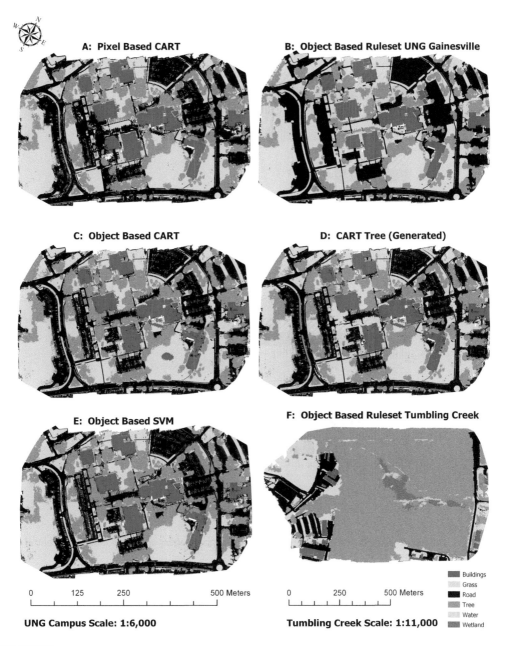

FIGURE 11.7
GEOBIA based classification of multispectral imagery of the UNG-Gainesville Campus using different algorithms [A through F].

The classification results are presented as maps in Figure 11.7 and the areas of the classes are presented in Figure 11.8. The ruleset based classification has had minimal editing done to remove the automobiles contained by the 'road' class; automobiles proved to be difficult to classify due to their large spectral variation (Figure 11.7B). The areas of the different land cover classes shown in Figure 11.8 are reasonably consistent. A ruleset based classification was also done for the Tumbling Creek Forest and Wetland (Figure 11.7F) and further

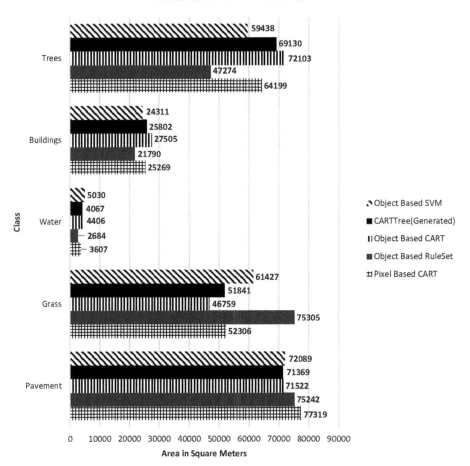

FIGURE 11.8
GEOBIA classification of October 2018 leaf using the multispectral imagery and the nDSM for the central urbanized area of UNG-Gainesville.

classification experiments will be conducted for this natural area. Further similar GEOBIA studies will be conducted to determine which of these classification algorithms will be the most robust choice for developing land cover maps for all of the UNG campuses.

A formal accuracy assessment has not been conducted as of yet and a larger systematic study of the quantitative accuracies of the classifications (Congalton and Green 2019) will be conducted in the future. An object based error analysis will be conducted as proposed by MacLean and Congalton (2012) and Ye (2018) as the next part of this project. A visual inspection and comparison with the imagery shows the maps to be largely viable and accurate with minimal manual editing of the classes. An object based quantitative error assessment will be conducted in detail as the exceptional spatial variation in the image and the shadows make it difficult to remove all sources of error in the classification.

The high spatial resolution landcover map, the DSM, and nDSM provide a good foundation to build geospatial applications like non-point source pollution, erosion, and sedimentation modeling. In particular, the land cover map is essential for thermal modeling of urban habitats and natural areas and is needed to create an emissivity map can be created for the different land cover types. Different emissivities for the land cover types determine the thermal power emission from these varying land cover surfaces.

11.5.4 Modeling Radiative Thermal Dynamics of the Urban Campus

The Stefan-Boltzmann law models the radiative power emitted of heated objects as follows:

$$P = \sigma Ae(T^4 - T_0^4) \, \text{Watts}/\text{m}^2 \tag{11.2}$$

where $\sigma = 5.670367 \times 10^{-8}$ W·m^{-2}·K^{-4} (Stefan-Boltzmann constant), A = area of GSD, e = emissivity, T = radiant surface temperature in K, and T_0 = ambient temperature in K.

The Zenmuse XT camera was mounted on a DJI Inspire 1 quadcopter and several areas in the urbanized portion of the UNG-Gainesville Campus were imaged to create both 2D thermal orthophoto mosaics and 3D thermal models by utilizing both Agisoft and Pix4D and the high overlap radiometric jpeg images.

The 2D orthophoto mosaics have a temperature associated with each ground pixel and are amenable to raster algebraic operations. The Stefan-Boltzmann law was applied to the thermal orthophoto mosaic to generate a 'thermal power-scape' in which the thermal power emitted from or absorbed by is represented. Two examples of thermal 'power-scapes' are shown in Figure 11.9. In the first case, right before dawn at 6 am and an ambient temperature of 10° C. Some surfaces are cooler than the ambient air temperature and are absorbing heat and many surfaces are warmer than the surroundings and emitting heat as shown by the temperature histogram (Figure 11.9A). In the second case for an adjacent area imaged later in the day at 9 am right after sunrise, all the surfaces warm up above the ambient air temperature of 18° C and begin emitting thermal energy, also shown by the shift in the histogram in Figure 11.9B.

This establishes a methodology to explore land surface-environment energy and power exchanges for both built and natural environments utilizing sUAS thermal imaging. These projects will be scaled up for larger study areas on campus to begin to explore the thermodynamic models of both urban habitats and forests.

11.5.5 Conducting Energy Audits of Buildings

The 3D thermal models developed using high overlap thermal imagery are very useful to examine and analyze the thermal radiative losses occurring from buildings in cold weather (Yang and Lin 2018). Regions of greater thermal radiative losses can be readily identified (Figure 11.5). As the 3D models are not rasterized it is not possible to perform algebraic operations on these with contemporary GIS software like ArcMap and ArcPro.

However, it is possible to work with the thermal imagery captured almost orthogonal to the building surfaces by manual piloting of the DJI Inspire 1 quadcopter to approximate the radiative emissions from the building. The FLIR Tools + software was utilized to measure the average temperatures of the building walls and windows in polygons drawn on the building surfaces as shown on the left portion of Figure 11.10. The areas of the building walls and windows were measured by utilizing the 3D building model available on Google Earth Pro, as shown on the right portion of Figure 11.10. The emissivity of the glass windows and wall were utilized to calculate the net radiative thermal power using the

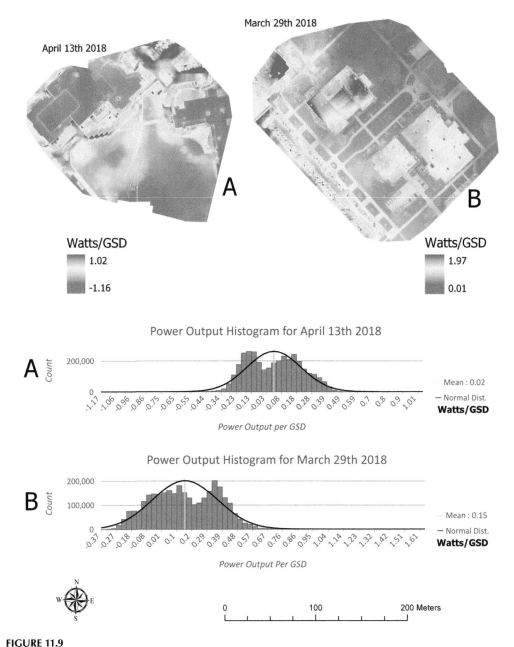

FIGURE 11.9
Early morning thermal 'power-scapes' with thermal power radiated per GSD (9.5 cm); the thermal imagery for Figure 11.9 A was collected at 6 am and for Figure 11.9 B at 9 am.

Stefan-Boltzmann law in an Excel spreadsheet. The results of the building thermal power loss estimation are summarized in Tables 11.1 and 11.2.

Ventilation heat loss accounts for much more of the heat transfer from buildings, and can be estimated as a fraction of the total building thermal power output: 'Ventilation and air infiltration into buildings represents a substantial energy demand that can account for between 25–50% of a building's total space heating (or cooling) demand' (Davies 2004; Liddament 1998). Assuming negligible conductive heat loss implies that the radiative heat

FIGURE 11.10
The average temperatures were measured using the thermal imagery (left) and FLIR Tools + software; the areas of the building windows and walls were measured using the 3D models in Google Earth Pro (right).

loss can account for 50–75% of building heat loss. Therefore, for the Nesbitt and Science Buildings, the estimated thermal outputs estimated can be considered to be 50–75% of the total power output for each building respectively.

Table 11.3 describes the upper and lower bounds of total power output for either building using the assumption that the heat loss due to convection is between 25–50% of the power output. This gives an sUAS measurement based estimated thermal power output between 100 kW and 150 kW for Nesbitt, and between 85.2 kW and 128 kW for the Science Building. The range of electrical energy costs, which can be used as a proxy for thermal energy transferred to or from the building, can be calculated using the average cost of electricity as $0.1 per kWhr; these are shown in Table 11.4.

The next step was to ascertain the validity of these results. The UNG facilities management informed us that the metering of the electricity consumed was done for sections of the campus as a whole such that these results could not be compared to the electrical bill.

TABLE 11.1

sUAS Based Thermal Power Output Estimate from the Nesbitt Building

Thermal Power Output Estimate from Nesbitt Building	
Total Window Area = 1459 m^2	Total Brick Facade Area = 3110 m^2
Emissivity $_{\varepsilon window}$ = 0.92	Emissivity $_{\varepsilon brick}$ = 0.8
Power Emission (Window) = 37.6 kW	Power Emission (Brick) = 37.5 kW
Total Thermal Radiative Power Emission = 75.1 kW	

TABLE 11.2

sUAS Based Thermal Power Output Estimate from the Science building

Thermal Power Output Estimate from Science Building	
Total Window Area = 337 m^2	Total Brick Facade Area = 2336 m^2
Emissivity $_{\varepsilon window}$ = 0.92	Emissivity $_{\varepsilon brick}$ = 0.8
Power Emission (Window) = 13.0 kW	Power Emission (Brick) = 50.9 kW
Total Thermal Radiative Power Emission = 64.0 kW	

TABLE 11.3

Estimation of Total Thermal Power Output of the Nesbitt and the Science Buildings

Building	Radiative Power: (kW)	(25%) Ventilated Power: (kW)	(50%) Ventilated Power: (kW)	(25%) Total Power Output: (kW)	(50%) Total Power Output: (kW)
Nesbitt	75.1	25.0	75.1	100	150
Science Building	63.9	21.3	63.9	85.2	128

TABLE 11.4

Comparing Thermal Power Emission Using Electricity Costs Per Month as a Proxy for Energy Costs Estimated Using Two Different Methods

Building	Area	Estimated Heating/Cooling Cost Per Month (Using $1/ft²/year)	sUAS Based Estimated Cost Per Month Lower Limit (Using $0.1/kwhr)	sUAS Based Estimated Cost Per Month Upper Limit (Using $0.1/kwhr)
Nesbitt	130,000 ft²	$10833	$7300 (100 kW emission)	$10950 (150 kW emission)
Science	60,000 ft²	$5000	$6206 (85 kW emission)	$9344 (128 kW emission)

However, the rule-of-thumb for UNG facilities management to estimate electrical utility expenses for heating and cooling a building is $1.0 per ft² per year (Moody 2019). This rule-of-thumb is utilized for planning new spaces and for developing reports on estimated energy consumption of buildings by UNG facilities management.

Knowing the internal square footage of the Nesbitt (130,000 ft²) and the Science (60,000 ft²) Buildings, the approximate electrical energy bill can be calculated and compared with the results obtained by sUAS thermal imaging. The results are summarized in Table 11.3, and we can see that the sUAS data based estimates of thermal power emitted via primarily both radiation and convection are in a range that is commensurate with the estimated costs based on the rule-of-thumb used by facilities management at the UNG-Gainesville Campus, which is $1.0 per ft² per year. The costs estimated from these two different methods fall in the same range and provide a validation of this basic approximation. Repeated sUAS based thermal measurements of a group of buildings whose metered total power consumption information is available will refine the methodology for sUAS based thermal energy audits. In particular, when the capability to generate pixelated 3D thermal models upon which raster algebraic operations can be conducted becomes available in image processing software, the thermal power emissions for buildings can be estimated more accurately. These methods are being experimented with utilizing MATLAB, Blender, and FLIR Tools + software and are giving promising results.

This has developed a workflow for approximating thermal emissions and efficiencies of campus buildings. This methodology is being further refined by analyzing the 3D thermal models with custom scripts on MATLAB software and will be scaled-up for energy audits of other buildings on all of the UNG campuses.

11.6 sUAS Data Positional Accuracy and Processing Issues

The imagery acquired from an sUAS platform has GNSS positional stamps accurate to about a planimetric RMSE of about 2.0 m. This means that repeated mapping flights will yield both 2D orthophoto mosaics and 3D models that are within about 1–2 meters of each other. The uncertainty in the GNSS derived Ground Control Points (GCP) is to the order of a 5.0 cm RMSE. Since the GSD sizes imaged by an sUAS are to the order of 5 to 10 cm, registration of data layers imaged at two different times to within a GSD, is difficult; the uncertainty in the data is to the order of the uncertainty in the GCPs. Highly accurate GCPs are necessary to georeference sUAS imagery correctly. Furthermore, to get multiple sUAS imagery bands in near perfect registration, all of this data has to be collected in the same flight and processed together to produce a multispectral mosaic and a DSM, DEM, and nDSM. This is absolutely necessary for developing GEOBIA applications for automated feature extraction and for performing viable raster algebra calculations. For this reason, imagery from a multispectral sensor was utilized for land cover classification in this project.

11.6.1 Development of a GCP Network

A Ground Control Point network across the urban and the forest portions of the UNG-Gainesville Campus was established using a Trimble Geo7x GNSS receiver with a Zephyr 2 external antenna for data collection step. All visible manhole covers had an 'X' pattern spray painted utilizing a stencil. Similarly, other points that are readily identifiable from the air and are fixed to the ground, like corners of pavements, were also captured as GCPs. A photograph of the student with the GNSS unit at the specific location was taken and was entered into a log to allow for unambiguous identification of these points at a later time. Differential correction of the GNSS data was performed the next day using Trimble's GPS Pathfinder office to achieve higher accuracies in the resulting GCP points to about 5.0 cm RMSE. This GCP network is still being developed at the UNG-Gainesville Campus and will be extended to the other UNG campuses over time. This GCP network will be used to georeference the sUAS derived land cover maps for use in further geospatial modeling applications.

11.7 sUAS Data Processing Issues

Effectively, the UNG-Gainesville Campus had two distinct study areas; one urban and another forested and both of an extent of about 0.5 km². Each one of these study areas required high overlap (85%) forward and sideways overlap with two mutually orthogonal flight lines. With multispectral imaging with the RedEdge MX camera, five bands were captured (B, G, R, NIR, red-edge). The resulting data volume is about 30,000 2 MB images for a total data collection of about 60 GB. Handling and processing such large datasets bring several issues with it including data indexing and download from the sUAS and performing image processing operations for which substantial computing power is needed. Cloud computing using Microsoft Azure proved to be the critical last step in the workflow in order to produce 2D multispectral orthophoto mosaics and 3D models that will be utilized for deriving land cover maps and other geospatial models.

11.7.1 Data Download and Indexing

Typically, sUAS image data is captured on an SD card and then downloaded onto a computer for processing. For capturing an area as large as 0.5 km² (124 acres or 50 hectares) there are several battery exchanges needed to complete the entire mission of high overlap mutually perpendicular flight lines. This can generate to the order of 30,000 images with up to 60 GB of data. The imagery is written onto an SD card in a file beginning with a numeral prefix like '001' until 1,000 images are captured at which time another directory with a prefix '002' is created in which another 1,000 images are captured and so on. The Micasense RedEdge camera is controlled with the AtlasFlight software, which requires to be reconnected to the sUAS upon switching out a depleted battery. In this case the images begin to be recorded in the same procedure as earlier, in a folder with a prefix of '001'. Sometimes this procedure to reconnect to the Micasense RedEdge camera wifi needs an sUAS reboot, in which there is a new directory with a prefix '001' created with subsequent file numbers repeated from the beginning once more.

At the end of the mission covering 0.5 km^2, there is a large number (roughly 25) of directories, some empty due to an sUAS reboot, and many images of different locations with the same file name. Therefore, a sequential indexing of the image data is essential such that as the data is copied from the SD card, it is all placed in one directory on the computer with each image having unique names.

A relatively simple way of performing this indexing is to rename the files in each directory with an alphabet successively, i.e. append an 'a' to the images in the first directory, append a 'b' to the images in the next directory, and so on; then all of these images are copied onto a single large directory on the computer. Given that the total data can have up to 30,000 or more 2 MB images, the copying of this data via an SD card reader and a USB port can take up to several hours. Furthermore, data transfer rates within the computer, in the local network, or to the Cloud also have a significant bearing on the time taken for implementing the image processing workflow. Rapid development of computing capabilities and the ubiquity of broadband promise to ameliorate these constraints.

11.7.2 The Necessity for Cloud Computing

The requirement of collecting high overlap multispectral imagery with mutually orthogonal flight lines generates high data volumes that are up to 60 GB with about 30,000 photos for imaging a 0.5 km^2 study area. It is a challenge to process such large datasets using Agisoft and Pix4D software and can take several days to generate orthophoto mosaics and DSMs and DEMs utilizing local workstations. In the workflow that involves photo alignment, building a dense point cloud, a mesh for 3D models, and a DSM before an orthophoto mosaic and a DEM is developed, there are several parameter choices that must be made at each step. This highlights the necessity for high performance computing that scalable cloud computing can provide. The UNG Information Technology (UNG-IT) Department with the support of the Office of Research and Engagement provided access to the Microsoft Azure cloud-computing platform to experiment with high performance computing.

Cloud computing has proved to be the most essential component of completing the workflow to convert the large volume sUAS data into actionable information. The on-demand cloud computing model requires users to understand that the cost of access is now dependent on actual usage of the resources. This fundamental mindset shifts from an 'always on' local model to a 'when needed' model. Computing costs accrued for the project are then recorded by the different resources used such as network/storage consumed, as well as compute cycles that are spent when the virtual machine is on. The support from UNG-IT has been critical in the development and deployment of the data-handling infrastructure needed. This involves the creation of network folders on the UNG computer network that synchronize with a network accessible folder on Microsoft Azure on the Cloud, for data access closest to the compute instance. This allows data to be accessed from a network folder on the Azure platform, followed by data processing using both Agisoft and Pix4D; the resulting geospatial products are then synchronized back to a folder in the UNG computer network. Azure allows for customization of hardware capabilities including additional cores for computing and inclusion of powerful graphics capabilities. These are known as Azure Virtual Machine (VM) options (Table 11.5). Some computing tests were run to compare the lowest Azure VM options available with local workstation based VM. This speeds up the computing time by a factor ranging from 3.5 to 5 for these experiments as shown in Table 11.6 (for Agisoft) and Table 11.7 (for Pix4D). Speedup of up to 25 times has been observed with experiments with higher end Azure VMs and more experiments are needed to quantify the computing gains realized. As sensors and sUAS capabilities miniaturize further, big datasets will increasingly become the norm in the future and

TABLE 11.5

Virtual Machines (VM) Used in the Experiment; Azure Cloud VM vs Local VM

Machine	Type	vCPU	Memory (GB)	Total GPU Memory (GB)	GPU
AgisoftVM (VMware)	Local	6	32	1	NVIDIA GRID K1
NV12	Cloud	12	114	48	NVIDIA Tesla M60 x2
NV24	Cloud	24	224	32	NVIDIA Tesla M60 x4

TABLE 11.6

Agisoft, 7830 Photos, Multispectral, 0.495 km^2 (~122.3 Acres)

Agisoft Test	Local (minutes)	Cloud NV12 (minutes)	Cloud NV24 (minutes)
Matching	292	17	9
Alignment	179	114	84
Depth Maps Generation	168	19	7
Dense Cloud Generation	54	32	21
Processing DEM	3	3	2
Processing Orthomosaic	34	25	12
Total	730	210	135

TABLE 11.7

Pix4d, 7830 Photos, Multispectral, 0.495 km^2 (~122.3 Acres)

Pix4d Test	Local (in minutes)	Cloud NV12 (in minutes)	Cloud NV24(in minutes)
DSM Generation	6	5	3
Orthomosiac Generation	145	62	41
DTM Generation	1	1	1
Point Cloud Densification	17	10	5
Reflectance Map Generation	238	102	69
Index Map Generation	2	2	2
Total	409	182	121

scalable cloud computing will have to be an essential part of the workflow in handling this increasingly higher spatial, radiometric, spectral, and temporal resolution data.

11.7.3 Comparing Local vs Cloud Computing for sUAS Applications

An interesting area of this study is identifying the most optimal configuration for the processing of sUAS projects and workflows. Achieving the highest efficiency of cost and time spent is dependent on the computing resources available. This goal is achieved by increasing the number of CPUs and GPUs inside the system. Acquiring additional resources to build a local system can require a costly expenditure upfront and to maintain in the long term. With advancements in cloud computing, an alternative on and off computing approach can optimize the project by utilizing cloud resources to increase resources on-demand when needed and powering off when not needed thus reducing the capital expenditure on computing resources.

The process in which we used to conduct our cloud based image processing optimization consisted of the seven basic steps outlined in Figure 11.11.

The overall time taken to process one project using low-end Azure VMs was experimentally determined. A plethora of cloud computing Azure VM configurations are available and can be chosen for the type and the scope of the project. The NV series in Microsoft Azure was chosen for our tests due to higher end CPU cores and access to high-end graphics processing units (Table 11.5).

Our test consisted of three VMs with different resource sizes using Agisoft Photoscan and Pix4d Mapper to compare applications in a single-node configuration processing the same multispectral imaging data. Results were collected from the reporting features in both applications tested and summarized in Table 11.6. Agisoft Photoscan demonstrated a speed up of 3.5 to 5 times when compared to current local resources available. A similar test was conducted for Pix4D Mapper (Table 11.7) and a minimum speedup by a factor of 3.4 was observed.

Our testing observed a speed-up with default configurations compared to our local test machine. Rapid advancements in cloud computing allow the research to leverage next generation hardware as it is made available as a sizing option. Further research in this area will be conducted to continue optimizing each step in the workflow.

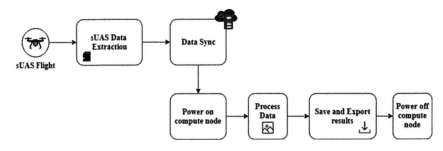

FIGURE 11.11
Basic sUAS data processing workflow.

11.7.4 Lessons Learned for Using Cloud Computing for sUAS Applications

The following is a summary of some lessons learned for effective deployment of cloud computing for sUAS applications:

Selecting an Appropriate GPU – When experimenting with different sUAS tools during the project, having a good understanding of graphics libraries is required when selecting the appropriate GPU card. A cloud provider can offer a variety of GPU options to switch between different features depending on the workload requirements. Identifying specific image processing operations that can utilize GPU Compute vs GPU Visualization workloads will reduce time and errors. Verifying operations that are GPU optimized can be done using computing resource monitoring tools. In our study, a software tool developed by graphics card maker NVIDIA called 'Nvidia-smi' was utilized during computing work periods to observe and verify GPU utilization.

File Transfer Latency – Placement of photos in storage location can impact the processing time of the project and should be considered before submitting the job. Choosing the fastest and lowest latency path to the project photos will greatly reduce processing time. During our testing, we found varying results when selecting virtual network drives as our project path, which can also introduce network latency. Further optimization in delivering fast network accessible storage comparable to local machine storage access is needed to avoid unnecessary copy operations of the data between locations.

Reducing Idle Time on VM – Cloud computing costs are accumulated by how long and how many resources used. For example, a powerful cloud VM used in our study costs around $3 an hour. Once a job is submitted, the user may move onto other tasks while waiting for the job to complete. Some safeguards can be added to avoid leaving a machine idle to prevent unnecessary cost. Some controls can be introduced such as a daily auto-shutdown policy, scripted batch processing with job completion email notification, or a machine shutdown sequence that executes after the job completes.

11.8 Issues in sUAS Higher Education

The rapidly emergent sUAS technology is a great opportunity not only to integrate physical and life science applications, but to build bridges to the social sciences and to social concerns as well. Inherently, sUAS applications are local and enrich the community in many ways.

Effective utilization of the sUAS based data to solve local ecological, agricultural, or urban problems is an opportunity to enrich both teaching and research experiences in the integrative sciences. There is a need for the development of sUAS academic programs at the K12 and undergraduate level that are rooted in STEM literacy. This is particularly important for STEM workforce development opportunities. There is an untapped pool of talent in the K12 and undergraduate pipelines that ultimately feed graduate and professional programs in geospatial science and technology. By its very nature, sUAS mapping has a focus on local mapping projects that generate very high spatial resolution data. This gives a strong impetus for geographically distributed sUAS programs in regional universities, public and private colleges, and K12 institutions that can then develop the expertise for sUAS mapping applications in their service regions. Whereas there is a large community of remote pilot license holders, a select few have proficiency in the mapping sciences with which the data collected is converted to geospatial products and actionable information. University programs that have a focus on the development of sUAS mapping applications are being developed in universities worldwide.

11.8.1 Current Geospatial Curricula at UNG-Institute for Environmental and Spatial Analysis

The Lewis F. Rogers Institute for Environmental and Spatial Analysis (IESA) at UNG has had a special focus on Geographic Information Science (GIS) education, research, and outreach since 1998. It has developed innovative curricula up to the baccalaureate level to promote programs and curricula that have its core GIS applications as related to environmental science, urban planning, natural resources, natural hazards, and socioeconomic issues. Students take introductory GIS and remote sensing courses early in their baccalaureate program, typically by their second year at UNG. This leaves two additional years of upper level coursework and projects in GIS applications for different content areas; this enables students to have experiences in synthesizing the GIS toolkit with advanced disciplinary subjects learned in the Bachelor's degree curriculum. The IESA program continues to grow since its inception 20 years ago and students graduate with about 30 baccalaureate degrees each year.

Beginning in 2002, the remote sensing coursework centered on analysis of multispectral imagery such as data from the Landsat program (30 m) and traditional aerial photography (1 m). Upper level elective courses in Digital Image Processing and Advanced Image Processing were added to expand the coursework related to remote sensing. An emphasis on GEOBIA techniques and multi-sensor fusion in the advanced courses now assumes particular importance in the automated classification of sUAS imagery. The period of 2013–2019 brought adjustments to the geospatial science and technology curricula with the current form being captured in Figure 11.12.

A special topics course in exploring sUAS mapping was offered in 2014 in which a faculty member and a student learned to operate an eBee Sensefly and conducted test flights over the Science Building over the UNG-Gainesville Campus. The results of these exploratory experiments were presented at conferences and published (Sharma and Hulsey 2014). This began a conversation among GIS faculty on how to integrate sUAS applications into the UNG STEM programs for research and teaching.

In the summer of 2017, a special topics course was offered focusing on using sUAS to produce a high-resolution orthophoto of a portion of the UNG-Gainesville Campus. The course required students to have completed several prerequisite courses in remote sensing and geospatial science and technology. In this course students were divided into groups for the purposes of completing steps in the orthophoto production process; from initial planning to data collection to the final production and assessment of the orthophoto. These initial sUAS centric curricular experiments began a process of sUAS projects to help clarify a vision for course and curricular development. The results of the sUAS projects summarized in this chapter have helped in developing effective sUAS workflows, the software application packages needed, the curricular concepts involved, and the programmatic

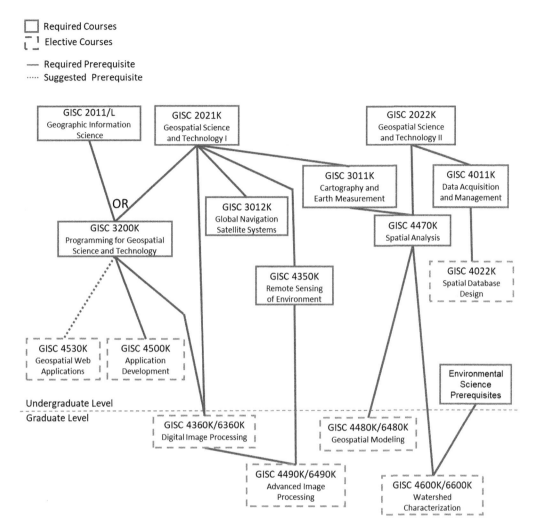

FIGURE 11.12
Foundational geospatial program structure for developing sUAS mapping curricula and courses.

structure that can support the development of a formal specialization in sUAS application development.

11.8.2 Future Curricular Directions

The rapidly decreasing cost of increasingly capable sUAS is increasing their use by geospatial professionals in the region served by UNG and this is influencing the development of novel sUAS curricula and programs. Future directions for sUAS curriculum within IESA include additional coursework that builds on the three-year Applying Geospatial and Engineering Technology (AGET) grant received from the National Science Foundation (NSF). The AGET project developed courses for an Associate of Science degree pathway in Geospatial Engineering Technology (GET) and for a certificate in land surveying. Courses are currently being developed centered around the use of sUAS technology for surveying and photogrammetry applications. Graduate level courses in remote sensing and digital image processing are being developed to extend current undergraduate level courses in the same areas. Graduate courses will focus on GEOBIA automated feature extraction, machine learning, SfM techniques, cloud computing, and other areas that build on or are used by sUAS applications.

11.9 Summary and Conclusions

The sUAS is emerging a platform for increasingly smaller and capable sensors that can make spectral and morphological measurements in study areas up to a few square kilometers with high spatial, temporal, and spectral resolution with increasingly high positional accuracy. The conversion of large datasets into actionable information provides a rich and tangible experience of the scientific process for students. Study areas within a campus are readily accessible and can be compared with their geospatial renditions and models by direct observation. Annual leaf-off and leaf-on multispectral mapping of all five of the UNG campuses will be a valuable research and teaching endeavor involving teams of faculty, students, and staff. The workflow to create a highly spatially and spectrally detailed 2D and 3D digital record of a university campus will inform planning, design, and implementation of campus sustainability. The 3D models developed can be 3D printed after conversion to a Stereo Lithography (STL) file format demonstrating that an sUAS is an effective 3D terrain scanner (Figure 11.13).

Thermal mapping of the urban areas of the UNG campuses at different times of the day will provide a valuable insight into the energy and power flows from the ground and building surfaces to and from the ambient environment. Thermal mapping for such applications is best done in cold weather when the temperature contrasts are greater. It will provide insights that will guide the designing of more energy efficient living spaces of the 21st century. Sustained sUAS operations will enable geospatial modeling applications like campus thermal audits, land cover mapping, non-point pollution, and hydrology. The generation of sUAS based thermal 3D models lays the foundation for developing thermodynamic models of both urban and forested areas on campus. This provides opportunity for undergraduate students from multiple disciplines like physics, mathematics, engineering, biology, and computer science to participate in research experiences based on physical measurements of the campus environment.

FIGURE 11.13
A 3D printed model of the Nesbitt Building.

The further refinement of sUAS technology will increasingly nurture local applications that enrich communities by enabling informed decision-making. sUAS will be an important part of real-time geospatial awareness in both urban and rural 'smart' spaces of the 21st century. Effective and geographically distributed sUAS operations will require the implementation of science and engineering principles in local management and governance. The conversion of sUAS data into information requires fundamental STEM literacy and at least a minimal knowledge of remote sensing and GIScience data handling skills. The generation of a geographically distributed workforce that can capture and analyze 'local' sUAS data for rapid decision-making is a great challenge for higher education. These types of sUAS applications will guide 'smart' spaces and near real-time geospatial awareness to enable agriculture, forestry, natural resource management, hydrology, disaster management, urban planning, public safety, infrastructure inspection, and historical preservation.

However, rooted sUAS mapping applications are in STEM knowledge; the social, economic, ecological, historical, and archeological local knowledge is essential for effective utilization of this technology. New sUAS mapping centric courses and programs are rapidly emerging in universities across the world; they will accelerate the integration of this technology into socio-economic processes worldwide. The ubiquity of broadband and scalable cloud computing will allow the capability to warehouse and process the large volume of data generated. An sUAS generates spectral and morphological data that is local and is experiential in that direct ground validation is possible. sUAS platforms are promoting a new era where geospatial science and technology becomes 'smart-and-local'. Increasing sUAS data collection internationally will give rise to best practices in sUAS applications and workflow design that will continue to shape this important technology and weave it even more intricately into the socio-economic fabric of the 21st century.

Acknowledgements

This work has primarily been supported by the UNG Foundation through the Eminent Scholar of Teaching and Learning endowed chair. In addition, a part of this work was funded by the AmericaView Consortium via GeorgiaView mini grants in 2017 and 2018. The curricular development in IESA was funded by the 'Applying Geospatial and Engineering Technology (AGET)' grant, number 170568, received from the National Science Foundation in the Summer of 2017. A special thanks to Parker Goins who is a UNG student and a licensed sUAS pilot who helped begin the sUAS flights on the UNG campus in 2018. Michael Mirolli was invaluable for his help in the chapter review process. Mark Leggiero has diligently worked on the energy audits using thermal imagery and 3D building thermal models.

References

AUVSI. (2014). The Economic Impact of Unmanned Aircraft Systems Integration in the United States. https://higherlogicdownload.s3.amazonaws.com/AUVSI/958c920a-7f9b-4ad2-9807-f9a4e95d1ef1/UploadedImages/New:Economic%20Report%202013%20Full.pdf. Last retrieved May 20, 2019.

Carrivick, J. L., Smith, M. W., Quincey, D. J. and Carver, S. J. (2013). Developments in Budget Remote Sensing for the Geosciences. https://onlinelibrary.wiley.com/doi/full/10.1111/gto.12015. Last retrieved May 16, 2019.

Clarke, R. and Moses, L. (2014). The Regulation of Civilian Drones' Impacts on Public Safety. www.sciencedirect.com/science/article/pii/S0267364914000594. Last retrieved May 16, 2019.

Congalton, Russell and Green, Kass. (2019). *Assessing the Accuracy of Remotely Sensed Data: Principles and Practices*, 3rd edition, Russell G. Congalton, Kass Green. Boca Raton, FL: CRC Press.

Davies, Morris. (2004). *Building Heat Transfer*, John Wiley and Sons, ISBN: 0-470-84731-X.

DJI X3. (2019). DJI Zenmuse X3 Camera Specifications. www.dji.com/zenmuse-x3/info. Last retrieved May 20, 2019.

DJI Zenmuse XT. (2019). DJI Zenmuse XT Thermal Camera Specifications. www.dji.com/zenmuse-xt/specs. Last retrieved May 20, 2019.

FAA. (2019). www.faa.gov/uas/educational_users/. Last retrieved May 8, 2019.

FAA. (2016). www.faa.gov/news/fact_sheets/news_story.cfm?newsId=20516. Last retrieved May 16, 2019.

Fonstad, Mark A., Dietrich, James T., Courville, Brittany C., Jensen, Jennifer L. and Carbonneau, Patrice E. (2013). Topographic Structure from Motion: A New Development in Photogrammetric Measurement (PDF). *Earth Surface Processes and Landforms*, 38(4): 421–430. ISSN 1096-9837.

Iqbal, Faheem, Lucieer, Arko, Barry, Karen and Wells, Reuben. (2017). Poppy Crop Height and Capsule Volume Estimation from a Single UAS Flight. *Remote Sensor*, 9(7), 647. doi:10.3390/rs9070647.

James, M. R. and Robson, S. (2012). Straightforward Reconstruction of 3D Surfaces and Topography with a Camera: Accuracy and Geoscience Application. *Journal of Geophysical Research: Earth Surface*, 117(F3): F03017. doi:10.1029/2011jf002289. ISSN 2156-2202.

Jenkins, D. and Vasigh, D. (2013). The Economic Impact of Unmanned Aircraft Systems Integration in the United States. https://robohub.org/_uploads/AUVSI_New:Economic_Report_2013_Full.pdf. Last retrieved May 23, 2019.

Liddament, M. W. (1998). Preface to Special Issue on Optimum Ventilation and Air Flow Control in Buildings. *Energy and Buildings*, 27(3), 221–222.

Lucieer, A., Jong, S. M. and Turner, D. (2014). Mapping Landslide Displacements Using Structure from Motion (SfM) and Image Correlation of Multi-temporal UAV Photography. https://journals.sagepub.com/doi/pdf/10.1177/0309133313515293. Last retrieved May 16, 2019.

MacLean, Meghan and Congalton, Russell. (2012). Proceedings of the ASPRS 2012 Annual Conference, Map Accuracy Assessment Issues When Using an Object Oriented Approach, Proceedings of the ASPRS 2012 Annual Conference Sacramento, California March 19–23, 2012 Sacramento, California, March 19–23, 2012.

Micasense. (2015). Micasense RedEdge Camera Datasheet. https://support.micasense.com/hc/en-us/articles/215261468-RedEdge-Data-Sheet-PDF. Last retrieved May 20, 2019.

Micheletti, Natan, Chandler, Jim H. and Lane, Stuart N. (2015). *Structure from Motion (SfM) Photogrammetry, Geomorphological Techniques*, British Society for Geomorphology, Chap. 2, Sec. 2.2 (2015).

Moody, William. (2019). Private Communication, William Moody, Director Facilities and Operations, UNG-Gainesville campus, May 10, 2019.

Mott, John H. (2016). Integration of Unmanned Aircraft Systems in Public Higher Education. *International Journal of Aviation Sciences*, 1(2), 196–205.

Niklaus, Frank, Vieider, Christian and Jakobsen, Henrik. (2008). MEMS-Based Uncooled Infrared Bolometer Arrays–A Review. *Proceedings of SPIE - The International Society for Optical Engineering*, 6838. doi:10.1117/12.755128.

Sharma, J. B. and Hulsey, D. (2014). Integrating the UAS in Undergraduate Teaching and Research – Opportunities and Challenges at University of North Georgia. *International Archives of the Photogrammetry, Remote Sensing and Spatial Information Sciences*, XL-1, 377–380, www.int-arch-photogramm-remote-sens-spatial-inf-sci.net/XL-1/377/2014/ (accessed August 10, 2019).

Stalin, J. L. and Gnanaprakasam, R. (2017). Volume Calculation from UAV Based DEM. *International Journal of Engineering Research and Technical Research*, V6(06). doi:10.17577/ijertv6is060076.

Verhoeven, G. (2011). Taking Computer Vision Aloft – Archaeological Three-Dimensional Reconstructions from Aerial Photographs with Photoscan. https://onlinelibrary.wiley.com/doi/full/10.1002/arp.399. Last retrieved May 16, 2019.

Wallace, L., Lucieer, A., Malenovský, Z., Turner, D. and Vopěnka, P. (2016). Assessment of Forest Structure Using Two UAV Techniques: A Comparison of Airborne Laser Scanning and Structure from Motion (SfM) Point Clouds. www.mdpi.com/1999-4907/7/3/62. Last retrieved May 16, 2019.

Westoby, M. J., Brasington, J., Glasser, N. F., Hambrey, M. J. and Reynolds, J. M. (2012). 'Structure-from-Motion' Photogrammetry: A Low-cost, Effective Tool for Geoscience Applications. *Geomorphology*, 179: 300–314. doi:10.1016/j.geomorph.2012.08.021.

Wing, M. G., Burnett, Jonathan, John, Brungardt and Josh, David. (2013). Eyes in the Sky: Remote Sensing Technology Development Using Small Unmanned Aircraft Systems. https://academic.oup.com/jof/article/111/5/341/4756650. Last retrieved May 16, 2019.

Yang, M., Su, T. and Lin, H. (2018). Fusion of Infrared Thermal Image and Visible Image for 3D Thermal Model Reconstruction Using Smartphone Sensors. www.ncbi.nlm.nih.gov/pmc/articles/PMC6069248/. Last retrieved May 16, 2019

Ye, S. (2018). A Review of Accuracy Assessment for Object-Based Image Analysis: From Per-pixel to Per-polygon Approaches. www.sciencedirect.com/science/article/abs/pii/S0924271618300947. Last retrieved May 16, 2019.

Zylka, A. (2014). Small Unmanned Aerial Systems (sUAS) for Volume Estimation. https://scholarworks.uvm.edu/hcoltheses/44/. Last retrieved May 16, 2019.

12

Flying UAVs in Constrained Environments: Best Practices for Flying within Complex Forest Canopies

Ramesh Sivanpillai, Gregory K. Brown, Brandon S. Gellis

CONTENTS

12.1 Introduction

Tropical forest canopies have been characterized as the last frontier in terrestrial biology with access, especially for high-canopy wet forests, identified as the primary reason (Wilson and Moffett, 1991; Nadkarni and Parker, 1994; Lowman and Wittman, 1995). These forest canopies host diverse communities of epiphytic plants that establish niche-space for high non-plant diversity (e.g., fungi, lichens, protozoa, invertebrates, vertebrates). Epiphytes are nonparasitic plants that grow on branch and trunk surfaces of trees, and trees hosting epiphytes known as phorophytes (Figure 12.1).

Epiphytism is an adaptation that places these plants in a favorable light environment for photosynthesis that would not be possible on the very light-limited forest floor. In wet tropical forest canopies, especially, vascular plant epiphytes such as orchids, bromeliads, and ferns, in concert with nonvascular plants such as mosses, can form arboreal mats of vegetation with an organically rich arboreal soil on the larger phorophyte branches. Accessing these epiphytic communities, both visually, and physically to study and collect, especially when they are 10–80 m above the forest floor, constitutes a major challenge for biologists interested in these epiphytic organisms.

Accessing the typically dense tropical canopy can be categorized into two types: *high-tech* and *low-tech* methods. The high-tech methods include building canopy towers,

FIGURE 12.1
Epiphytes growing on the branches and tree trunks in a wet secondary forest in Costa Rica. (Photo: Brown, 2018).

scaffolding, walkways or use of cranes, hot air balloons, and canopy rafts. These high-tech methods are expensive and might not be suitable for all forested environments. For example, transporting construction materials inside a forest environment might not only be challenging, but could also damage the trees and other vegetation. Low-tech methods are ground-based and include the use binoculars and clipper-poles, or involve climbing individual trees. Low-tech methods tend to the less expensive. Barker (1997) and Lowman et al. (2012) provide detailed descriptions of the various high- and low-tech methods for accessing tree canopies along with their limitations.

Climbing individual trees, if feasible, is preferred by many botanists who study vascular plant epiphytes such as bromeliads, orchids, gesneriads, and ferns. By tree climbing, the researcher will obtain a closer, better view of the epiphytic vegetation, and may get close enough to handle, sample, or collect target-organisms. Tree climbing techniques needed for such observations and collections have improved over time (Barker, 1997; Dial et al., 2004; Picart et al., 2014), yet despite these improvements, tree climbing is inherently dangerous and physically demanding, arduous work. Furthermore, after the effort and risk, the researcher might not find the specific epiphytic species on the targeted branch, resulting in a waste of time and resources. This problem gets bigger as the individual trees and canopy increase in height, and climbing becomes increasingly more "blind" (i.e., can't see from the ground the epiphytic composition on the tops of branches), resulting in the researcher's descent from the initial target tree to re-establish a new climbing target based on the visual information obtained (e.g., higher or lower branch, or different adjacent tree).

12.1.1 Unmanned Aerial Vehicles for Monitoring Natural and Built Environments

Unmanned Aerial Vehicles (UAVs), or drones, are widely used to collect data for monitoring and mapping natural and built environments. Drones allow users to collect aerial images and data (e.g., LiDAR point cloud) for their study area from more than one vantage point. Numerous studies have reported the utility of UAVs for collecting data on crops, rangelands, forests, and other environments (Rango et al., 2009; Rango and Laliberte, 2010; Watts et al., 2012). With advances in image processing software, data collected by drones can be processed for generating virtual landscapes by wrapping the imagery data over 3D surface models of the terrain (Zarnowski et al. 2015; Themistocleous et al. 2015).

UAVs are available in rotary and fixed-wing models. The rotary UAV models are capable of hovering over the target, and are available as quad-, hexa-, and octocopters, depending on the number of rotors. Similarly, fixed-wing models vary in size and payload capacity. Some fixed-wing models can be launched by tossing them into the air, while others require specialized launching gear, and both require significant open space for takeoff. Rotary-wing models do require an open volume of air space ideally located over more or less level space (Figure 12.2), but can also be launched from an elevated platform (Sajithvariyar et al., 2019), under exceptional conditions, with special precautions. Users select the particular UAV design and model based on specific data collection requirements, but when there is limited to no open space users have to carefully evaluate the risks before launching the drone towards the canopy.

Drones are capable of carrying a wide array of sensor types: true-color, infrared, thermal (FLIR), and LiDAR, and they can be custom built to suit specific monitoring or data collection needs. For example, applications in crop dusting (precision farming), powerline inspection (infrastructure monitoring), require very accurate and precise positional data. For such applications users have built custom Global Positioning Systems (GPS) receivers that are capable of achieving very high levels of accuracy instead of using the one provided by the drone manufacturer.

Availability of low-cost drones has resulted in numerous monitoring and mapping applications in forestry (Zahawi et al., 2015; Zhang et al., 2016), peatland (Palace et al., 2018), wetland (Bertacchi et al., 2019), intertidal seagrass (Konar and Iken, 2018), oil-palm plantation (Khokthong et al., 2019), and rangeland vegetation (Tay et al., 2018). Studies

FIGURE 12.2
Quadcopter UAV is launched off of a landing pad in a tropical forest at La Selva Biological Station, Costa Rica. Rotary wing UAVs require an open volume of air space in order to be launched vertically towards the canopy. Pilot and spotters have to pay close attention to wind drift at both surface and canopy levels to minimize the chances for the drone to crash into trees.

have reported the utility of UAVs for their applications in spatial ecology (Anderson and Gaston, 2013), plant ecology (Cruzan et al., 2016), forestry (Tang and Shao, 2015), wildlife research (Christie et al., 2016), and conservation of protected areas (Lopez and Mulero-Pazmany, 2019).

UAVs are used for collecting data about plants and animals in difficult to access environments (Table 12.1), and findings from most of these studies conclude that UAVs have utility for collecting useful data (Weissensteiner et al., 2015; Johnston et al., 2017; Rush et al., 2018), and a few recommend that they can be a part of the sampling protocol (Bonnin et al., 2018; Gooday et al., 2018). Flexibility of collecting data coupled with their low acquisition cost are often cited reasons for their widespread use.

Drones are also increasingly used for coordinating response activities following disasters such as wildfires (Tang and Shao, 2015; Allison et al., 2016), floods (Loftis et al., 2017; Rabta et al., 2018), landslides (Antonino, 2016; Qiui, 2016; Song et al., 2019), and accidents involving automobiles and trains (Tang et al., 2016; Milas et al., 2018).

Operating a research drone often requires a team of two (i.e., pilot and spotter), or more, depending on the application, and the individuals involved must have the necessary qualifications and permits. The pilot is in charge of takeoff, flying, and landing the drone using a hand-held remote controller unit, and also controls the position of the camera or sensor while in flight. The spotter is responsible to maintain the line-of-sight with the drone while in flight. The pilot also has to follow instructions from the spotter if the drone flies out of the pilot's line-of-sight. Most remote control units are digitally connected to a video display and recording device, and a third crew member often monitors the video feed and provides instruction to the pilot in terms of positioning the drone and sensor for certain types of missions.

Most countries require the pilot to obtain certain credentials prior to flying a UAV. Certain countries have additional restrictions for flying drones over a public gathering, in wilderness areas, or in restricted airspace (e.g., near airports and defense installations). Protocols in most countries also require a crew member to conduct pre-flight and post-flight safety inspections and report any accidents that might have occurred during the flight. Operator(s) are required to follow all rules and protocols established for their study area before, during, and after each drone mission. Failure to follow these can lead to fines and/or other punitive actions.

TABLE 12.1

Select Studies that have Reported the Use of Unmanned Aerial Vehicles for Monitoring Plants and Animals Found in Not Easily Accessible Environments

Study	Authors
1. Nesting status of canopy-breeding birds	Weissensteiner et al. (2015)
2. Nest counts of breeding Glossy Ibis populations	Afan et al. (2018)
3. Classifying bird nest sites	Kamm and Reed (2019)
4. Estimating bird population	Lyons et al. (2019)
5. Counting Lesser Black-backed Gulls	Rush et al. (2018)
6. Detecting Chimpanzee nest	Bonnin et al. (2018)
7. Radio tracking wild bird populations	Tremblay et al. (2017)
8. Counting crocodiles in a National Park	Thapa et al. (2018)
9. Counting marine mammals ashore	Gooday et al. (2018)
10. Estimating rocky shore mussel demography	Gomes et al. (2018)
11. Assessing gray seal colony	Johnston et al. (2017)
12. Enumerating cetaceans	Angliss et al. (2018)

12.1.2 Drones for Imaging Epiphytes

With the availability of good quality drones at an affordable cost, and advances in image capturing hardware and processing software, there is considerable interest among botanists and canopy biologists to use drones for collecting data on vascular plant epiphytes. These researchers could view drones as a panacea to overcome difficulties associated with climbing and descending trees in tropical environments for epiphyte reconnaissance. Using an appropriate copter-drone, data can be collected on epiphytes growing on high canopy surfaces that are visually inaccessible from the ground (Figures 12.3 and 12.4).

FIGURE 12.3
Bromeliad arboreal mats consisting of three genera, *Tillandsia*, *Guzmania*, and *Aechmea* are visible in this photo acquired 40 m above-ground level by the camera onboard a DJI Spark™ (La Selva Biological Station, Costa Rica, 2018).

FIGURE 12.4
Werauhia kupperiana (Bromeliaceae, bromeliad family) with characteristic leaf pigmentation pattern for this species, and dehisced capsules (mature fruits), visible in this photo acquired 50 m above-ground level by the camera onboard a DJI Mavic Pro™ (near Barbilla National Park, Costa Rica, 2018).

The UAV systems best suited for canopy flights are relatively small, and can be easily carried (e.g., in a backpack) into isolated field sites. Given these conditions, botanists could assume that they only have to select a target tree, or trees, unpack the drone, affix the rotors and battery, launch the drone from the forest floor into the canopy, and begin collecting imagery/data on the epiphytic species of interest, or other attributes of the arboreal vegetation mat. However, operating these drones within the complex 3-dimensional volume of a tropical forest poses numerous challenges.

12.2 Best Practices for Flying Drones within Complex Environments

Our team has been conducting experimental and proof-of-concept flights to study epiphytic plants, mostly from the flowering plant family Bromeliaceae, commonly known as bromeliads (Lehmitz et al., 2018; Wirsching et al., 2018). Three different quadcopter models (DJI Phantom 3 Advanced™, DJI Mavic Pro™, DJI Spark™) have been used, and flight experiences to examine epiphytes range from within a controlled interior environment (Williams Conservatory, University of Wyoming), to tree canopies at Marie Selby Botanical Gardens, Sarasota, Florida (USA), to flights launched within different tropical forest types in Costa Rica. Based on these flight experiences we have developed a list of best practices that can be helpful for others interested in vertical launching of drones within complex environments such as a tropical forest.

12.2.1 Crew: Qualifications, Training, and Preparation

- Assembling a qualified, practiced crew is critical. For high-canopy and distant flights (>50 m from pilot to UAV), a crew of four is deemed minimal, and consists of a pilot, image acquisition expert, and two spotters. One, or more individuals within this four-person crew must have expertise in the UAV hardware and the UAV system software, or a fifth member, the flight engineer with these skills, must be added. One of the spotters should be an expert in the specific plant group being studied (i.e., bromeliads, orchids, etc.), or have expertise in tropical canopy ecology. Ideally, the second spotter should be cross-trained in both the system hardware and software being utilized in the project, and have pilot experience.

- It is important to establish a common and concise language (or terminology) for UAV operations that is understood by all crew members. While the drone is in the air, all crew members have to be engaged in conversation and every attempt must be made to streamline the terms and phrases used during a mission.

- For botanical applications, all crew members must have a general, basic understanding of gross plant morphology, e.g., inflorescence, rosette, floral bract, distal etc.

- It is important for at least one of the crew members to have extensive knowledge of, and experience in, UAV technology, including the systems used for communication between the drone and image/video device. Either this person, or another crew member must be trained to download and save digital files from the camera/sensor's memory devices.

- Pilot practice and mastery in controlling the drone during the imaging process, especially while collecting videos, is essential. Pilots must be able to navigate the drone close to the target plant in a slow and steady manner. Rapid movements (shudders, abrupt motions) can greatly reduce video quality. Some experienced drone pilots recommend at least 200 hours of practice before one can gain reasonably good piloting skills.

- Pilots must practice precise vertical takeoff and landings in order to launch the drones in the field with limited open space.

- One or more crew members must be fully familiar with the photo-application software and knowledgeable to manipulate this (e.g., f-stop, ISO), and not rely only on the auto-exposure function of the camera.

12.2.2 Planning and Preparation for Fieldwork

- It is advisable to generate a list of equipment and accessories, i.e., connector cables that are needed for a given mission. Prior to the fieldwork, it is imperative to test all equipment including back up and redundant systems.

- Complete necessary software updates and check their compatibility between the drone, camera, and the visual systems. Crew member assigned to this task must maintain a log of software installs along with compatibility requirements if the crew has to revert to an earlier version or installation after reaching the field.

- Binoculars are an essential part of the inventory. It is necessary to have at least one pair for pre-flight reconnaissance, and also during flights. Binoculars are especially important for spotters when the drone reaches a certain height and/or distance from the launch location. This linear distance is variable, depending on the UAV size, but in most cases, binoculars become essential for spotting at 40–50 m.

- Include extender cord to connect the micro USB in the controller to appropriate display device (phone or tablet).

- Have several extra propeller blades in the inventory. Despite best efforts, drones can come into contact with leaves and small branches that may damage blades, or possibly cause a crash.

- If feasible, it is advisable to have a second drone of the same make and model as a back-up, in case of a fatal crash of the primary vehicle. Include a small set of tools that are appropriate for the UAV hardware being used.

- Additional batteries are essential, as is a means to recharge these (i.e., automobile charger) in the field. This will reduce the down time during each mission.

- It is recommended to have a glare-guard (hood or canopy) for viewing devices (phone or tablet) in bright light.

- A portable takeoff and landing pad is essential for short understory vegetation.

- Pair the UAV with image device (phone, tablet) before leaving a Wi-Fi area.

- Establish the functional minimal focal-distance of the sensor system in different lighting conditions prior to conducting fieldwork.

- It is highly recommended to carry plenty of memory devices such as SD or microSD cards. During our flights with the DJI Mavic™ in default mode, we observed that one 32 GB memory card could be used for imaging for the duration of 3 batteries (or approximately 1 hour of flight time). Whereas a 16 GB memory card lasted for the duration of 1.5 batteries. However, this length could vary with other makes and drone models.

- It is advisable to have external storage devices with 1 or 2 Terabyte storage capacity for backing up the data instead of using the "cloud." Often in forested and other remote locations it is difficult to have internet access, so having capability to download data from memory cards to an external backup device can be valuable.

- Almost all standard first-aid kits contain medication and accessories for dealing with minor situations. When acquiring data in forested or wooded environments, the crew could encounter poisonous plants, insects, and animals. We recommend all crew members are aware of the common risks associated with the study area's flora and fauna and take additional medications and accessories.

12.2.3 Issues for Under Canopy sUAS Data Acquisition

- Cameras onboard most UAVs are capable of acquiring and storing image frames in formats such as RAW, TIFF, and JPG. Files stored in RAW format are larger in size in comparison to files stored in JPG format. JPG files can be displayed on most computers and other electronic devices. However, images stored in RAW format retain more color depth or range of values, in comparison to other formats. For applications that require identification of detailed plant parts, we recommend storing image frames in RAW format.

- Start an individual flight with wide angle shot(s) before close-up image capture.

- Turn off the obstacle avoidance behavior option on the UAV.

- End individual flight as soon as the battery sensor shows 10% of power left. Do not rely on the UAV's auto-return (Home) function due to low or no GPS signal strength inside the forest canopy.

- Do not assume wireless connections between the controller and display device will work automatically all the time; go with hard connections for hardware.

- Comment relevant to DJI Mavic: Propellers can have some limited interactions with vegetation, e.g., small epiphytic cacti (*Rhipsalis* sp.), dangling aerial roots, leaves, and lignified twigs <1 cm dia. (estimated diameter).

- For very high missions, it is advantageous to have one spotter with binoculars be able to lay on the ground with an object (e.g., backpack) to elevate head at a comfortable angle for tracking the UAV.

- In bright light conditions employ a shading hood or canopy to reduce screen-glare and improve viewing image quality.

- Include a set of heavy work-gloves, preferably made of leather, in case of hand-held takeoffs, or if hand-grab landings become required.

- In windy conditions approach target plant, if possible, from the down-wind direction.

- When flying to investigate isolated trees that are more than 40 to 50 m away from the control site, pilot should approach the tree from either the right or left side, and not directly at the trunk/canopy. The off-line approach helps spotter(s) communicate estimated distances between the target and UAV.

- Consider use of a strong laser pointer (green) to help indicate locations or specific parts of target plant. This would only be useful for scenarios where the distance between mission team and UAV is probably <50 m distant, and light conditions are not too bright.
- It is best for viable image capture to position the UAV so that the target plant is fore-lit.
- Prior to attempts for any close-up photos of a target plant, or plants, have the UAV do a slow horizontal-plane "orbit" (360° fly around) of the canopy environment in video mode.
- Do not capture video of takeoffs and landings; this is a waste of storage memory.
- Take abundant photos and videos. Post-flight analyses of images frequently reveal details not seen or noticed during flight image/video capture.
- Based on our experience, it is not advisable to use a First-Person-View (FPV) system while flying the drone underneath, or within, the forest canopy. Pilot and a spotter must maintain direct visual contact with the drone during the flight. If FPV is used, the pilot must be fully competent with this system. This probably involves, at least, another 10–15 hours of practice prior to fieldwork in addition to an estimated 100–200 hours of the basic pilot training. When using FPV, do not back the UAV up, instead, rotate the UAV 180° and go forward. Finally, for image/data gathering with FPV, do not have exposure control set at maximum.

12.2.4 Post-Flight Inspection

- At the end of a flight mission, inspect the UAV for any minor and major structural damages that could have happened during flight. Parts such as rotors and air intake valves below the rotors have to be checked carefully to see if any small insects or plant parts are caught in them.
- Prior to the next flight mission ensure that the battery voltage is above the value recommended by the manufacturer.
- At the end of the day, detailed post-flight inspection must be conducted as per the guidelines provided by both the UAV manufacturer and the oversight authority in the country or region where the aircraft is operated. This could include entering pertinent flight information in notebooks and uploading digital log files to websites.
- Any deviation from normal UAV flight responses or aberrant vehicle motions and behaviors must be recorded and attended to at the earliest opportunity to minimize any future risks to missions and crews.

12.2.5 Camera Issues and Lighting Conditions

- For close-up photography, a higher quality camera, one capable of macrophotography, is needed. For the three systems tested, camera focal length was the key limitation for close-up photography. Both Mavic Pro & Spark can be maneuvered to within a few centimeters of a target plant, or specific plant part, but images were out-of-focus.
- Pay attention to background and foreground lighting conditions, especially in full sun conditions. "Sun-glare," the light reflected from most smooth, glabrous, and

often shiny tropical tree leaves in full sunlight, can result in badly over-exposed images. Overcast sky conditions generally resulted in better conditions for UAV image capture.

12.2.6 Data Management

- Take careful, detailed field notes about flight locations, field conditions, and any distinctive qualities or outcomes of each flight mission.
- If possible, at the end of the field-day, evaluate, label, and file images while mission details are still fresh in the minds of all crew members.
- Huge amounts of data (images) can be collected in a short amount of time. Develop systematic labelling protocol for files. We developed RSEPII (Rating Scale for Epiphytic Plant Image Interpretation) to catalogue images for easy retrieval based on their file name.

12.3 Conclusions

The described best practices and recommendations are based on flying three different quadcopter models in progressively more complex environments, with proof of concept flights taking place in both primary and secondary forests in Costa Rica. Researchers and users who have experience with other UAV makes and models might add more recommendations to the ones listed and described by us.

Based on the Costa Rica study we can concur with previous studies that use of UAVs resulted in considerable cost and time savings for collecting data on epiphytes. This expedition, lasting only five days, resulted in the acquisition of more than 2000 photos. From these images, about 60 percent contained useable images of epiphytic species growing in several Costa Rican forests. Traditional data collection methods such as the use of binoculars and tree-climbing would have yielded only a very small fraction of the data volume generated by flying UAVs. Based on the experimental studies conducted by our team using three small UAVs, we recommend the use of small rotary-wing UAVs for the reconnaissance of, and photographic study and documentation of, vascular plant epiphytes in high-canopy tropical forests.

Botanists can add UAVs as part of their tool kit and not rely solely on traditional methods such as scanning the canopy with binoculars or climbing individual trees. However, we cannot overemphasize the importance of trained pilot(s) with several hours of flight experience in complex, tight space. As each mission has to be flown on manual mode, i.e., without flight planning missions, the pilot has to remain on constant alert for any unexpected activity from the drone and the environment.

We also recognize that UAS is a rapidly advancing field with the introduction of new hardware and software for flight planning and processing data. As newer drones with advanced technologies are introduced, few or some of the practices listed here might become obsolete. For instance, if the ability for the operator to control the camera's focal length is introduced in a new drone, our recommendation pertaining to positioning the drone/camera within close proximity of the target will not be applicable. Also, newer technologies that rely on visual cognition, machine learning, and pattern recognition need to be prototyped and tested. This technology will not be relying on GPS signals for proximity awareness, which will alleviate issues associated with flying the drones inside the forest canopy. Currently obtaining signals from the required number of satellites inside a forest canopy is a major limitation. However, many of the recommendations pertaining to crew training and practice, flight time procedures, and data management practices will continue to be useful for at least a few years to come.

References

Afan, I., Manez, M., Diaz-Delgado, R. (2018). Drone monitoring of breeding waterbird populations: The case of Glossy Ibis. *Drones* 2(4):42. doi:10.3390/drones2040042.

Allison, R.S., Johnston, J.M., Craig, G., Jennings, S. (2016). Airborne optical and thermal remote sensing for wildfire detection and monitoring. *Sensors* 16:1310. doi:10.3390/s16081310.

Anderson, K., Gaston, K.J. (2013). Lightweight unmanned aerial vehicles will revolutionize spatial ecology. *Frontiers in Ecology and the Environment* 11(3):138–146.

Angliss, R.P., Ferguson, M.C., Hall, P., Helker, V., Kennedy, A., Sformo, T. (2018). Comparing manned to unmanned aerial surveys for cetacean monitoring in the Arctic: Methods and operational results. *Journal of Unmanned Vehicle Systems* 6(3):109–127.

Antonino, D. (2016). Drones: New tools for natural risk mitigation and disaster response. *Current Science* 110(6):958–959.

Barker, M. (1997). An update on low-tech methods for forest canopy access and on sampling a forest canopy. *Selbyana* 18(1):61–71.

Bertacchi, A., Giannini, V., Franco, C.D., Silvestri, N. (2019). Using unmanned aerial vehicles for vegetation mapping and identification of species in wetlands. *Landscape and Ecological Engineering* 15(2):231–240.

Bonnin, N., Van Andel, A.C., Kerby, J.T., Piel, A.K., Pintea, L., Wich, S.A. (2018). Assessment of Chimpanzee nest detectability in drone-acquired images. *Drones* 2:17. doi:10.3390/drones2020017.

Christie, K.S., Gilbert, S.L., Brown, C.L., Hatfield, M., Hanson, L. (2016). Unmanned aircraft systems in wildlife research: Current and future applications of a transformative technology. *Frontiers in Ecology and the Environment* 14(5):241–251.

Cruzan, M.B., Weinstein, B.G., Grasty, M.R., Kohrn, B.F., Hendrickson, E.C., Arredondo, T.M., Thompson, P.G. (2016). Small unmanned aerial vehicles (micro-UAVs, drones) in plant ecology. *Applications in Plant Sciences* 4(9):1600041.

Dial, R.J., Sillett, S.C., Antoine, M.E., Spickler, J.C. (2004). Methods for horizontal movement through forest canopies. *Selbyana* 25(1):151–163.

Gomes, I., Peteiro, L., Bueno-Pardo, J., Albuquerque, R., Pere-Jorge, S., Oliveria, E.R., Alves, F.L., Queiroga, H. (2018). What's a picture really worth? On the use of drone aerial imagery to estimate intertidal rocky shore mussel demographic parameters. *Estuarine, Coastal and Shelf Science* 213:185–198.

Gooday, O.J., Key, N., Goldstien, S., Zawar-Reza, P. (2018). An assessment of thermal-image acquisition with an unmanned aerial vehicle (UAV) for direct counts of coastal marine mammals ashore. *Journal of Unmanned Vehicle Systems* 6(2):100–108.

Johnston, D.W., Dale, J., Murray, K.T., Josephson, E., Newton, E., Wood, S. (2017). Comparing occupied and unoccupied aircraft surveys of wildlife populations: Assessing the gray seal (*Halichoerus grypus*) breeding colony on Muskeget Island, USA. *Journal of Unmanned Vehicle Systems* 5(4):178–191.

Kamm, M., Reed, J.M. (2019). Use of visible spectrum sUAS photography for land cover classification at nest sites of a declining bird species (*Falco sparverius*). *Remote Sensing in Ecology and Conservation*. doi:10.1002/rse2.104.

Khokthong, W., Zemp, D.C., Irawan, B., Sundawati, L., Kreft, H., Holscher, D. (2019). Drone-based assessment of canopy cover for analyzing tree mortality in an oil palm agroforest. *Frontiers in Forests and Global Change* 2:12. doi:10.3389/ffgc.2019.00012.

Konar, B., Iken, K. (2018). The use of unmanned aerial vehicle imagery in intertidal monitoring. *Deep Sea Research Part II* 147:79–86. doi:10.1016/j.dsr2.2017.04.010.

Lehmitz, M., Wirsching, E.M., Brown, G.K., Sivanpillai, R. (2018). Utility of close range UAV photos for identifying epiphytic plants. *Proceedings of the 2018 ASPRS Annual Conference*, 5–7 February. Denver, CO (USA).

Loftis, J.D., Wang, H., Forrest, D., Rhee, S., Nguyen, C. (2017). Emerging flood model validation frameworks for street-level inundation modeling with StormSense. *Proceedings of the 2nd International Workshop on Science of Smart City Operations and Platforms Engineering (SCOPE 2017)*, pp. 13–18, 18–21 April, Pittsburg, PA (USA). doi:10.1145/3063386.3063764.

Lopez, J.J., Mulero-Pazmany, M. (2019). Drones for conservation in protected areas: Present and future. *Drones* 3:10. doi:10.3390/drones3010010.

Lowman, M.D., Wittman, P.K. (1995). The last biological frontier? Advances in research on forest canopies. *New Phytologist* 19(4):161–165.

Lowman, M.D., Schowalter, T.D., Franklin, Jerry F. (2012). *Methods in Forest Canopy Research*. Berkeley, CA: University of California Press.

Lyons, M.B., Brandis, K.J., Murray, N.J., Wilshire, J.H., McCann, J.A., Kingsford, R.T., Callaghan, C.T. (2019). Monitoring large and complex wildlife aggregations with drones. *Methods in Ecology and Evolution*. doi:10.1111/2041-210X.13194.

Milas, A.S., Cracknell, A.P., Warner, T.A. (2018). Drones – The third generation source of remote sensing data. *International Journal of Remote Sensing* 39(21):7125–7137.

Nadkarni, N.M., Parker, G.G. (1994). A profile of forest canopy science and scientists – who we are, what we want to know, and obstacles we face: Results of an international survey. *Selbyana* 15(2):38–50.

Palace, M., Herrick, C., DelGreco, J., Finnell, D., Garnello, A.J., McCalley, C., McArther, K., Sullivan, F., Varner, R.K. (2018). Determining subarctic peatland vegetation using an unmanned aerial system (UAS). *Remote Sensing* 10:1498. doi:10.3390/rs10091498.

Picart, L., Forget, P-M, D'Haese, C.A., Daugeron, C., Beni,S., Bounzel, R., Kergresse, E., Legendre, F., Murienne, J., Guilbert, E. (2014). The Cafotrop Method: An improved rope-climbing method for access and movement in the canopy to study biodiversity. *Ecotropica* 20:45–52.

Qui, J. (2016). Killer landslides: The lasting legacy of Nepal's quake. *Nature* 532:428–431. doi:10.1038/532428a.

Rabta, B., Wankmuller, C., Reiner, G. (2018). A drone fleet model for last-mile distribution in disaster relief operations. *International Journal of Disaster Risk Reduction* 28:107–112.

Rango, A., Laliberte, A. (2010). Impact of flight regulations on effective use of unmanned aircraft systems for natural resources applications. *Journal of Applied Remote Sensing* 4(1):043539.

Rango, A., Laliberte, A., Herrick, J.E., Winters, C., Havstad, K., Steele, C., Browning, D. (2009). Unmanned aerial vehicle-based remote sensing for rangeland assessment, monitoring, and management. *Journal of Applied Remote Sensing* 3:033542.

Rush, G.P., Clarke, L.E., Stone, M., Wood, M.J. (2018). Can drones count gulls? Minimal disturbance and semiautomated image processing with an unmanned aerial vehicle for colony-nesting seabirds. *Ecology and Evolution* 8:12322–12334. doi:10.1002/ece3.4495.

Sajithvariyar, V.V., Sowmya, V., Gopalakrishnan, E.A., Soman, K.P., Bupathy, P., Sivanpillai, R., Brown, G.K. (2019). Opportunities and challenges of launching UAVs within wooded areas. *Proceedings of the 2019 ASPRS Annual Conference*, 27–31 January, Denver, CO (USA).

Song, K., Wang, F., Dai, Z., Lio, A., Osaka, O., Sakata, S. (2019). Geological characteristics of landslides triggered by the 2016 Kumamoto earthquake in Mt. Aso volcano, Japan. *Bulletin of Engineering Geology and the Environment* 78(1):167–176.

Tang, L., Shao, G., (2015). Drone remote sensing for forestry research and practices. *Journal of Forestry Research* 26(4):791–797.

Tay, J.Y.L., Erfmeier, A, Kalwij, J.M. (2018). Reaching new heights: Can drones replace current methods to study plant population dynamics? *Plant Ecology* 219 (10):1139–1150.

Tang, Z., Nie, Y., Chang, J., Zhang, J., Liu, F. (2016). Photo-based automatic 3D reconstruction of train accident scenes. *Proceedings of the Institution of Mechanical Engineers, Part F: Journal of Rail and Rapid Transit* 232(1):144–158. doi:10.1177/0954409716662089.

Thapa, G.J., Thapa, K., Thapa, R., Jhawali, S.R., Wich, S.A., Poudyal, L.P., Karki, S. (2018). Counting crocodiles from the sky: Monitoring the critically endangered gharial (*Gavialis gangeticus*) population with an unmanned aerial vehicle (UAV). *Journal of Unmanned Vehicle Systems* 6(2):71–82.

Themistocleous, K., Ioannides, M., Agapiou, A., Hadjimitsis, D.G. (2015). The methodology of documenting cultural heritage sites using photogrammetry, UAV and 3D printing techniques: The case study of Asinou Church in Cyprus. Proceedings of SPIE Third International Conference on Remote Sensing and Geoinformation of the Environment (RSCy2015). 9535, 953510. doi: 10.1117/12.2195626.

Tremblay, J.A., Desrochers, A., Aubry, Y., Pace, P., Bird, D.M. (2017). A low-cost technique for radio-tracking wildlife using a small standard unmanned aerial vehicle. *Journal of Unmanned Vehicle Systems* 5(3):102–108.

Watts, A.C., Ambrosia, V.G., Hinkley, E.A. (2012). Unmanned aircraft systems in remote sensing and scientific research: Classification and considerations of use. *Remote Sensing* 4:1671–1692.

Weissensteiner, M.H., J.W. Poelstra, J.B.W. Wolf. (2015). Low-budget ready-to-fly unmanned aerial vehicles: An effective tool for evaluating the nesting status of canopy-breeding bird species. *Journal of Avian Biology* 46:425–430.

Wilson, E.O., Moffett, M.W. (1991). Rain forest canopy: The high frontier. *National Geographic* 180(6):79–107.

Wirsching, E.M., Lehmitz, M., Sivanpillai, R., Brown, G.K. (2018). Determining optimal target distances for sUAV while imaging epiphytes. *Proceedings of the 2018 ASPRS Annual Conference*, 5–7 February, Denver, CO (USA).

Zahawi, R.A., Dandois, J.P., Holl, K.D., Nadwodny, D., Reid, J.L. (2015). Using lightweight unmanned aerial vehicles to monitor tropical forest recovery. *Biological Conservation* 186:287–295.

Zarnowski, A., Banaszek, A., Banaszel, S. (2015). Application of technical measures and software in constructing photorealistic 3D models of historical building using ground-based and aerial (UAV) digital images. *Reports on Geodesy and Geoinformatics* 99:54–63. doi:10.2478/rgg-2015-0012.

Zhang, J., Hu, J., Lian, J., Fan, Z., Ouyang, X., Ye, W. (2016). Seeing the forest from drones: Testing the potential of lightweight drones as a tool for long-term forest monitoring. *Biological Conservation* 198:60–69.

Index

9781032475158